ELECTRIC POWER ECONOMICS

SELECTED WORKS

About the Author

Mohan Munasinghe

Mohan Munasinghe was born in Sri Lanka, and has earned post-graduate degrees in engineering, physics and economics from Cambridge University, U.K., the Massachusetts Institute of Technology, USA, McGill University, Canada, and Concordia University, Canada.

He has received a number of international honours and prizes for outstanding research work and applications, and is the author of over 30 books and monographs, as well as several hundred technical papers on electric power, energy, environment, economics and information technology. He serves also on the editorial boards of several scholarly journals.

He is presently Division Chief for Environmental Policy and Research at the World Bank, Washington DC, and until recently served as Division Chief for Energy and Infrastructure Operations. During his 15 years with this institution, Professor Munasinghe has had field experience in most developing countries, and worked on electric power, environmental, petroleum, water, telecommunications, transport, urban and multi-purpose projects. From 1982 to 1986, he was the Senior Energy Advisor to the President of Sri Lanka, the Chairman of the Computer and Information Technology Council of Sri Lanka, and the President of the Sri Lanka Energy Managers Association (SLEMA), while on leave of absence from the World Bank.

He has been active in a number of international think-tanks in the areas of energy, information technology and the environment -- recently serving as Chairman, Energy Research Group, IDRC-United Nations University, and Chairman, Committee on Computers and Informatics, Third World Academy of Sciences, Trieste, as well as member of several expert advisory panels of the National Academy of Sciences, USA. He is Visiting Professor, Institute for Technology Policy in Development, State University of New York, Stonybrook, NY. He has taught also in several Sri Lankan, Canadian and other US universities, and conducted international training seminars for many senior developing country officials.

Finally, he maintains a lively interest in sports, having formerly represented Sri Lanka (Junior Davis Cup) and Cambridge University in lawn tennis, and won several titles.

ELECTRIC POWER ECONOMICS

SELECTED WORKS

Mohan Munasinghe

Foreword by Marcel Boiteux

Introduction by Gabriel Sanchez-Sierra

Butterworths

London Boston Singapore Sydney Toronto Wellington

PART OF REED INTERNATIONAL P.L.C.

First published 1990

British Library Cataloguing in Publication Data

Munasinghe, Mohan 1945–
Electric power economics: selected works.
1. Electricity supply industries
I. Title
338.4′762131

ISBN 0-408-05622-3

Library of Congress Cataloging-in-Publication Data

Munasinghe, Mohan, 1945–
Electric power economics: selected works/
Mohan Munasinghe; foreword by Marcel Boiteux;
introduction by Gabriel Sanchez-Sierra.
p. cm.
Includes bibliographical references.
ISBN 0-408-05622-3:
1. Electric utilities—Developing countries. 2. Electric power consumption—Developing countries. I. Title.
HD9685.D452M86 1989
333.79′32′09172′4—dc20 89-25255
CIP

Printed and bound by Hartnolls Ltd, Bodmin, Cornwall

To Anusha and Ranjiva

This collection of essays and its companion publication entitled *Energy Analysis and Policy* are being issued to commemorate the receipt of two awards by Mohan Munasinghe, for:

(a) exceptional contributions to energy literature and research, from the International Association for Energy Economics, USA (1987);
(b) outstanding research and applications in power and energy system analysis and pricing, from the Latin America and Caribbean Conference on Electric Power and Energy (1988).

These twin volumes are published with the compliments and felicitations of the following:

Bulletin of Indonesian Economic Studies
Energy Economics
Energy Journal (publication of the International Association for Energy Economics)
Energy Policy
Energy Systems and Policy
IEE Review (publication of the Institution of Electrical Engineers, UK)
IEEE Transactions on Power Apparatus and Systems (publication of the Institute of Electrical and Electronic Engineers, USA)
International Journal of Ambient Energy
International Journal of Electrical Power and Energy Systems
Journal of Consumer Research
Journal of Development Economics
Journal of Energy and Development
Natural Resources Forum
Natural Resources Journal
Pacific and Asian Journal of Energy
Pakistan Economic and Social Review
Power Engineering Journal (publication of the Institution of Electrical Engineers, UK)
Proceedings of the IEEE (publication of the Institute of Electrical and Electronic Engineers, USA)
Rand Journal of Economics (formerly *Bell Journal of Economics*)
SLEMA Journal (publication of the Sri Lanka Energy Managers Association)

CONTENTS

FIGURES

TABLES

FOREWORD

Welcome to the latest work of Mohan Munasinghe!

The twin honours for outstanding research and applications, awarded to him by the International Association for Energy Economics (in 1987) and the Latin American and Caribbean Conference on Electric Power and Energy (in 1988), are evidence of the sustained interest that his work has stimulated among the international scientific community, in the areas of both energy and electric power. The profound insights and experiences of the author are amply demonstrated by the diversity and richness of the themes contained in this particular volume of collected papers on the economics of electricity, produced during the course of the last two decades.

In fact, Mohan who has all the attributes of the perfect scholar, and acquired an impressive array of degrees, also has succeeded in complementing his intellectual achievements with solid practical applications. His experience in the field includes over 15 years at the World Bank, where he is currently a Division Chief, as well as 4 years in his native land -- Sri Lanka -- serving as Senior Energy Advisor to the President.

Over the years, he has made original research contributions in practically every major area of electricity economics including pricing, demand forecasting, investment planning, power system reliability, outage costs and rural electrification. On the topic of marginal cost pricing, in particular, he has been a leader in implementing efficient pricing policies, establishing rigorous procedures for tariff setting, and bringing about regulatory reforms.

His work carries all the more weight because his analysis and interpretations draw upon case studies based on the many projects that he has dealt with in the real world. Professor Munasinghe's ideas especially enrich our knowledge of the developing countries, in which the electric power sector faces more formidable problems than in the industrialised world.

This latest and most stimulating publication consolidates an already long series of his past writings on the subject. It will help to further our understanding of fundamental concepts in electricity economics, and guide the future development of the power sector -- one of the most important in any modern economy.

Marcel Boiteux
President, World Energy Conference and
Honorary President, Electricite De France
Paris, France

ACKNOWLEDGEMENTS

I am deeply indebted to all the colleagues and friends, too numerous to mention here, who have contributed so generously to my intellectual development over the years. Special mention should be made of those who helped to co-author the various works that provide the basis for several chapters in this volume, as indicated in the relevant footnotes.

I am most grateful and priviledged also, to have the Foreword contributed by the doyen of modern marginal cost pricing and distinguished elder statesman of the power sector, Marcel Boiteaux, and the Introduction contributed by another leading authority in the field, Gabriel Sanchez-Sierra. Finally, many thanks are owed especially to Chitrupa Fernando, as well as to Mark Bernstein, Jayant Sathaye, Susantha Perera and the staff of Butterworths, for assistance in reviewing, compiling and editing this volume.

INTRODUCTION[1]

When Mohan Munasinghe first suggested that I prepare the introduction to this collection of essays, I readily agreed to undertake what was a seemingly easy task. After all, through my involvement in the energy sector, I had followed his work, participated in some of the same events and even worked with him at the World Bank, particularly on my own country, Colombia. With such a wealth of first-hand knowledge regarding Mohan's professional contributions, I was certain that I could draft several pages with no difficulty whatsoever. I soon realized , however, that I had lost sight of a very important fact; this publication is one of the latest in a long list of books numbering approximately thirty, aside from which my friend and colleague has authored more than one hundred technical papers. Given the glowing commentaries made in previous critiques of his work, no superlative had been left unsaid.

I therefore decided to focus on the man behind the scholar and, basically, to respond to two key questions -- what kind of professional training has laid the groundwork for Mohan Munasinghe's reputation as one of the leading authorities on energy economics, policy and planning; and what personal, rather than professional, attributes have given his work a "human perspective"?

Professor Munasinghe has received degrees in Physics, Engineering and Economics, from Cambridge University, the Massachusetts Institute of Technology, McGill University and Concordia University. Hence, his technical analyses offer the advantage of a broad-based, multi-disciplinary treatment, which is essential in dealing suitably with the area of energy.

Mohan can accurately be described as "a scholar who lives in the real world." Throughout his academic and professional career, he has been deeply concerned about the problems he has studied, and he has been actively involved in the search for appropriate solutions. He has left the walls of the classroom behind and seen his theoretical knowledge in action in practical applications that have not only benefited the energy community, but have also contributed to social development in Third World countries around the globe.

Perhaps one of his most gratifying experiences at the personal level was the four-year period (1982-1986) during which he served as the Senior Energy Advisor to the President of his native Sri Lanka, and personally supervised the country's energy sector. However, Mohan's field experience has certainly not been limited to his homeland. Over his more than 15 years at the World Bank, he has collaborated with numerous developing countries, in development. In that context, he has been able to strike an excellent balance between the philosophy of an international organization and his own Third World-oriented views. Mohan has never lost his feeling for his roots.

Through such varied efforts, Mohan has achieved international recognition and acclaim. In fact, the present volume, *Electric Power Economics*, and its companion publication, *Energy Analysis and Policy*, follow on the two major awards mentioned

[1] The opinions expressed in this introduction are the exclusive responsibility of the author and are not expressed on behalf of the Latin American Energy Organization (OLADE) or any of its 26 Member Countries.

earlier in the text. A brief glance at the index of each of these two volumes, and at the list of supporting organizations, will give the readers an idea of the depth and breadth of Mohan's work and the worldwide recognition it has received. At the same time, the papers included in this volume do not provide merely a sampling of Professor Munasinghe's research interests over the past decade; they provide insight into what have been the key issues for energy economists and policymakers worldwide, especially those working in the power sector.

The first essay included in this volume is entitled "Power Sector Issues and Policy in Developing Countries". It is based on the author's keynote address presented in 1988, at the Fifth Latin American and Caribbean Regional Conference on Electricity Economics and Pricing, where he was honored for his outstanding theoretical work and practical applications in the areas of power and energy system analysis and pricing.

This paper provides a valuable overview of some of the major financial, technical and manpower-related constraints which hamper the expansion of power supplies in developing countries, and it assesses some of the options for dealing with these problems. The conclusions point out a series of additional analyses needed before arriving at any final plan for power expansion, and offer some concrete measures to be undertaken in the meantime.

The next two essays in this collection are comprehensive overview papers on the theory and application of electric power system investment planning and pricing, especially prepared for the *Proceedings of the IEEE* -- the principal publication of the largest international professional society of engineers. The second essay, "Engineering-Economic Analysis of Electric Power Systems," was originally published in 1984. It discusses advances and developments in this area and provides a new theoretical and analytical framework for power system planning, in which future reliability levels and prices are optimized. In the traditional framework, reliability levels and prices were set arbitrarily. The basic economic principle underlying the methodology described in this paper is that decisionmakers should seek to provide electricity services in a manner which maximizes the net benefits of consumption for society as a whole. Several well documented case studies illustrate the practical application of the theoretical models presented earlier.

The third essay is on "Principles of Modern Electricity Pricing", the original version of which was published in 1981. However, this topic is as timely now as it was at the beginning of this decade, given the fact that tariffs are one of the most important issues in power utilities around the world. Professor Munasinghe discusses the traditional approach used in pricing policy and the factors which have arisen in recent times calling for a re-orientation in this approach. Specifically, the paper discusses the many objectives of power pricing, not all of which are mutually consistent, and offers a pricing framework based on long-run marginal costs (LRMC) that permits pragmatic trade-offs among alternative goals.

The fourth essay was first presented in 1981, at the USAID-sponsored Energy Management Training Program for senior officials from developing countries. By discussing "Electricity Tariffs in Burma," it illustrates the situation of a power-deficit developing country whose large amounts of untapped hydroelectric resources will be useful for the long term, but whose short-term efforts must be geared to optimizing its current energy configuration and formulating on-target plans for the future. This situation occurs in many Third World Countries, including Latin America, where the structure of energy consumption has not been in line with the mix of the country energy resource endowment, and where planning and pricing approaches have often been ineffectively oriented. Specifically, this article reviews the existing electric power tariff structure and the current procedure for setting tariffs, and then analyzes a proposed new tariff structure based on a transitional version of LRMC. Finally, it points out some of the important issues which needed to be addressed in future tariff revisions.

In "Electricity Demand Analysis and Forecasting," the fifth essay written in 1982, Professor Munasinghe provides an introduction to methods and models used in analyzing and forecasting electricity demand. He also identifies some of the main problems which may arise during practical implementation. Finally, he compares the demand functions described herein with others offered by the energy economics literature.

The sixth essay, "The Economic Analysis of Power Projects: Case Study of Rural Electrification," was originally presented in the January 1988 issue of *Energy Economics*. Although no one questions the fact that rural electrification could provide the bases for economic growth and improvements in the quality of rural life, just how effective it is as a driving force for rural development has been the subject of much debate. In light of the stumbling-blocks that do indeed exist for rural electrification efforts, this article focuses on the criteria which should be used in assessing the technical and economic feasibility of undertaking such a project. Once again, the theory involved is effectively supported by a case study, this time in Malaysia.

The seventh essay is the outgrowth of some of Professor Munasinghe's most recent research and study. It is based on a paper originally published in the January 1989 issue of *The Energy Journal*. This contribution identifies the basic concepts and key issues involved in "Reliability of Electricity Supply, Outage Costs and Value of Service." Through a literature survey, the paper provides an overview of important aspects included in the definition, measurement and application of outage costs. Finally, it comments on how the introduction of some of the principal demand side issues into electric power systems can influence regulatory criteria and practices.

The eighth essay, "Economic Criteria for Optimizing Power System Reliability Levels," first published in the *Bell Journal of Economics* in 1979, describes some of the early and original contributions that Mohan has made in the area of optimal reliability and outage costs. This paper is a fine example of how the design of a distribution system may be optimized so as to maximize the net benefits of electricity consumption for society. In this approach, outage costs are measured considering electricity as an intermediate product which is used to produce goods and services in various sectors of the economy. A case study in the small Brazilian city of Cascavel is used to illustrate the application of the optimization methodology.

The ninth essay, "Costs Incurred by Residential Electricity Consumers Due to Power Failures" first appeared in the *Journal of Consumer Research* in 1980. It also grew out of the search for appropriate standards of reliability for electricity supply. This paper recognizes the need to analyze demand-side and supply-side social costs and benefits for industrial, commercial and residential consumers but concentrates exclusively on household consumption. On the basis of a survey conducted in Cascavel, Brazil, the article highlights the loss of time as the principal outage cost for a household, and assesses the marginal economic value of such losses. The appendix explains the method used to derive outage costs and the questionnaire for the Residential Consumer Survey.

The tenth essay, entitled "Rural Electrification: International Experience and Policy in Indonesia," originally appeared in the *Bulletin of Indonesian Economic Studies* in August 1988. Here, Professor Munasinghe reviews the quantitative and qualitative benefits of this controversial topic. Then, he presents a case study of rural electrification policy analysis in Indonesia, as an illustration of some of the principal policy issues. Since he considers that "The success or failure of rural electrification activities in a developing country invariably depends on the extent to which the relevant issues have been systematically analyzed and addressed," he enumerates six points as prerequisites for an effective rural electrification effort, based on the experience of Indonesia and other developing countries.

"Optimization of Electric Power Distribution Systems," the final essay, was prepared in the mid-1980's, and includes extracts from Munasinghe's classic reference volume, *Rural Electrification for Development*. It addresses a very timely topic -- loss reduction as one of the main ways of achieving efficiency in the electric power sector. Technical losses take place during the generation, transmission and distribution stages of a power system, and non-technical losses also occur. This particular analysis focuses on the reduction of energy and peak power technical losses in distribution networks, where up to three fourths of total system losses occur.

The article analyzes the two principal reasons for current high loss levels, summarizes the basic procedures for determining economic loss levels for all distribution system components, and even provides a check-list for evaluating the status of engineering standards and practices in rural power distribution systems in developing countries. It provides technical and economic data from selected rural electrification projects during the 1975-84 period and, finally, lists seven measures that could aid low- and middle-income developing countries in reducing overall rural electrification costs significantly.

These eleven papers and the fourteen others included in the companion volume, *Energy Analysis and Policy*, will undoubtedly furnish the reader with a sound overview of the major issues being dealt with in the power and energy area. However, in addition to their scholarly contribution, may I recommend that these essays be read keeping in mind the man behind them -- with the understanding that Mohan is not merely attempting to provide a compendium of technical information, but also trying to get across fundamental principles in a new and flexible approach to dealing with Third World development problems.

Finally, I would like to thank my friend Mohan for having paid me the honour of asking me to prepare the introduction to this set of essays which I consider to be one of the most important contributions to the understanding of the critical issues in power economics in the last two decades. Over the years, it has been a great personal pleasure to have been able to share ideas on planning and development with Mohan. I have seen many of these ideas mature in real world applications, and observed their soundness proven time and time again in practical programs such as those currently being implemented by the Latin American Energy Organization (OLADE).

Gabriel Sanchez-Sierra
Executive Secretary
Latin American Energy Organization (OLADE)
Quito, Ecuador.

Chapter 1

POWER SECTOR ISSUES AND POLICY IN DEVELOPING COUNTRIES[1]

1.1 Introduction

Electric power is a vital pre-requisite in any modern economy. Even in the least developed countries, the availability of a reliable power supply at reasonable cost is important for economic growth and development. The critical issue today is that the demand for electric power in developing countries keeps expanding faster than the growth of supply, which is constrained by financial, technical and manpower related impediments. This paper reviews some of these difficulties and various options that are available to address them at the present time.

A recent review of power sector performance in developing countries since the early 1960's underscores the central theme of the need for greater efficiency in the sector (Munasinghe et al. 1988). The issue is not that power is unimportant to modern economies, or that this capital intensive sector will not continue to merit significant resources in the future, but rather that developments over the past few years have highlighted the need to improve efficiency in the production and use of electricity. Thus, both governments and donors would be well advised to scrutinize more carefully continued requests for resource flows to the sector, and to search actively for methods of increasing technical, financial and managerial effectiveness.

The developing countries have made significant gains in terms of access to, and per capita consumption of electricity over the last few decades. Other evidence also suggests that in many countries the power sector tends to be better organized and perform better than other sectors of the economy. However, during periods of high growth, power utilities have had to weather oil price increases and high inflation and have been hampered in their efforts to attain financial targets because governments have been slow in responding to changing conditions and in granting tariff increases. Thus, an urgent need has arisen to arrest deteriorating trends, given that power investments absorb as much as half of all public investments in some countries, and often are the cause of severe debt-related and macroeconomic stresses.

The evidence available tends to confirm that the social compact on which power utilities are based and the assumptions under which they operate within developing economies need to be rescrutinized. Over the past decade, the decline in sector performance has been paralleled by a shift towards large monolithic government-controlled electric utilities. This centralizing trend is based on several reasons, such as: the high growth of the power sector, economies of scale in planning and operations, opportunities for improving coordination and efficiency, reduced reserve margins and reliability gains, the need to undertake larger and longer term investments, and pressures for nationalization and elimination of foreign ownership.

[1] Edited version of award acceptance keynote address presented at the Fifth Latin American and Caribbean Conference on Electric Power and Energy, held in Caracas, Venezuela, in September 1988.

Although some of this rationale for concentration is still valid, there is a growing consensus about the urgent need for greater efficiency and organizational reform in the power sector.

Power Sector Status in Developing Countries

Despite the many indicators of deteriorating performance in the electric power sector, it is evident that access to service has increased considerably in terms of both average per capita kilowatt hour generation and percentage of population served. In 51 countries surveyed by the World Bank, generation increased at an annual average of 7%, from 196 kWh/capita in 1968 to 529 kWh/capita in 1982. Correspondingly, the average rate of per capita GDP growth was only 2%. The annual average growth rate of connections for 29 power projects for which data was available was 9%. Even in countries with little or negative economic growth, the majority had more than 5% annual growth in generation and better than 3% annual growth in connections.

Such improvements in access have required sustained high rates of investment and expansion in total power sector assets as high as 15% annually in many countries. The growth of underlying demand has placed an even greater economic burden, with many utilities often being forced to scale down their programs due to budget and other constraints. Furthermore, investments in the past decades have been predominantly weighted toward generation, with a consequent underfunding of the distribution side. Thus, many developing country power systems now suffer from unbalanced investments, with overcapacity in generation coexisting with serious underinvestment in distribution.

One consequence of this unbalanced investment is the high level of technical losses recorded in many developing country power systems. High loss levels drive up supply costs and increase financial burdens. They are also indicative of a poor quality of service since substandard distribution networks which lead to losses are also responsible for voltage fluctuations and power outages. Frequently, investments on the distribution side that are relatively small compared to supply side investments, suffice to achieve significant energy savings by reduction of technical losses. There are also considerable non-technical losses in third world electricity systems. These occur due to outright theft and inadequate metering, billing or collection. For the countries reviewed by the World Bank, total network losses during the 1967-78 period averaged 13%, whereas a reasonable technoeconomic norm would be less than 8%. It is also disconcerting to note that in most cases, losses have either progressively increased or remained unchanged, despite the heavy power sector investments. Service quality has shown a similar adverse trend.

While an accurate assessment of supply costs is not available, it is quite apparent that tariffs have not kept up with costs even in financial terms. On average the operating ratio (ie. total costs as a fraction of total revenue) for over 300 projects reviewed by the World Bank deteriorated from 0.65 in the 1966-73 period to 0.80 during 1980-85. The trend of operating ratio is shown in Fig.1.1. Despite increasing emphasis on economic efficiency pricing placed by external agencies, only relatively modest progress has been achieved globally, even though marginal cost based tariff structures have been implemented in several countries.

The shortfall in revenue which is indicated by this adverse trend in the operating ratio, has largely contributed to the poor financial performance of the LDC power sector. Cost escalation has also been responsible for this decline. In addition to the operating ratio, sector financial performance has been monitored using several other key utility financial ratios such as rate of return on assets, self financing ratio and days receivable. Rates of return have declined progressively, from 9.2% in 1966-73 to 7.9% in 1974-79 and 6.0% in 1980-85. The trend in rate of return is shown in Fig.1.2. The self-financing ratio (which measures the amount of investment requirements financed by internal cash generation) also worsened over time, starting at an average of 25% in 1966-73 but falling to 17% by 1980-85. Days receivable (which reflects the utility's ability to collect its bills and act against delinquent customers) increased from 77 days during 1966-73 to 108 in the 1970s, and to 112 days in the 1980-85 period.

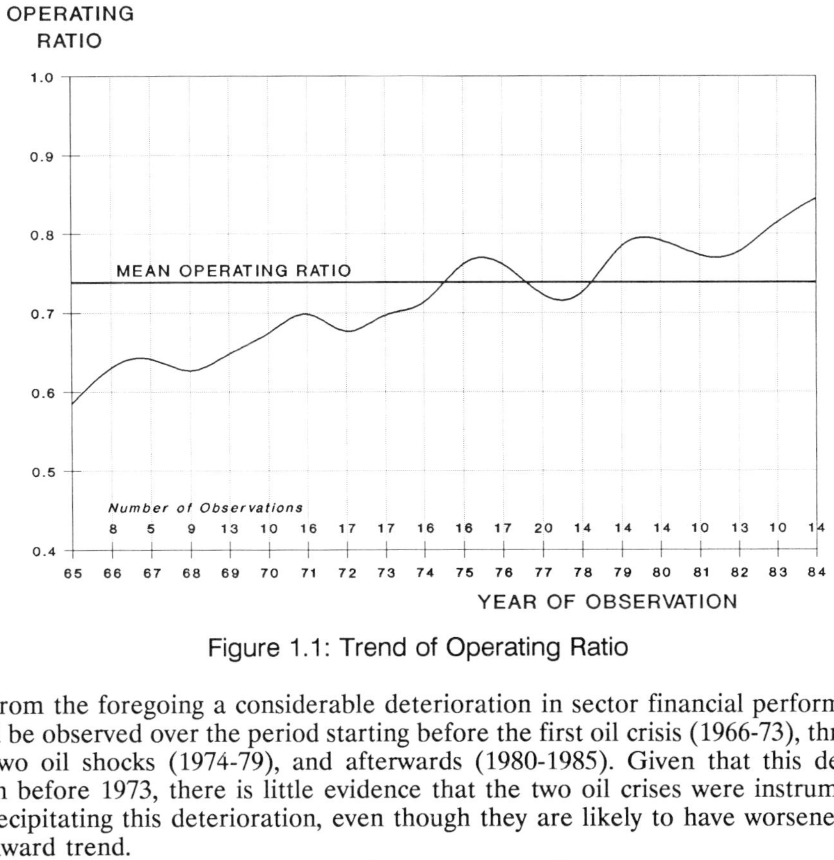

Figure 1.1: Trend of Operating Ratio

From the foregoing a considerable deterioration in sector financial performance could be observed over the period starting before the first oil crisis (1966-73), through the two oil shocks (1974-79), and afterwards (1980-1985). Given that this decline began before 1973, there is little evidence that the two oil crises were instrumental in precipitating this deterioration, even though they are likely to have worsened the downward trend.

A widely prevalent phenomenon in some developing country power sectors has been that of over-optimism in demand forecasting. Indeed, this phenomenon has significantly contributed to the growing economic burden. Such over-optimism can be partly attributed to the unexpected slowdown following the twin oil shocks of 1973-74 and 1979-80, where the energy price increases and recessionary conditions were more severe than anticipated.

Institutional performance, reviewed on the basis of such indicators as the number of consumers per employee, adequacy of maintenance and general utility efficiency, has stagnated at a relatively unsatisfactory level over the past two decades. Part of this can be attributed to the high growth rates in demand, frequently calling for a doubling of the plant in service in seven to ten years. The requirements for maintenance and efficient operations have increased commensurately and have strained the available management and manpower resources. Furthermore, rapidly changing external conditions, such as foreign exchange shortages and economic volatility, make the managerial task increasingly difficult.

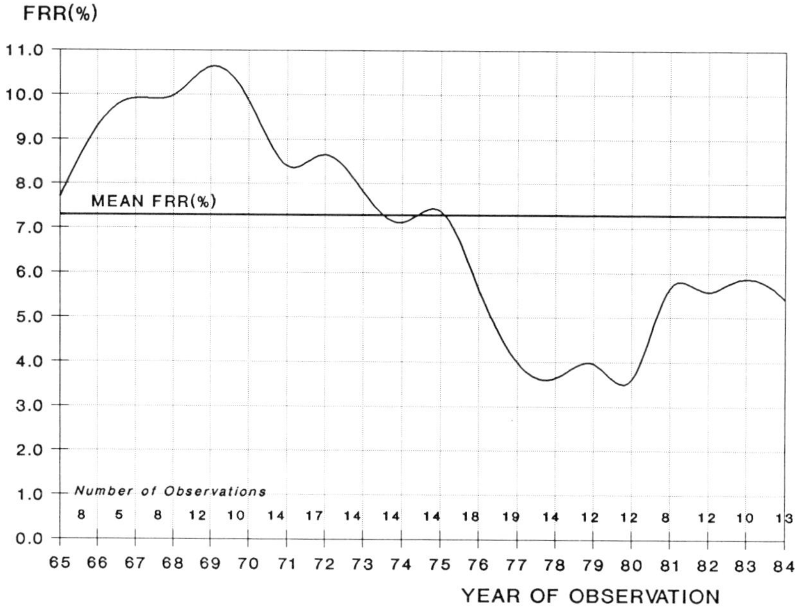

Figure 1.2: Trend of Rate of Return on Assets

The sociopolitical pressures to increase public sector employment are reflected in the number of consumers per employee. Excluding bulk users, a sample of third world utilities has shown that about two thirds have fewer than 100 connections per employee. This is a low figure, even in the context of an unskilled labour-surplus economy, especially since many utility workers are skilled and practically all should be at least semi-skilled. Only a small fraction (10%) of power utilities had more than 150 consumers per employee. Adequacy of maintenance is more difficult to measure quantitatively. Survey based qualitative assessments of maintenance practices indicate that almost half of the utilities surveyed fell into the poor to very poor category. Qualitative assessment of general utility efficiency has indicated similarly weak results.

1.2 Future Prospects and Critical Issues

Future Prospects

The need for greater productive efficiency in the power sector arises from the investment burden, especially in the case of countries with serious debt problems (e.g., major Latin American countries like Argentina, Brazil and Mexico). Table 1.1 presents a cross section of recent energy sector investments in developing countries, with electric power accounting for about three quarters of the total. On average, developing country energy investments still account for about 25 percent of total public capital expenditures. At the same time, oil importers are spending an average of 15-20 percent of export earnings on petroleum imports, and serious fuelwood shortages and deforestation problems continue unabated. High power and energy related expenditures has contributed significantly to the indebtedness of many countries (especially in Latin America) during the last decade, with correspondingly grave macroeconomic consequences. Figure 1.3 indicates how developing country debt as a fraction of GNP has risen alarmingly in the 1975-1987 period, from an average of 15% in 1975 to about 35% in 1987. In the highly indebted countries of Latin America, the ratio of debt to GNP had increased to an average of 50% by 1987.

Table 1.1

Estimated Annual Energy Investment as a Percent of Annual Total Public Investment During the Early 1980's

Over 40%	30-40%	20-30%	10-20%	0-10%
Argentina	Equador	Botswana	Benin	Ethiopia
Brazil	India	China	Egypt	
Colombia	Pakistan	Costa Rica	Ghana	
Korea	Philippines	Liberia	Jamaica	
Mexico	Turkey	Nepal	Morocco	
			Nigeria	
			Sudan	

Source: World Bank data.

Despite some anomalies, the link between energy demand and GDP is well established. Electric power, in particular, has a vital role to play in the development process, with future prospects for economic growth being closely linked to the provision of adequate and reliable power supplies. Figure 1.4 is an indicative example of this relationship between electricity use and income in a sample of 14 developing countries. A more systematic analysis of World Bank and UN data over the last two or three decades indicates that the ratio of percent growth rates (or elasticity) of power system capacity to GDP is about 1.4, in the developing countries.

The World Bank's most recent projections of economic growth during the period 1988-95 are in the range of 3.7 to 4.6% annually (World Bank 1989). Assuming an annual economic growth rate of 4.2% for the next decade (based on a medium scenario), and a value of 1.4 for the income elasticity of electricity demand, the installed power capacity should grow at about 6.4% annually. This implies that the developing countries will require additional electric system capacity of around 430 gigawatts (GW) during the next decade -- starting from a base of about 560 GW in 1988. This is a substantial increase in capacity, by any standard.

The power sector investment needs corresponding to these indicative projections are also very large. Based on World Bank data, the average cost of installed generation, transmission and distribution capacity is approximately $1900 per kW (in constant 1988 dollars). Thus even under the medium economic growth scenario, the required investment over the period 1989-98 would be around $820 billion or an average of over $80 billion annually. In comparison, the present annual rate of investment in developing countries is only around $50 billion, a rate which is already proving difficult to maintain. In order to avoid this impending crisis, there is an urgent need for the developing countries to explore and implement effective demand and supply side options -- based on fundamental principles of economic and technological efficiency.

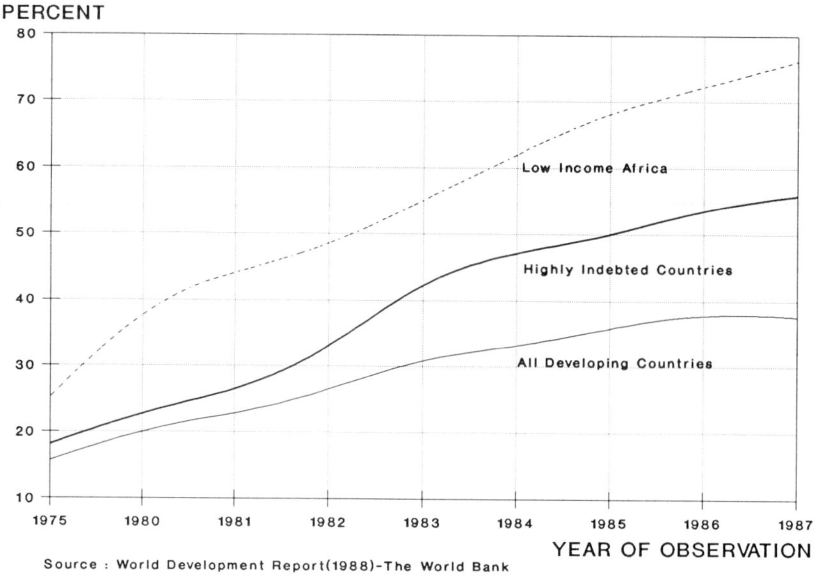

Figure 1.3: Debt in LDC's 1975-1987

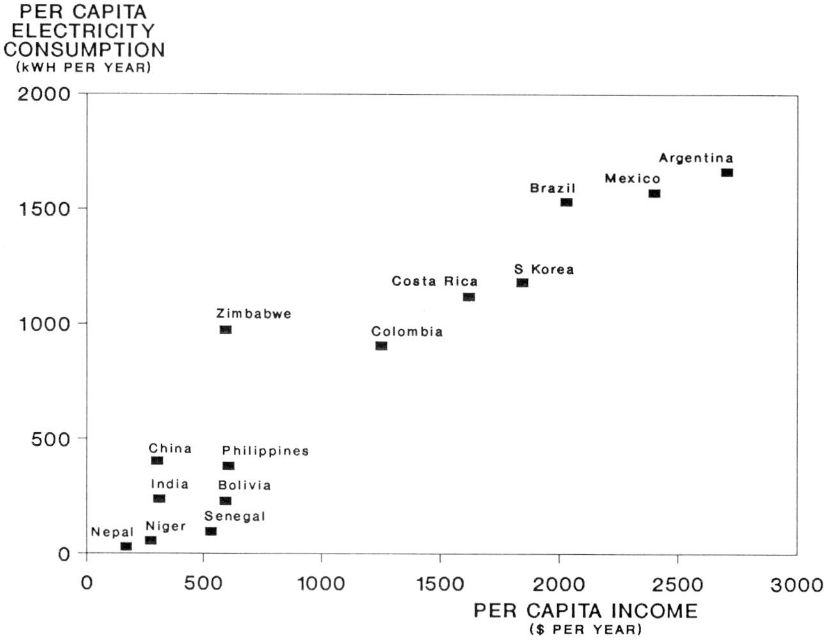

Figure 1.4: Electricity Intensity in LDC's (1982)

Issues and Constraints

Although the criteria underlying government policy in the power sector rarely have been spelled out explicitly, several of them nevertheless are discernible. First, most developing countries feel that power is an engine for growth and modernization that should be sustained through centrally directed investments and activity. Second, electric utilities are considered a tool for addressing social equity and employment issues and improving the quality of life. Third, the sector is sometimes perceived as a vehicle for raising resources and taxing away surpluses (although significant implementation problems are acknowledged).

While the availability of electric power has brought many benefits, the scarcity of resources underscores the need to focus attention on sector weaknesses and improve the efficiency of resource use. In some cases the failure of the existing arrangements between national governments, power producers and consumers is not due to shortcomings of the policy criteria themselves, but to the imprecise interpretation and application of these objectives. Thus, the desire for growth and modernization mentioned above has resulted in unquestioned funding of power needs, continuous central government subsidies to the sector, nonoptimal or unbalanced investments, and lack of productive efficiency and incentives to maintain technical and financial discipline. Social equity and employment objectives have led to excessive subsidies to consumers, inefficient pricing, and inadequate resource mobilization. Attempts to tax sector surpluses have created highly skewed price structures, cross subsidies, and incorrect price signals. Altogether, this environment has encouraged a casual attitude towards official interference, resulting in excessive and counterproductive government intervention in nearly every aspect of utility activities.

The foregoing considerations and desperate circumstances of many developing country power utilities have generated pressures for new approaches. In particular, there appears to be considerable interest in the scope for more decentralization and greater private participation, as one means of improving power utility performance and relieving developing country governments of the crippling economic burden of financing the chronic deficits of these state-owned enterprises. The history of the utility sector in industrialized countries indicates periods of fierce internal debate and introspection, followed by significant policy reforms to adapt to changing external circumstances. The rapidly evolving environment faced by developing country utilities today suggests that it would be both appropriate and healthy to carry out such self-examination, rather than adhering rigidly to an outdated framework.

Enhanced private participation in the power sector is likely to be more successful when it is one element in a broader economic package involving policy reforms in the trading, industrial and financial sectors, as well as deregulation and liberalization of the electric power business (see Section 1.4 for details). Such private participation has to do with the infusion of private capital and the transfer of ownership to private hands as well as with the introduction of private sector management methods in order to improve efficiency of power supply enterprises. However, it is naive to expect either the transfer of ownership or management or both to private hands, to solve all the problems of the power sector. Thus, in many countries it has become one of several methods aimed at restructuring public sector enterprises.

A new outlook for the power sector should take several key factors into account. First, conditions within the utility must provide incentives for technical, financial and managerial efficiency. Remedial measures must address a number of problems that have plagued these institutions, including: weak planning, inefficient operation and inadequate maintenance, high technical and non-technical losses, low quality of supply and frequent power failures, inability to raise prices to meet revenue requirements, poor management, excessive staffing and low salaries, poor staff morale and performance, undue government interference, and so on. For many utilities these problems have persisted over time despite efforts to identify and correct the causes through consultant studies and institutional development programs.

Second, the national framework (within which the utility functions), might be restructured to improve performance. Key aspects to be considered include clear-cut government policy guidelines for utility management, delegation of authority to implement agreed policies and corresponding accountability, a rational regulatory framework, government non-interference in daily sector activities, and reforms in the financial climate and access to capital.

Third, dealing with exogenous factors outside the policymakers' control requires that decisions be made with due allowance for uncertainty. Thus, unpredictable changes in environmental constraints, demand forecasts, inflation, exchange rates, interest rates, and fuel prices, require a new mindset and scenario-oriented approach to decisionmaking, that is quite different from the deterministic methods that were more useful in the past. It also calls for a greater emphasis on flexibility in the investment program and the use of technologies that can adapt with reasonable speed to the changing environment. Planners and decisionmakers would also need to consider strategies of hedging the risk brought about by volatility in international markets.

Finally, it is becoming increasingly evident that effective policy must be determined by using a holistic framework that fully accounts for key macroeconomic and intersectoral linkages as well as energy-environmental interactions -- an approach that is more comprehensive than the narrower, intrasector analysis used earlier.

As indicated above, the current situation in the LDC power sector will be difficult to sustain in the long run and continued sector growth in many countries will hasten the deterioration rather than lead to improvements. Unless there is an improvement in the quality of service to accompany the expansion of service to new customers, there will be increasing difficulties in persuading customers to pay higher tariffs. Governments are very sensitive to the objections of customers (generally the politically influential urban minorities) who resist paying higher prices for poor quality service which shows little sign of improvement. Inability to generate higher revenues result in the underfunding of all activities and deterioration of service levels, thus reinforcing the vicious cycle.

It seems apparent from the above that there is a strong argument in favor of putting greater emphasis on efficiency and restructuring issues rather than concentrating solely on continued power sector expansion. More attention needs to be given to improving the quality of service and reducing losses through rehabilitation and reinforcement of power systems. Given the present state of many power utilities, it may not be possible to achieve rehabilitation and service improvements simultaneously with adequate system expansion. Furthermore, in many countries there is a backlog of maintenance requirements in addition to the need to improve operations. Finally, environmental protection measures also may increase the future costs of power supply (Tietenberg 1988). These needs already place a burden on institutions which do not have sufficient qualified manpower to meet the requirements adequately for both expanding the system and operating existing systems. Given such constraints, in some cases this approach may well lead to a slower expansion of supply to new customers than would otherwise be planned by governments and utilities.

1.3 Analytical Framework

Since it is crucial to relieve developing country governments of the crippling burden of financing the chronic deficits of state owned electric enterprises, this has increased pressures for power sector reform. It has also highlighted the need to study power-energy- macroeconomic linkages more thoroughly, because effective national power and energy strategies cannot be developed in isolation.

Advantages of an Integrated Approach

A better understanding of economy-wide linkages is useful, whatever the prevailing political system. It will help decisionmakers in formulating policies and providing market signals and information to economic agents that encourage more efficient energy production and use. We summarize below a hierarchical framework for integrated national energy planning, policy analysis and supply-demand management, that also identifies the role of the power subsector in a broader context (see Figure 2.1 in Chapter 2)[2].

At the highest and most aggregate level, it must be clearly recognized that the energy sector is part of the whole economy. Therefore, energy planning requires analysis of the links between the energy sector and the rest of the economy. Such links include the energy needs of user sectors, input requirements of the energy sector, and impact on the economy of policies concerning energy prices and availability. The second level of the integrated approach treats the energy sector as a separate entity composed of sub-sectors such as electricity, petroleum products and so on. This permits detailed analysis, with special emphasis on interactions among the different energy sub-sectors, substitution possibilities, and the resolution of any resulting policy conflicts. The third and most disaggregate level pertains to analysis within each of the energy subsectors. It is at this lowest hierarchical level that most of the detailed energy resource evaluation, planning and implementation of projects is carried out by line institutions (both public and private). In practice, however, the three levels of INEP merge and overlap considerably. Energy-environmental linkages also should be incorporated into the analysis, as far as possible[3].

This integrated framework facilitates policymaking and does not imply rigid centralized planning. Thus, such a process should result in the development of a flexible and constantly updated energy strategy designed to meet national goals. This national energy strategy (of which a power investment program and pricing policy are important elements), may be implemented through a set of energy supply and demand management policies and programs that make effective use of decentralized market forces and incentives.

To achieve the desired national goals the policy instruments available to third world governments, for optimal power and energy management include: (a) physical controls; (b) technical methods; (c) direct investments or investment-inducing policies; (d) education and promotion; and (e) pricing, taxes, subsidies and other financial incentives. Since these tools are interrelated, their use should be closely coordinated

---

[2] See the companion volume: *Energy Analysis and Policy*, Chapters 2 and 4, for a detailed discussion of the Integrated National Energy Planning (INEP) concept and its application.

[3] See the companion volume: *Energy Analysis and Policy*, Chapter 1, for a more detailed discussion of energy use and environmental impacts.

for maximum effect. In view of their importance, we pay particular attention below, to investment and pricing policy.

The chief constraints that limit effective policy formulation and implementation are: (a) poor institutional framework and inadequate incentives for efficient management; (b) insufficient manpower and other resources; (c) weak analytical tools; (d) inadequate policy instruments; and (e) other constraints such as low incomes and market distortions.

Next, we examine specific power subsector issues, within the wider integrated framework. It is convenient to begin by recalling that economic efficiency requires both: (a) efficient production of electricity, by ensuring least-cost supply through optimal long run investment planning as well as short run power system operation and management; and (b) efficient electricity use, by providing price signals that ensure optimal consumption patterns and resource allocation.

Efficient Supply and End Use

Efficiency in electricity supply requires that the power system is optimally planned and operated. This is important, even from a practical point of view, since an inefficient power utility that is routinely permitted to pass on excessively high costs to consumers, under the umbrella of marginal cost pricing (as described in the next section), will have very little incentive to reduce costs and produce more efficiently. There are also solid analytical reasons for insisting on supply efficiency as a prerequisite to LRMC based pricing. We begin by studying optimal investment planning below, and then discuss operating efficiency later.

Optimal Reliability and Traditional Least-Cost Planning

While the close relationship between optimal investment and pricing policies have been generally recognized for some time (Boiteux 1949), theoretical work in recent decades has focussed on the optimal conditions for price and capacity levels that must be simultaneously satisfied to maximize the net social benefits of electricity consumption. In this context, determining the optimal capacity level is equivalent to establishing the optimal level of reliability, since capacity additions do improve reliability.

The complex analysis underlying the joint optimality conditions may be summarized simply as follows. The optimal price is the marginal cost of supply. Simultaneously, the optimal reliability (capacity) level is defined by the point at which the marginal cost of increasing reliability is exactly equal to the corresponding marginal gain to consumers by reduction of power outages, poor voltage and frequency variations. In conventional economic analysis, the short-run marginal cost (SRMC) is the cost of meeting additional electricity consumption, with capacity fixed, while the long-run marginal cost (LRMC) is the cost of providing an increase in consumption (sustained indefinitely into the future) in a situation where optimal capacity adjustments are possible. When the system is optimally planned and operated (ie., capacity and reliability are optimal), SRMC and LRMC coincide. However, if the system plan is sub-optimal, significant deviations between SRMC and LRMC will have to be resolved within the pricing policy framework. Finally, if there are substantial outage costs outside the peak period, then the optimal marginal capacity costs may be allocated among the different rating periods (ie., peak, intermediate and off-peak) in proportion to the corresponding marginal outage costs.

Electricity tariffs are often not readily subject to change (except under special conditions, e.g., when spot pricing is feasible), and simultaneous optimization of price and capacity is a theoretical ideal. Thus from a practical point of view, it may be appropriate (or necessary) to uncouple the joint price and reliability optimization conditions by attempting to "optimize" reliability in the presence of fixed or given tariffs, at least on the first round. Once the optimal reliability and investment levels (subject to given prices as defined above) are determined, tariffs can be revised to reflect any changes in the marginal cost of supplying electricity implied by the new reliability level. Using this new level of tariffs and resulting demand, reliability and capacity can be reoptimized iteratively. We note that application of the marginal cost pricing rule has been attempted in several countries (most notably France), and while interpretations vary among practitioners, the approach has gained wide acceptability. The optimal reliability rule is more difficult to apply. This is mainly because shortage costs are not easy to estimate, although considerable progress in this area has been made in recent years.

It must be emphasized that the reliability (or supply quality) optimization model strengthens rather than replaces the conventional least-cost criterion. In the conventional approach to power system design and planning, costs are minimized subject to supplying the load at some (arbitrarily) given standard of supply. With the new approach, it is possible to determine an optimal reliability level, by using a social cost-benefit approach (based on a national rather than a utility specific viewpoint), to evaluate the inherent trade-off between the increase in power system supply costs required to achieve a higher level of reliability, and the corresponding decline in consumer shortage costs. It would be possible then to design the power system to meet the forecast load, subject to the new reliability requirement, using the traditional least-cost planning techniques. This would permit the application of existing sophisticated least cost system planning models.

We will wrap up the above discussion with a simplified static presentation of the reliability optimization model (see Chapter 2 for a more detailed dynamic version which includes reliability expectation and intertemporal effects). Consider a simple expression for the net benefits (NB) of electricity consumption:

$$NB(D,R) = TB(D) - SC(D,R) - OC(D,R)$$

where:

TB = total benefits of power consumption if there were no outage.
SC = supply costs (ie. system costs).
OC = outage costs (ie. costs to consumers of supply shortages).
D = demand
R = reliability.

In the traditional approach to system planning, both D and R are exogenously fixed, and therefore NB is maximized when SC is minimized (which is least cost system expansion planning). However, if R is treated as variable:

$$\frac{d(NB)}{dR} = \frac{\partial(SC+OC)}{\partial R} + \frac{\partial(TB-SC-OC)}{\partial D} \cdot \frac{\partial D}{\partial R} = 0$$

is the necessary first order maximization condition.

Assuming $\partial D/\partial R = 0$ yields: $\partial(SC)/\partial R = -\partial(OC)/\partial R$

Therefore, reliability should be increased by adding to capacity until the above condition is satisfied. An alternative way of expressing this result is that since TB is independent of R, NB is maximized when total costs: TC = (SC + OC) are minimized. The above criterion is one which effectively subsumes the traditional system planning rule of minimizing only the system costs.

Recognition that reliability should not be a fixed parameter in power systems planning, but needs to be determined endogenously on the basis of consumer outage costs provides significant potential for enhancing economic efficiency in power systems. The recent literature has made significant strides in understanding and addressing these issues (see Chapter 7). One of the earliest attempts to use outage costs in optimizing the planning of a distribution network was the case of Cascavel, Brazil (Munasinghe 1979). Several utilities in both developed and developing countries have adopted this approach, including Sweden, Norway, France, Italy, Brazil and Columbia.

Based on the fact that consumer outage costs vary significantly, recent proposals for rate structures have considered the possibility of unbundling the quality attributes of electric power supply, thus allowing the customer to select between different levels of quality (*The Energy Journal* 1988). The primary determinant of quality in this case would be reliability of the power supply. Thus, in developing countries, power utilities could use these concepts to offer "priority service" -- which provides supply quality to different consumers according to their willingness to pay -- thereby efficiently allocating the scarce supply. These ideas are related to the "spot pricing" approach discussed in the next section.

In recent years, a great deal of attention has been focused on the energy-environmental linkage. The process of extracting energy resources and converting them into end-use energy forms (such as electric power), often imposes a heavy environmental burden. In the industrialized countries, the environmental consequences of many decades of dependence on fossil fuels are only now becoming apparent in the form of acid rain, air and water pollution, and the disruptive impacts of the growing carbon dioxide concentrations in the earth's atmosphere (Tietenberg 1988, OECD 1985).

Thus, explicit consideration of environmental effects is now well within the purview of the power sector planner, who can influence environmental impacts by choice of supply technology, fuel type, and mitigating measures adopted. Such considerations become particularly urgent in light of the future projections for developing country power sector growth discussed in Section 1.2. If these forecasts hold, total emissions from developing countries (although still a fraction of developed country emissions) could almost double over the next 10 years, unless state-of-the-art technologies are used to limit them (USAID 1988). The above analytical framework for investment choice and policy may be readily modified to take account of environmental effects, by incorporating an environmental cost term in the expression for the net benefits of electricity consumption. Then, the decision rule for maximizing net benefit would be the minimization of total costs -- now consisting of the system and outage costs together with the environmental cost term. Thus, one of the main challenges facing power system analysts is the more accurate identification and quantitative estimation of such environmental costs (and benefits). The use of appropriate regulatory measures and financial tools such as pollution taxes will also play an increasingly important role in the future.

The above analysis has been discussed in a deterministic framework. It is worth reiterating that the present decisionmaking climate is characterized by high levels of uncertainty with regard to trade and economic conditions, energy prices, future demand, interest rates, environmental constraints, and technological change. Consequently, there is much greater scope in developing countries to use models which identify risk-diversifying energy policy options that are robust over a wide range of exogenous scenarios, rather than cost minimizing, deterministic but risky solutions that were more appropriate for the narrower band of eventualities which existed in the past. Methods for dealing with uncertainty in the power sector include sensitivity analysis and Monte-Carlo or probabilistic simulation. These methods are aimed at providing the planner with some insight into possible future outcomes, so that his plans can be designed to cope with a broad range of scenarios.

Optimal Power System Operation and End Use Efficiency

Beyond the need to ensure optimal investment planning (as discussed above), supply efficiency also requires the optimal operation of power systems -- implying that plant performance, operating and maintenance procedures, loss levels, etc. are optimized. Efficient supply management and loss reduction (see Chapters 2 and 11 for details), together with efficient electricity end use, load management and pricing (described in the next section), constitute an attractive package of policy options for most power utilities in both the developing and developed countries.

As one indicative example of the state-of-the-art in supply and demand management possibilities, we summarize the results of a recent Swedish study (Bodlund et al. 1989). The power sector in Sweden, currently supplied in equal shares by hydro and nuclear generation, faces the following severe restrictions: (a) hydro expansion limited by environmental constraints; (b) mandatory phasing out of all nuclear units by 2010; and (c) no increase permitted above present CO_2 emission levels. The demand for electricity derived services is projected to increase by 50%, from 1987 to 2010. If end use efficiency remains unchanged, then under this "frozen efficiency" scenario, the electrical load would also increase by 50%, from 129 TWh in 1987 to 195 TWh in 2010, at an average annual growth rate of 1.8%.

The *same* output of electrical services could be provided, but with steadily declining electricity input needs and load levels -- based on the increasingly energy efficient scenarios A, B, C and D, shown in Figure 1.5. The corresponding loads in 2010 would be 140, 111, 96 and 88 TWh, respectively. Only options C and D permit the load to be met after all the nuclear plants are retired in 2010. Figure 1.6 indicates that the total costs of energy supply also fall progressively under the scenarios A through C (some of the costs are undefined for scenario D). In addition, there are three supply scenarios based on different selection rules for generation -- the supply costs rise steadily as we move through the economic despatch, natural gas/biomass, and environmental despatch options. These costs exclude taxes and subsidies, and are based on a 6% discount rate, 1987 world oil and coal prices, and coal equivalent gas prices for steam power generation.

Pricing Policy

In the past, electric power pricing policy in most countries was determined mainly on the basis of financial or accounting criteria, for example, raising sufficient sales revenues to meet operating expenses and debt service requirements, while providing a reasonable contribution towards capital needed for future power system expansion.

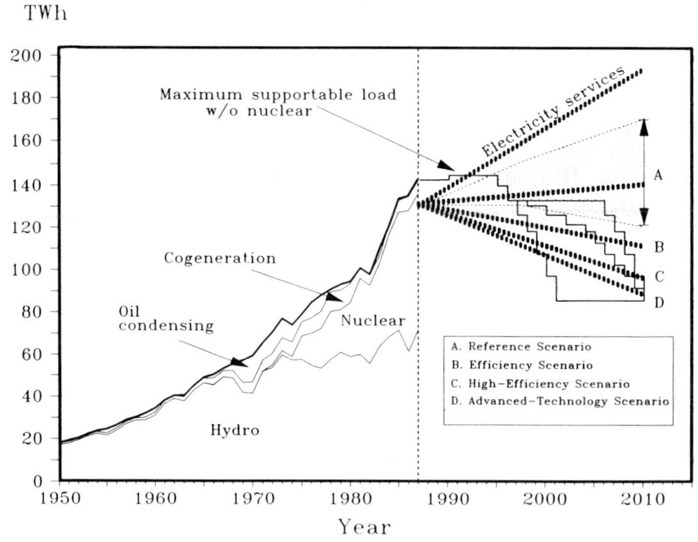

Figure 1.5: Generation Capability and Load Under Various Scenarios

Notes:
1. Electricity services = end use efficiencies frozen at 1987 levels.
2. Scenario A = normal penetration of energy efficient end use technologies (vertical arrows show range of uncertainty).
3. Scenario B = high penetration of energy efficient end use technologies that are cost effective and commercialised.
4. Scenario C = same as scenario B, but includes uncommercialised newly developed technologies.
5. Scenario D = same as scenario C, but includes advanced technologies still in the R&D stage.
6. Downward stepped curves indicate generation capability with different options for phasing out 12 nuclear plants.

Source: Bodlund et al (1989)

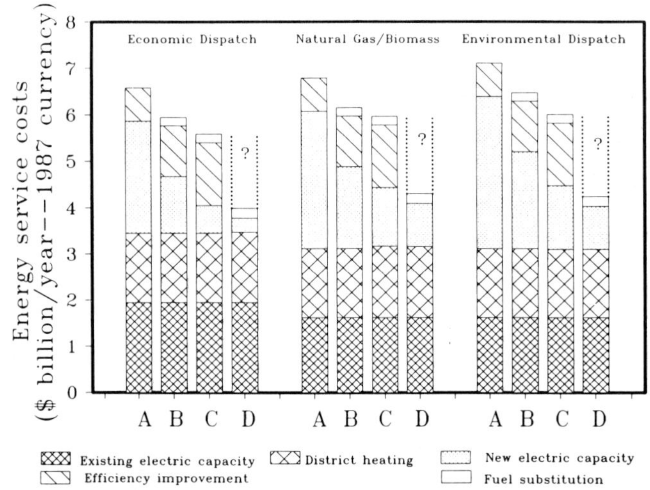

Figure 1.6: Total Energy Supply Costs Under Various Scenarios

Notes:
1. Scenarios A to D are the same as in Figure 1.5, but efficiency improvement costs are unknown for scenario D.
2. Economic despatch = traditional least cost generation expansion and operation.
3. Natural gas/biomass = intensive use of gas and biomass, with coal use banned (to limit CO_2 emissions).
4. Environmental despatch = generation expansion and operation in order of increasing CO_2 emissions per kWh produced.

Source: Bodlund et al (1989)

However, there has recently been increasing emphasis on the use of economic principles in order to produce and consume electric power efficiently, while conserving scarce resources, especially in the developing countries. In particular, a great deal of attention has been paid to the use of marginal cost pricing policies.

A comprehensive approach to power pricing recognizes the existence of several objectives or criteria, not all of which are mutually consistent. First, national economic resources must be allocated efficiently not only among different sectors of the economy, but within the electric power sector itself. Second, certain principles relating to fairness and equity must be satisfied, including: (a) fair allocation of costs among consumers according to the burdens they impose on the system, (b) assurance of a reasonable degree of price stability, and (c) provision of a minimum level of service to low income consumers. Third, power prices should help raise sufficient revenues to meet the financial requirements of the sector. Fourth, the power tariff structure must be simple enough to facilitate metering and billing customers. Finally, other economic and political requirements also exist, such as subsidized electricity supply to certain sectors or geographic areas.

Since the above criteria are often in conflict with one another, it is necessary to accept certain trade-offs between them. The marginal cost concept and the relationship between LRMC and SRMC were summarized earlier. The marginal cost approach to price-setting has both the analytical rigor and inherent flexibility to provide a tariff structure that is responsive to these basic objectives. In the first stage of calculating marginal costs, the economic (first-best) efficiency objectives of tariff-setting are satisfied because the method of calculation is based on future economic resource costs (rather than sunk costs) and also incorporates economic considerations such as shadow prices and externalities. The structuring of marginal costs permits an efficient and fair allocation of the tariff burden on consumers. In the second stage of developing marginal cost based tariffs, deviations from strict marginal costs are considered in order to meet important financial and other social, economic (second- best), and political criteria. This second stage of adjusting strict marginal costs is generally as important as the first-stage calculation, especially in the developing country context.

The basic theory of marginal cost pricing may be exposited by the simple supply-demand diagram shown in Figure 1.7. Let DD' be the demand curve (which determines the kWh of electricity demanded per year, at any given average price level), while SS' is the supply curve (represented by the marginal cost MC of supplying additional units of output).

At the price P, and demand Q, the total benefit of consumption is represented by the consumers willingness to pay, ie. the area under the demand curve DAQO. The cost of supplying the output is the area under supply curve SBQO. Therefore, the net benefit, or total benefit minus supply cost, is given by area DABS. Clearly, the maximum net benefit DCS is achieved when price equals marginal cost at the optimal market clearing point C, ie. (P_o, Q_o).

By formulating the problem algebraically, it is possible to derive the more sophisticated result involving the rule for economically efficient pricing during periods of peak demand that infringe on the power system capacity limit (see Chapter 3 for a more detailed treatment). In mathematical terms, the net benefit of electricity use is given by:

$$NB = \int_0^Q [P(q) - MC(q)].dq$$

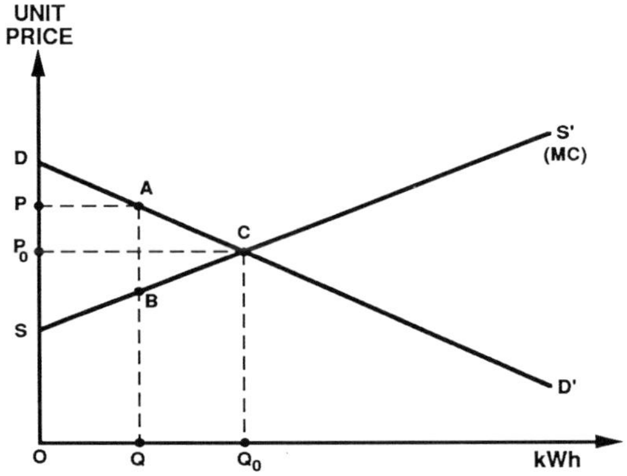

Figure 1.7: Supply and Demand for Electricity

where P(q) and MC(q) are the equations of the demand and supply (marginal cost) curves respectively. Maximizing NB subject to the capacity constraint $Q < \underline{Q}$ (where $\underline{Q}$ is the capacity limit), is formally equivalent to unconstrained maximization of the Lagrangian:

$$L = NB - m(Q - \underline{Q})$$

The first order necessary condition is:

$$\partial L/\partial Q = P - MC - m = 0$$

MC may be interpreted as the short-run marginal cost of supplying more output with capacity fixed, while m has the usual meaning of the marginal cost of adding to capacity. Therefore, during off-peak periods, $P = MC$ (since $Q<\underline{Q}$ and $m=0$). Similarly, during peak periods, $P = MC + m$ (since $Q=\underline{Q}$ and $m>0$).

The marginal cost approach provides an explicit framework for analyzing system costs and setting tariffs. If departures from the strict marginal costs are required for non-economic reasons, then the economic efficiency cost of these deviations may be estimated, even on a rough basis, by comparing the impact of the modified tariff structure relative to (benchmark) strict marginal costs. Furthermore, since the cost structure may be studied in considerable detail during the marginal cost calculations, this analysis helps pinpoint weaknesses and inefficiencies in the various parts of the power system, such as over-investment, unbalanced investment, or excessive losses at the generation, transmission, and distribution levels in different geographic areas. This aspect is particularly useful in improving system-expansion planning.

Any marginal cost based tariff is a compromise among many different objectives. Therefore, there is no "ideal" tariff. By using the marginal cost approach, it is possible to revise and improve the tariff on a consistent and ongoing basis and thereby approach the optimal price over a period of several years, without subjecting long-standing consumers to "unfair" shocks in the form of large abrupt price changes.

Recent advances in low cost metering and switching equipment have made it possible to consider (where appropriate), more sophisticated approaches to supply-demand balancing such as spot pricing and load control. Spot pricing of electric power is an effort to exploit the full potential of the marginal cost pricing framework by the use of modern communication and computation capabilities (Schweppe et al. 1988, Berrie 1987). This scheme calls for the variation of electricity price at frequent intervals in real time (for instance, each hour) in keeping with the marginal cost changes in the system. These prices are communicated to the consumer who is then expected to respond by load adjustment. While the implementation of spot pricing schemes entails an added cost in communication, metering and computation, the potential savings to both the power producer and the consumer can often be significant. Producers benefit by achieving some of their demand management objectives such as peak shaving and load shifting. Consumers benefit by being able to select service levels according to their individual needs and by a reduction in total cost.

There may be significant potential for the application of the concepts of spot pricing the more sophisticated in developing countries, in which utilities are subject to high variations in supply availability. Spot pricing is an efficient means of rationing power supplies. Also, the rapid growth of power systems in developing countries provides opportunities for incorporating system design innovations and new approaches to producer-consumer interaction.

1.4 Options for Improving Power Sector Performance

Enterprise Efficiency - Scope for Greater Autonomy and Management Reforms

Although many difficulties have plagued developing country power utilities, probably the most pervasive has been excessive government interference in organizational and operational matters. Such interference has adversely affected least-cost procurement and investment decisions, hampered attempts to raise prices to efficient levels, mandated low salaries tied to civil service levels, and promoted excessive staffing. This in turn has resulted in inadequate management, the loss of experienced staff due to uncompetitive employment conditions and poor job satisfaction, weak planning and demand forecasting, inefficient operation and maintenance, high losses, and poor financial monitoring, controls and revenue collection.

In order to address these difficulties, an important principle must be recognised - that given the complexity of energy problems and the scarcity of resources and managerial talent in developing countries, each set of issues should be dealt with by that level of decisionmaking and management best suited to analyzing the difficulty and implementing the solution. This hierarchical approach corresponds closely to the INEP concept summarized earlier in this paper. Thus, political decisionmakers, senior government officials and ministry level staff should focus on critical macroeconomic and energy sector strategy and policy, in order to determine global expectations of power utility performance. The senior management of a power company appropriately guided and buffered by an independent board of directors, would then conduct their daily operations free from government interference, to meet the overall national policy objectives and targets within regulatory guidelines. As far as possible, the utility management should be assured of continuity at the top, even in the face of political changes. While the enterprise is provided wider autonomy, it would now become more accountable in terms of performance measured against an agreed set of specific objectives and monitored indicators. The increased responsibility of senior management of the power enterprise would require more regular consultation with government and especially consumer representatives, to discuss performance problems and successes. This rarely occurs in many developing countries. Major changes in enterprise management may be required, to mirror changes in the utility's external environment discussed above. The enterprise's organizational structure and procedures may be inadequate. Once again, the fundamental principle that will help to address these problems is delegation of authority. In many developing country power utilities, the senior management attempts to deal with all problems, and trivial issues often get more attention than critical ones.

Provided that middle managers could be adequately trained, senior managers could (by appropriate delegation of tasks) free themselves to deal with higher level policy. The middle management would account for their performance through an agreed set of performance indicators, while obtaining greater responsibility and latitude to make decisions. This process would then be repeated down to the lowest working levels. Obviously, staff training and education at all levels and stages of career would play a critical role in ensuring the success of such an approach.

Restructuring, Private Participation and Decentralization Options

The natural monopoly characteristics of some power enterprise functions and willingness to manipulate these enterprises for general policy purposes, are in many countries accepted as sufficient reasons for maintaining large centralized public sector organisations. Nevertheless, given the observed problems inherent in stimulating management of developing country state enterprises to be cost conscious, innovative, and responsive to consumer needs, there may be a need for more fundamental change. It could be worthwhile to trade off some of the perceived economies of scale in energy enterprises for other organisational structures which provide greater built-in incentives for management efficiency and responsiveness to consumers. In particular, varying the forms of ownership and regulation in the power sector should be considered.

Options for private and cooperative ownership of energy enterprises could include both local and foreign participation as well as joint ventures. As long as a given regulatory framework prevails, it can be argued that the form of ownership (private and public) would not by itself affect operating efficiency. The main point is that, to the extent possible, the introduction of competitive market forces should be encouraged. Governments may wish to divest themselves of either all or part of some government owned enterprises, or organisational structures. Thus, they might provide an environment in which governments, enterprise managements, and energy consumers are all better off. A first step towards decentralization could be for government-owned power enterprises to competitively contract out activities or functions better handled by others. Many companies already subcontract various construction related activities. Some portions of the billing and collection process, or routine maintenance, can also be subcontracted. Among the advantages of such arrangements have been lower costs and greater programming flexibility.

While several potential opportunities for private participation in the LDC power sector have been identified and are being developed around the world, these have not been without significant impediments (Sullivan 1988). Frequently, these projects are hindered at the outset by a policy and institutional framework that discourages private sector involvement. Since the power sector occupies a central position among publicly owned enterprises, changing this framework in a manner conducive to private participation may prove to be extremely difficult. Furthermore, LDC utilities often operate in a climate which is economically weak and politically unstable. This increases the risk of expropriation and frequent disruptions. Furthermore, even in circumstances where the private enterprise was initially encouraged and approved, the likelihood of subsequent adverse regulatory changes and other contractual breaches would be greatly enhanced. Finally, the LDC power sector is characterized by extraordinary technical uncertainties such as fuel shortages, lack of skilled personnel and a weak supporting infrastructure.

There are also opportunities for decentralization on a spatial basis. For example, larger countries may have independent regional power grids. Power distribution companies could be separate by municipality, with perhaps limited overlap in some fringe franchise areas, and have the right to purchase from various suppliers, when feasible. If private participation were allowed, one advantage might be that at least the large power consumers could also be legitimate shareholders who would be concerned not only with service efficiency but also with the financial viability of the company.

Power generation also has potential for efficiency improvements through divestiture. While the bulk power transmission and distribution functions might be regarded as having more natural monopoly common carrier type characteristics, this is not so with generation. In fact, there is substantial scope for competition in power generation with independent (perhaps foreign-owned enclave) producers selling to a central grid, as in the case of large industrial cogeneration. For example, in the US the Public Utilities Regulatory Policies Act of 1978 (PURPA) specifically encourages small privately owned suppliers to generate electricity in various ways for sale to the public grid. As a result there are now a large number of small companies producing electricity presumably at costs below those incurred by traditional large utilities. Similar laws have been passed elsewhere and are beginning to have an impact.

In fact, with appropriate legislation, innovative contractual arrangements and the strengthening of utility capability to manage such new schemes, greater scope for cogeneration and free-standing generation might be encouraged. Larger countries might find small entrepreneurs ready to invest in small hydro or similar generation facilities. The advantage to the power company would be a deemphasis on large lumpy capital-intensive projects together with the fact that the cogeneration and free-standing generation companies would put up all or part of the capital and be paid only out of revenues from power sold at guaranteed prices. For larger enclave generation facilities (perhaps peat, coal or nuclear), the concept would be that a foreign investor will finance and build the plant, operate and maintain it for an agreed period, and be repaid out of power sold at guaranteed prices convertible into foreign currency.

Finally, with the proper institutional structure, developing countries could take advantage of many technological innovations that could lead to increased power sector efficiency. Cogeneration -- ie. the simultaneous production of power and heat -- is an option that is frequently available in developing countries but not fully utilized. On the generation side, several advancements in gas turbines have been developed in recent years with overall conversion efficiencies nearing 50 percent. In energy conversion too, several advances have been reported (such as innovative coal combustion technologies) with superior environmental and technical characteristics. All of these technological options have imminent application potential in developing countries. Some promising new power technologies are described in OTA (1985) and Fulkerson et al. (1988).

Financial Options

Developing countries are turning increasingly to more innovative financing options, most of which have been used in the industrialised countries. Some of the financial instruments that are now being studied in third world nations include: (1) non-recourse and limited recourse financing (or project specific financing); (2) leasing of individual pieces of equipment or whole plants, by local or foreign investors; (3) private ownership or operation of generation and distribution facilities; (4) countertrade, involving barter type exchange of specific export goods for energy imports; (5) developing financial instruments to finance local costs, often involving the creation of new financial intermediaries; (6) revenue bonds, with yields tied to enterprise profitability; (7) tax-exempt bonds; and (8) sale of electricity futures that seek more stable longer term electricity price contracts.

The Multilateral Investment Guarantee Agency (MIGA), recently created by the World Bank, could also play a key role. It will seek to promote the flow of international capital to developing countries, by providing guarantees (on a fee basis) against the following non- commercial forms of risk: (1) transfer risk, arising from host government restrictions against convertibility and transfer of foreign exchange; (2) loss risk, resulting from legislative or administrative action (or omission) of the host government that leads to loss of ownership, control or benefits; (3) contract repudiation risk, when the outside investor has no recourse to an adequate forum, faces undue delays, or is unable to enforce a favorable judgement; (4) war and civil disturbance risk.

1.5 Conclusions

Despite gains in providing service to more electricity consumers during the last several decades, there has been a steady deterioration in the performance of many power utilities in the developing countries. Because of the continued growth of future demand for electric power, these companies will face increasing pressures. To summarize, the improvement of the LDC power sector would involve attention being paid to the following areas:

1. Integrated national energy planning and policy analysis - strengthen available analytical tools and practical methodology to study key power sector-energy sector-environmental-macroeconomic links and determine a suitable national strategy based on economic efficiency.
2. Improvement of productive and managerial efficiency, with special reference to maintenance, rehabilitation and distribution network investments, in order to improve losses and the quality of service.
3. Innovative financing and restructuring options, including regulatory reforms, more decentralization and greater private participation.
4. Energy-environmental analysis - develop a framework for power and energy sector policy in relation to environmental and natural resource management based on sound economic principles.
5. Investment planning under uncertain conditions - identify and strengthen analytical tools and approaches to determine robust investment policy decisions in changing external circumstances.
6. Improved demand management including efficient power pricing and end use.

More specific questions to be studied include:

1. Causal links between the external environment of energy sector institutions and their performance.

2. Identification of specific reasons for declining utility performance and their relative importance.
3. Planned versus actual investment programs and reasons for differences.
4. The distribution of investments among generation, transmission and distribution such that each area is funded to maintain optimal system efficiency and balance.
5. Comparison of marginal costs of supply and prices.

References

Berrie, T.W., "Power System Planning under Electricity Spot Pricing", *IEEE Transactions on Power Systems*, Vol.PWRS-2, No.3, August 1987.

Bodlund, B., et al., "Technology Options for the Swedish Power Sector", in Johansson, T.B., et al.(eds), *Electricity*, Lund University Press, Stockholm, Sweden, 1989.

Boiteux, M., "La Tarification des Demandes en Pointe", *Revue Generale de l'Electricite*, vol.58, 1949.

Energy Journal, Special Issue on Electricity Reliability, 1988.

Fulkerson, W., et al., *Energy Technology R&D*, Oak Ridge National Lab., Tenn., USA, 1988.

Munasinghe, M., *The Economics of Power System Reliability and Planning*, Johns Hopkins Univ. Press, Baltimore MD, 1979.

Munasinghe, M., "Practical Application of Integrated National Energy Planning (INEP) Using Microcomputers", *Natural Resources Forum*, vol 10, no.1, Feb 1986, pp. 17-38.

Munasinghe, M., J. Gilling and M. Mason, *A Review of World Bank Lending For Electric Power*, Energy Series Paper No. 2, The World Bank, Wash. DC, March 1988.

Office of Technology Assessment (OTA), *New Electric Power Technologies*, US Congress, Washington DC, July 1985.

Organization for Economic Cooperation and Development (OECD), *Environmental Effects of Electricity Generation*, Paris, 1985.

Schweppe, Fred C., M. Caramanis, R. Tabors and R. Bohn, *Spot Pricing of Electricity*, Kluwer Publishing, New York, 1988.

Sullivan, J.B., "The A.I.D. Experience with Independent Power Generation", paper presented at the Central American and Caribbean Workshop on Electric Power, San Jose, Costa Rica, August 1988.

Tietenberg, T., *Environmental and Natural Resource Economics*, Scott Foresman Co., Glenville, IL, 1988.

U.S.Agency for International Development (USAID), *Power Shortages in Developing Countries*, Washington, DC, March 1988.

The World Bank, *World Development Report*, Wash. DC, August 1989.

Chapter 2

ENGINEERING - ECONOMIC ANALYSIS OF ELECTRIC POWER SYSTEMS[1]

2.1 Introduction and Overview

Today's societies require increasing amounts of energy for residential, industrial, commercial, agricultural and transport uses. These energy needs are met by the commercial energy sources including the short-term, depletable fossil fuel supplies -- petroleum, coal, and natural gas -- as well as the longer run, renewable sources such as hydroelectric, biomass, solar, geothermal, wind, and tidal power (Munasinghe and Schramm 1983).

The rapid increases in energy prices in the 1970's have indicated that the era of cheap and abundant energy, especially oil, is over. The cost of energy is now becoming increasingly significant relative to the cost of other factor inputs such as capital, labor, and land. The critical dependence of modern economies on energy in various forms, especially electricity, underlines the need for effective development and use of scarce energy resources. At a more general level, the intricate links between energy and the different sectors of the economy, as well as the interactions among the various forms of energy, need to be carefully examined within a disaggregative but integrated framework. Within the electricity power sector itself, the application of rational economic analysis will improve the efficiency of investment decisions and pricing policy. Even small improvements can have a significant beneficial impact, given the large investment needs and consumer expenditure on electricity.

In 1978, world electricity generation was about 7617 terawatt-hours (TWh) of which about 73 percent was generated from fossil fuels, and the remainder was primary electricity (hydro, nuclear, and geothermal). The average world per capita consumption was 1803 kWh. The electricity generated was over 30 percent of all commercial energy produced. The importance of electricity is even more pronounced when it is realized that in many uses, especially for motive power, there is no practical economical substitute.

These figures conceal many geographic disparities. Thus in 1978, the industrialized countries produced about 7 times as much electricity as the developing countries, although their population was only about one third as large. Therefore, per-capita consumption of electricity in the industrialized economies was over 20 times greater than in the third world. The corresponding ratio of gross national product per capita which was over 12 to 1, underlines the close correlation between electricity use and income levels. At the same time, access to electricity in the industrialized world is over 90 percent, whereas it is under 30 percent in the developing countries.

The developing countries' average annual growth rate of electricity consumption has exceeded 8 percent in the 1973-1980 period, and is more than twice the developed

[1] Edited version of a paper published in *Proceedings of the IEEE*, Vol. 72, No. 4, April 1984, pp. 424-61.

countries' average growth rate. As development continues, the third world's future demand for electricity will tend to grow faster than the world's average. In the past, the convenience and relatively low price of electricity in most developing countries have helped to promote electricity use. As electricity and other energy prices are likely to rise during the next decade, however, the growth of demand is unlikely to accelerate further.

During the next decade, installed generation and distribution capability will grow at almost 3 to 4 percent per annum (p.a.) worldwide, but will roughly double in the developing countries requiring investments exceeding US$65 billion a year on average. The real cost of a kilowatt will continue to rise for several reasons. These include the shift to more costly coal and nuclear plants following the oil crisis, the reduced availability of cheaply exploitable hydroelectric sources, and the inability to realize further significant economies of scale -- particularly as systems expand into regions of lower consumer density such as small towns in rural areas. While the third world average growth rate of overall generation will be 6.5 percent p.a., the share of oil-fired generation in total generation is expected to decrease significantly, from 26 percent in 1980 to 7 percent in 1995. Although nuclear and geothermal generation will grow dramatically, their contribution in 1995 will still be relatively small because of their small starting base in 1980.

The large investments in, and rising cost of power have highlighted the need for increased economic efficiency in the electricity sector, and this paper seeks to explain some of the recent analytical techniques that would facilitate such improvements. Traditionally, major emphasis has been placed on improving technical and financial efficiency through least-cost, long-range planning for system expansion, and by optimizing short- and medium-term system operation and providing better management of utilities. These considerations on the "supply side" of the supply-demand equation have been tackled with a remarkable degree of success, usually by engineers using technically and financially oriented solutions. More recently, power economists have focussed attention on the objectives of economic efficiency in the national context. These developments have had a growing and significant effect on the "demand side," chiefly through the application of appropriate tariff policies based on the principles of marginal cost pricing. On the supply side as well, the acceptance of economic arguments is reflected in the use of economic opportunity costs. Such costs are represented by shadow prices that are the national costs of economic resources, instead of purely financial or accounting costs more relevant to the utility alone.

Section 2.2 contains a description of the economic-engineering principles underlying optimal investment planning and pricing in the electricity sector. The empirical modifications required to apply the models (developed in Section 2.2) to the real world are discussed in Section 2.3, within a practical policy-oriented framework. In Section 2.4, three case studies are presented, involving optimal investment planning and reliability, system loss optimization, and tariff setting, which illustrate the methodologies developed earlier. The conclusions are given in Section 2.5, followed by a mathematical appendix and references.

2.2 Engineering–Economic Modeling Framework

An engineering-economic modelling framework which is suitable for rational decisionmaking in the electric power sector is developed in this section. Different aspects of this framework have been developed over the last three decades, which when synthesized coherently as described below, form a powerful set of paradigms and models suitable for determining practical investment and pricing policies in the power utility area. The general approach stresses the national viewpoint, in which scarce economic resources should be used to maximize the net benefits of electricity consumption to society as a whole, subject to various constraints. On the investment side, it may be shown that this is often equivalent to minimizing the economic resource cost of supplying electricity services.

Electricity, the Energy Sector, and the Macroeconomy

Specific models used in the analysis of electric power system may be better understood by first clarifying the relationships among the power sector, the aggregate energy sector, and the rest of the economy (Munasinghe and Schramm 1983). In brief, as shown hierarchically in Figure 2.1, the electricity sector is a part of the energy sector, which, in turn, is a portion of the macroeconomy. Therefore, successful electricity modeling requires careful analysis at all three levels.

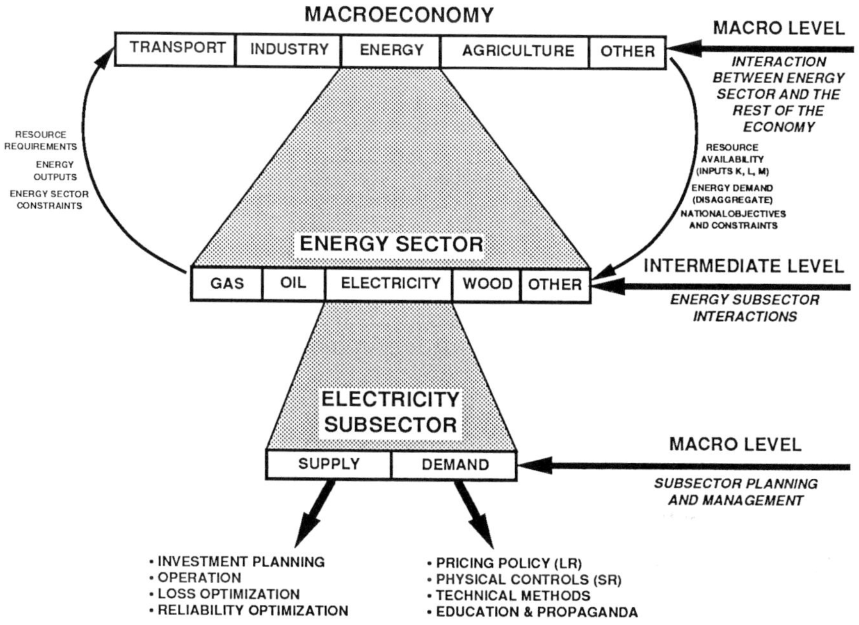

Figure 2.1: Hierarchy of Interactions and the Role of Supply and Demand Management in the Electricity Subsector

Since the energy sector is recognized as a part of the macroeconomy at the highest and most aggregate hierarchical level, electricity and energy policy development requires analysis of the links between the energy sector and the rest of the economy. Such links include the input requirements of the energy sector such as capital, labour, and raw materials; as well as energy outputs such as electricity, petroleum products, coal, gas, woodfuel, and so on; and the impact on the economy of policies concerning availability of energy supply, pricing, etc., in relation to national objectives.

The second hierarchical level in Figure 2.1 treats the energy sector as a separate entity composed of subsectors such as electricity, petroleum products, coal, gas and so on. This permits detailed analysis of the energy sector with special emphasis on interactions among the different energy subsectors, and the resolution of any resulting policy conflicts, e.g., competition among oil, coal, and gas for electricity generation; or electricity, oil and gas for home heating; or wood fuel and kerosene for cooking, especially in developing countries (Munasinghe and Schramm 1983). The third and most disaggregative level pertains to planning within each of the energy subsectors. Thus for example, the electricity subsector must determine its own long-run investment program and pricing policies. Although electricity has been defined as a subsector of the energy sector, elsewhere in this paper the former is referred to as the electricity sector, for convenience.

Energy management is an essential part of overall national economic management, and should be carried out and implemented in close coordination with the latter. It is important to realize that lack of data, time, and manpower resources will generally preclude the analysis of a full economy-wide model when electricity and energy-related decisions are made. However, this holistic approach or general equilibrium analysis is conceptually important. For example, the efficient shadow price of a given resource may be represented by the change in value of aggregate national consumption or output, due to a small change in the availability of that resource (Munasinghe 1979).

In practice, the partial approach shown in Figure 2.1 may be used, where key linkages and resource flows between the energy sector and the rest of the economy, as well as interactions among different energy subsectors, are selectively identified and analyzed, using appropriate shadow prices such as the opportunity cost of capital, shadow wage rate, and marginal opportunity cost for different fuels (Munasinghe 1980). In practice, surprisingly valuable results may be obtained from relatively simple models and assumptions.

Power sector management may be conceptually broken down into supply and demand management. While each of these two areas comprises a number of aspects (see Figure 2.1), the two principal policy decisions to be taken by power utilities relate to investments in and the pricing of electricity, respectively. In countries where the government is the main supplier of electricity services, it can intervene directly in the sector. Successful policy modeling in relation to the framework discussed above requires the following items. First, the national and corporate policy objectives must be clearly defined. Second, the policy models must explicitly define and quantify the most important technical-economic relationships within the electricity sector and interactions with the outside world. Third, other social and political constraints that are difficult to quantify must be systematically accounted for. Finally, the results should be readily translatable into straightforward policy options, and the methodology should be practical and applicable even when data are poor.

Basic Engineering-Economic Principles for Optimizing Investment and Pricing Policy

From the engineering-economic viewpoint, the basic objective of modeling is to determine a set of policies that will maximize the net benefits of electricity consumption to society as a whole. This also corresponds to the most efficient use of scarce economic resources, and maximization of output or GDP. In this sector, we will focus on the national economic viewpoint. As discussed later, there are several other important objectives such as meeting the basic energy needs of poor consumers, independence from foreign sources, raising financial resources for future investments, et cetera, which will also influence both pricing and investment policy.

The investment decision has traditionally been treated within the framework of the least cost system expansion plan. In recent times sophisticated system planning models and techniques have been developed, based on the criterion of minimizing the cost of supplying a given long- range demand forecast, at some acceptable reliability level or quality of supply (Anderson 1972, Sullivan 1977, Endrenyi 1978). The optimal size, mix, and timing of new capacity additions are treated in this way, and related models also provide for optimal (least-cost) operation of the system.

The theoretical foundations of optimal electricity pricing date back as far as the pathbreaking efforts of Dupuit and, subsequently, Hotelling; Ruggles provides a comprehensive review of work in this area up to the 1940's (Dupuit 1932, Hotelling 1938, Ruggles 1949-50). The development of the theory, especially for application in the electric power sector, received a strong impetus from the work of Boiteux, Steiner, and others, from the 1950's onwards (Boiteux 1949, Steiner 1957, Boiteux and Stasi 1974, Williamson 1966, Turvey 1968). Recent work has led to more sophisticated investment models which permit determination of marginal costs, developments in peak load pricing, consideration of the effects of uncertainty, and the costs of power shortages or outage costs suffered by consumers due to unreliable power supply, and so on (Chapters 7-9). Phenomena such as the "Averch-Johnson" effect have been studied in the United States, where private utilities may adopt overly capital-intensive technologies to inflate their rate bases and avoid rate-of-return regulation (Crew and Kleindorfer 1979).

While the close relationship between optimal investment and pricing policies have been recognized for some time, these links were systematically analyzed only in some of the more recent studies. By explicitly incorporating effects due to the stochasticity of supply and demand, and introducing the notion of shortage or outage costs in welfare maximizing models of electricity consumption, it has been shown that the optimal conditions for price and capacity levels must be simultaneously satisfied (Turvey and Anderson 1977, Crew and Kleindorfer 1979). In this context, determining the optimal capacity level is equivalent to establishing the optimal level of reliability (or lower consumer outage costs) and vice versa. Problems relating to the dichotomy of having to choose between short- and long-run marginal costs, and the correct allocation of capacity costs among peak, shoulder, and off- peak consumers have been illuminated in recent work.

We may summarize this complex analysis in simple terms, as follows: The optimal price is the marginal cost of supply. Simultaneously, the optimal reliability (capacity) level is defined as the point at which the marginal cost of increasing reliability is exactly equal to the corresponding reduction in marginal outage costs of consumers. Furthermore, let us define the short-run marginal cost (SRMC) as the cost of meeting additional electricity consumption, with capacity fixed, while the long-run marginal cost (LRMC) is the cost of providing an increase in consumption (sustained indefinitely into the future) in a situation where optimal capacity adjustments are possible. When the system is optimally planned and operated (i.e., capacity and reliability are optimal), SRMC and LRMC coincide. However, if the system plan is suboptimal, significant deviations between SRMC and LRMC will have to be resolved within the pricing policy framework. Finally, if there are substantial outage costs outside the peak period, then the optimal marginal capacity costs may be allocated among the different rating periods (i.e., peak, intermediate, and off-peak) in proportion to the corresponding marginal outage costs.

For practical purposes, the joint optimal price and reliability conditions are used in an essentially uncoupled form. Thus the optimal price is determined at some target reliability level and optimal reliability is set assuming fixed price. By iteration, the solutions may be made mutually consistent. The optimal reliability rule is not easy to apply, especially because outage costs are difficult to estimate (see Chapters 7 to 9). However, application of the marginal cost pricing rule has been attempted in several countries (most notably, France), and while interpretations vary among practitioners, the approach is gaining wide acceptability (see Chapter 3).

Basic Investment Optimizing Framework

A framework which determines the optimal investment (or reliability) level for a power system is summarized below (see Appendix A2.1 for a more rigorous exposition). It is shown that this methodology in which the reliability level is a variable to be optimized, subsumes the traditional least-cost investment planning approach of meeting a given demand forecast at a fixed or target reliability level. The optimization of power system losses, another vital area, is also discussed.

Reliability and Capacity Optimization

The model presented is static (or one period), deterministic, and also has several other simplifying assumptions which improve the clarity of the presentation but do not affect the principal results to be derived. The essence of the new optimizing methodology described here is that reliability or supply quality, R, should be treated as a generalized variable that is to be optimized. The quality of supply affecting customers may be defined very broadly in terms of a set of measurements including frequency and duration of interruptions, and the extent of voltage and frequency variations.

Changes in R have several important economic consequences. First, as supply quality increases, the costs of building, operating, and maintaining the system (SC) also increase. On the other hand as R rises, the shortage costs and inconvenience suffered by consumers due to disruption in supply including outages, low voltages, etc., (QC) will decrease. An improvement in supply quality will also raise the consumers' expectation regarding the future level of reliability, R*, and is likely to induce increased electricity demand which provides additional net benefits of consumption. Changes in R* may also affect QC, as consumers adapt their behaviour patterns to reduce their shortage costs. Thus by increasing R, it would be possible to trade off the higher system costs against the decrease in QC and increased net benefits of induced demand.

Let the demand for electricity in the area of service of a power system be given by

$$D = D(p,R^*,Y,Z)$$

where p is the price per unit, R* is the reliability of service quality consumers expect to receive, Y is a variable which captures the level of economic activity (e.g., income), and Z is a vector of other relevant explanatory variables. For purposes of optimization, we assume Y and Z are exogenously given.

The net benefits of electricity consumption may be written

$$NB(D,R) = TB(D) - SC(D,R) - QC(D,R,R^*) \qquad (2.1)$$

where TB is the total benefit of consumption. TB as well as SC and QC are functions of D, while the functional dependence of SC and QC on R and R* has been explained earlier.

Before optimizing this model, we recognize that electricity tariffs are often not readily subject to change, and simultaneous optimization of price and capacity is a theoretical ideal. Thus from a practical point of view, it may be appropriate (or necessary) to uncouple the joint price and reliability in the presence of fixed or given tariffs, at least on the first round. Once optimal reliability (subject to given prices as defined above) is determined, tariffs can be revised to reflect any changes in the marginal cost of supplying electricity implied by the new reliability level. Using this new level of tariffs and resulting demand, reliability can be reoptimized iteratively, as described later.

Allowing only R to vary, the first-order (necessary)) condition for maximization of net benefits is

$$dNB/dR = (\partial D/\partial R^*).(\partial R^*/\partial R).[\partial(TB-SC)/\partial D] - \partial(SC+QC)/\partial R - (\partial QC/\partial R)$$
$$= 0$$

The derivative $\partial R^*/\partial R$ represents the change in reliability expectation due to variations in R, and may appear unrealistic in the single-period static case presented here. However, it is conceptually the counterpart of terms such as $(\partial R^*_t/\partial R_{t-t'})$ for t' = 1, 2, . . ., S; which appear in the dynamic model and indicate the influence of actual reliability level changes in previous years, on the quality of supply expected today (see Appendix A2.1). We may rewrite the above as

$$(dSC/dR) = -(dQC/dR) + (dTB/dR) \qquad (2.2)$$

where

$$(dSC/dR) = (\partial SC/\partial R) + (\partial R^*/\partial R).(\partial D/\partial R^*).(\partial SC/\partial D)$$

$$(dQC/dR) = (\partial QC/\partial R) + (\partial R^*/\partial R).[(\partial D/\partial R^*).(\partial QC/\partial D) + (\partial QC/\partial R^*)]$$

and

$$(dTB/dR) = (\partial R^*/\partial R).(\partial D/\partial R^*).(\partial TB/\partial D)$$

The term (dSC/dR) on the left-hand side of Eq.(2.2) represents the total change in supply costs due to variations in the actual supply quality, and consists of two components. The first part (∂SC/∂R) is the direct effect of R on SC, while the second component captures the indirect effect via the chain of interactions in which R affects R*, which in turn affects the demand, which finally causes a change in supply costs.

The term (dQC/dR) is the total change in shortage costs with respect to reliability, and has three components, among which the first two may be interpreted analogously to the corresponding components of (dSC/dR). The last part (∂R*/∂R).(∂QC/∂R*) represents the change in QC due to changes in reliability expectation R*, which are themselves caused by variations in R.

Finally, the term (dTB/dR) denotes the change in total benefits caused by the induced demand changes arising from variations in R*, themselves caused by changes in R. We will refer to this term as the change in total benefits due to induced demand. TB and QC must be specified carefully to avoid overlap. The effect of R on TB is based on the fact that consumers will alter their demand level D because their reliability expectation R* has changed. In other words, their demand curve has shifted because their perception of shortage costs has altered, and this change in TB essentially reflects the long-term change in expected shortage costs. Therefore, the term QC in Eq.(2.1) should be interpreted as the remaining short-term unavoidable shortage costs, arising from the difference between the actual and expected reliability levels.

Equation (2.2) implies that to maximize the net benefits of electricity consumption, the quality of supply should be increased up to the point where the marginal increase in system costs is equal to the marginal decrease in the costs of poor supply quality, plus the increase in the benefits of induced demand. These marginal costs and benefits include both the direct effects of changes in R, and the indirect effects due to induced variations in R* and D.

We note that (dSC/dR) > 0, since all the components such as ∂SC/∂R, ∂R*/∂R, ∂D/∂R* and ∂SC/∂D would be positive. Similarly (dTB/dR) > 0, because all its components are also positive. Finally, we might expect (dQC/dR) < 0, assuming that the dominant component ∂QC/∂R is always negative (although the sign of some of the other components may be positive or indeterminate).

These results may be interpreted more clearly using Figure 2.2, where SC and QC are shown (by solid lines) as monotonically increasing and decreasing functions of quality of supply (R), respectively. The total costs curve is defined by TC = SC + QC. TB is also an increasing function of R due to the effects of induced demand. As shown in the figure, the optimal value of quality of supply (Rm) where the marginality conditions in Eq.(2.2) are satisfied, occurs when the TB and TC curves have exactly equal slopes (S_3 and S_4 respectively). At this point net benefits, NB = TB - TC, are maximized. The new economic criterion for power system optimization represented by Eq.(2.2) is general enough to be applicable to all levels of power system planning, that is, for generation, transmission, and distribution.

It is useful to analyze the way in which the above approach relates to, and subsumes the traditional economic criteria used in power system planning. First, suppose R = R* and the effects due to induced demand are ignored, i.e., ∂D/∂R = 0. Then Eq.(2.2) may be rewritten

$$\partial SC/\partial R = -\partial QC/\partial R \qquad (2.3)$$

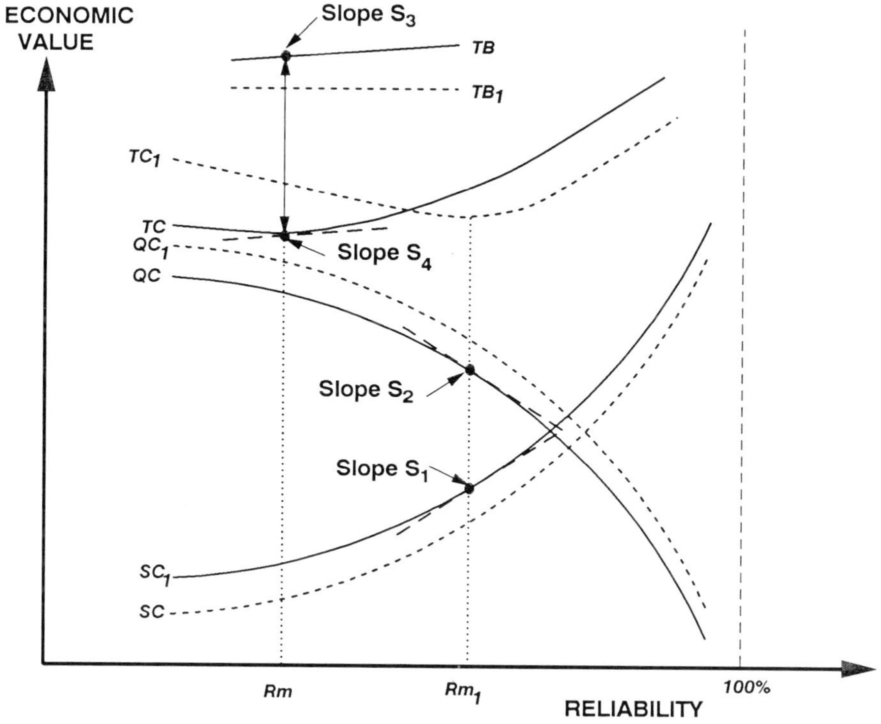

Figure 2.2: Optimal Reliability Levels

This more restricted marginality condition may be interpreted using Figure 2.2, in which the broken curves SC_1 and QC_1 are both shown slightly displaced to the right because the ignored components in (dSC/dR) and (dQC/dR) are both assumed to be positive and would tend to increase the slopes of SC_1 and QC_1, respectively. In other words, Eq.(2.3) implies that the optimal reliability level Rm_1 occurs when the slope of SC_1 (S_1) is equal to the negative slope of QC_1 (S_2). This point is also the minimum of the (broken) total costs curve: $TC_1 = SC_1 + QC_1$. Equivalently, since TB_1 is now independent of R, from Eq.(2.1), minimizing TC_1 is equivalent to maximizing net benefits. In general, Rm and Rm_1 would not coincide.

Next, suppose that both the demand D and the target reliability level R are assumed to be fixed. Then, maximizing NB reduces to simply minimizing the costs of supply SC for exogenously given D and R, which is the conventional and most commonly encountered planning criterion, i.e., least-cost system expansion. Clearly, the least-cost rule is most appropriate for a private utility which is concerned only with its own costs. In contrast, both Eqs.(2.2) and (2.3) imply that not only SC but also the costs and benefits of consumption should be included in the planning criterion. This is more comprehensive and represents the viewpoint of society as a whole. Finally, Eq.(2.2) is more general than Eq.(2.3), because the effects of induced demand have been included.

In this formulation, the reliability measure is defined in a very generalized way. Therefore, the selection of the optimal plan and the associated reliability level is made on the basis of economic cost-benefit analysis, and is quite independent of the actual index of reliability. However, from a practical point of view, it is important to develop reliability indices which not only characterize future system performance in a satisfactory manner, but are also meaningful at the consumer level, and could be easily used to determine the costs of shortages incurred by users. From this viewpoint, load-point indices of the frequency and duration type, referred to individual consumers on a disaggregate basis, would be the most convenient measures to use. It is also important to know the times when outages occur. Since outage costs are generally a nonlinear function of outage duration, ideally, the probability distribution of duration should be computed. However, in practice, a knowledge of the mean duration at specific times may be sufficient, e.g., during the periods of peak, intermediate, and off-peak demand.

Next, let us relax the assumption regarding the original fixed price p. Consider the situation where the stream of system supply costs SC associated with the optimal expansion plan i = m, on the first round, necessitate significant changes in the assumptions regarding the evolution of prices p which were themselves used to determine the initial demand forecast. For example, the use of a marginal cost pricing rule, or some simple financial requirements such as an adequate rate of return on fixed assets, may require previously unforeseen changes in future electricity tariffs, to compensate for the new supply costs. Such a shift in prices would directly affect load growth. Furthermore, the new target reliability levels implied by the first round optimal expansion path i = m, may themselves affect reliability expectation, and thus have a secondary impact on demand and shortage costs -- (see Eqs.2.1 and 2.2).

In such a situation, the impact of the new sets of price and expected reliability levels on the demand forecast and shortage cost estimates, should be considered when iterating through the model again. To clarify this result, the simplified model results (i.e., assuming $\partial D/\partial R = 0$) shown by the broken lines in Figure 2.2, have been redrawn in solid lines in Figure 2.3. Then the iterative changes in demand and shortage cost forecasts, would in general, shift the whole total cost (TC_2) curve as shown by the broken line in Figure 2.3, leading to a new optimal reliability Rm_2. In this fashion, the direct and indirect feedback effects of reliability on demand may be considered iteratively until a set of self-consistent price, demand, and reliability levels was determined.

Loss Optimization

The worldwide scarcity of energy resources and the increasing costs of energy supply have highlighted the importance of energy conservation and elimination of waste by both producers and users of energy. Power system loss reduction is one of the principal ways for achieving this in the electricity power sector.

In the process of delivering electricity to consumers, losses are incurred at the generation, transmission, and distribution stages of a power system. Generation losses may be improved by improving the efficiency of plant and reducing station use, e.g., using new technologies like combined-cycle thermal plants, replacing old boilers and generally uprating old thermal generators, using higher efficiency designs in new hydro installations or replacing older turbines, et cetera. Leaving generation aside (where acceptable norms for losses vary according to the mix of plant), recent work indicates that average energy losses in the power delivery system, i.e., transmission and distribution, should normally be below 10 percent of gross generation, while economically optimal loss levels may be as low as 5 percent.

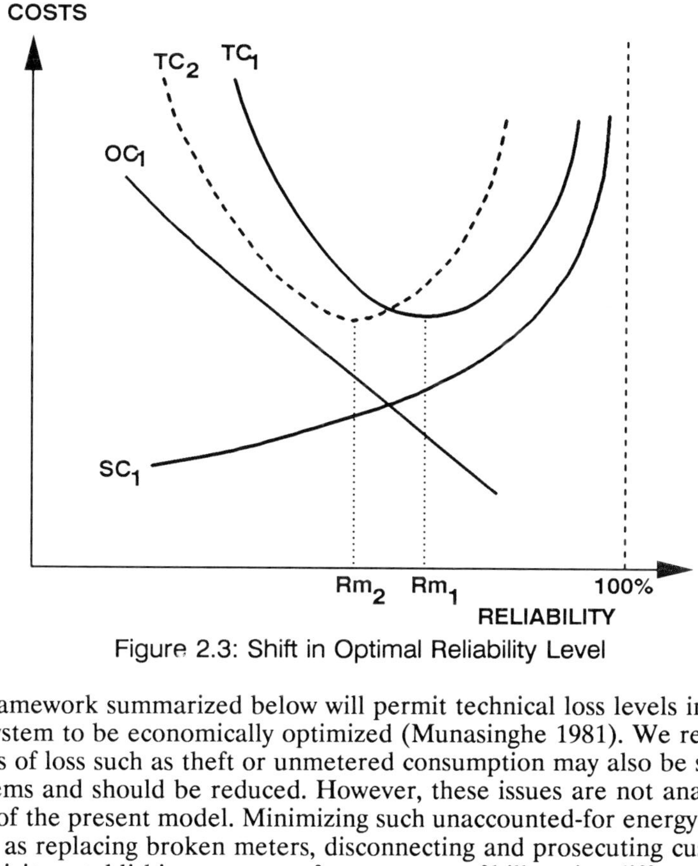

Figure 2.3: Shift in Optimal Reliability Level

The framework summarized below will permit technical loss levels in any part of a power system to be economically optimized (Munasinghe 1981). We recognize that other types of loss such as theft or unmetered consumption may also be significant in some systems and should be reduced. However, these issues are not analyzed within the scope of the present model. Minimizing such unaccounted-for energy may involve steps such as replacing broken meters, disconnecting and prosecuting customers who steal electricity, establishing a system of prepayment of bills, using different individuals to do the meter reading and billing (as an anticorruption measure), and so on.

The essence of the optimization model described below is the tradeoff between power system costs and the resultant decrease in the cost of losses. While system costs are relatively easy to measure in terms of the economic value of physical inputs like capital, labour and fuels, the value of losses is more difficult to establish. Therefore, after discussing loss optimization, we establish below how physical losses in a power system may be valued in economic terms.

We develop the model in terms of optimizing distribution losses, but clearly, the same analysis is applicable to any other part of the power system. Consider the electric power distribution system shown in Figure 2.4. An amount QI of electrical energy is supplied as input by the bulk power system of which an amount L is dissipated as losses and the remainder QO is delivered to consumers.

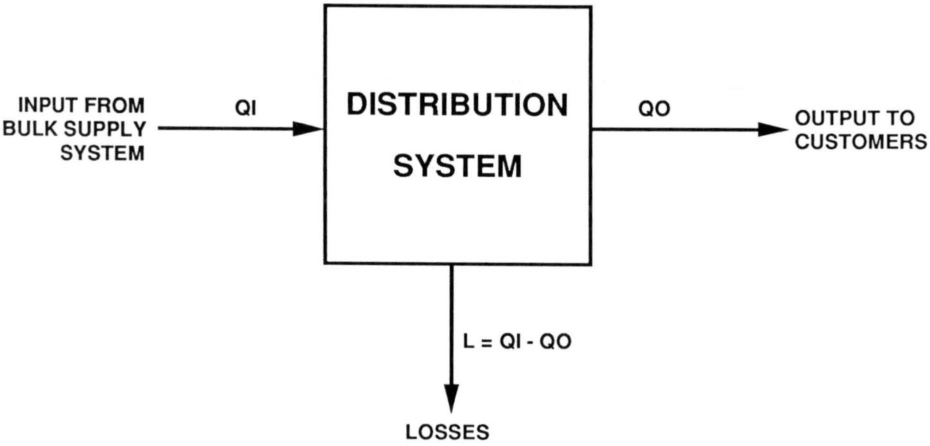

Figure 2.4: Simplified Representation of Losses in a Distribution System

This static analysis is deliberately simplified for clarity of presentation. Thus the quantities QI, QO, and L would actually be disaggregate (e.g., peak kilowatts, peak, shoulder, and off-peak kilowatt-hours, etc.), with each component being valued separately. Similarly, the calculation would have to be performed in present value terms over a long period (e.g., 20 years) to allow for the full life-cycle of components, and dynamic load growth. We also ignore quality of supply and outage cost considerations, explicitly included in Eq.(2.1). All these aspects are incorporated rigorously into the more detailed analysis in Chapter 11.

The net benefits (NB) of electricity consumption from the social viewpoint are given by

$$NB = TB - SC$$

TB is the total benefit of consumption and SC is the supply cost. TB depends on the amount of electricity consumed, i.e., TB(Q). We may break SC down into two principal components

$$SC = BSC + DSC$$

BSC is the bulk supply cost and DSC is the distribution system cost (investment, O&M, etc.)

We use VQI, the value of input electricity (QI), as a measure of BSC. Then

$$SC = VQI + DSC$$

and

$$NB = TB - (VQI + DSC)$$

Suppose we continue to supply QO to consumers but are able to incrementally reduce distribution losses by an amount dL, by improving the network. But QI = QO + L and since we have just assumed that QO is constant, dQI = dL. Therefore, QI decreases by the same amount as losses L, and VQI also falls. TB is unchanged since QO is the same.

The change in net benefits is given by

$$dNB = dVQI - dDSC = dVL - dDSC$$

where VL is the change in value of losses which is assumed to be negative. (Note that dVL = dVQI, although VQI is much greater than VL.) In other words

Increase in Net Benefits = (Decrease in Value of Losses)

- (Increase in Distribution System Costs)

Therefore, net benefits to society would increase if the reduction in value of losses exceeds the increase in the distribution costs. Thus an operational criterion which distribution planners should use is that loss reduction measures should continue up to the point where a marginal increment in distribution costs will be exactly counterbalanced by the decrease in value of losses. Equivalently, we can argue that net supply cost: NSC = VL + DSC should be minimized, to maximize NB.

These relationships are summarized in Figure 2.5, which shows that VL increases and DSC falls as L increases, while the economically optimal loss level L^* occurs when NSC (the sum of VL and DSC), is a minimum. At L^*, the slopes of VL and DSC are equal and opposite.

The Economic Value of Kilowatt and Kilowatt-Hour Losses: In most engineering studies that have been done so far, emphasis has been placed on applying accounting principles to loss evaluation, rather than economic principles. Although concepts such as present worth of annual revenue requirements, leveled annual costs, and equivalent investment costs are used, there is no application of economic theory in the above procedure (Nickel 1980).

The principal point we make is that both kilowatt and kilowatt- hour distribution losses at various time periods should be correctly valued at the long-run marginal costs (LRMC) of supply from the bulk supply system. This issue is not clearly addressed in the work of some previous authors who often suggest lower values for these losses, especially kilowatt losses (Johnson 1976, Baldwin et al. 1961). The valuation of kilowatt-hour energy losses does not pose major problems. If distribution losses decrease at any given moment, then the bulk supply LRMC of energy at different time (e.g., peak, shoulder, off-peak, or by season of the year) provide a measure of the value of kilowatt-hours lost in the distribution system.

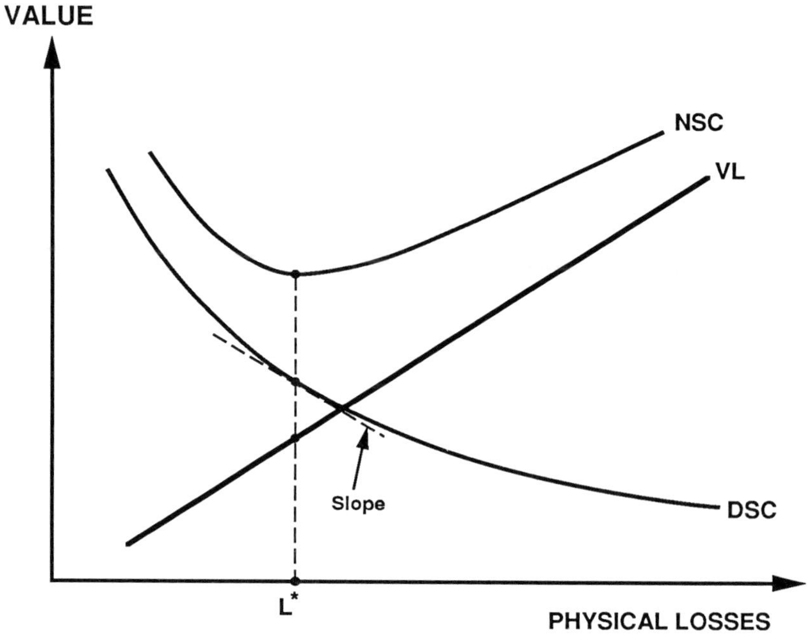

Figure 2.5: Optimal Economic Loss Level

However, when distribution system improvements are made, the greatest change usually occurs with respect to kilowatt losses during the peak period. Although the distribution feeder peaks and the bulk system peak may not overlap, any reduction in kilowatt losses during the bulk system peak will lead to a saving in generation and transmission (G&T) capacity. Even if G&T investments are not actually deferred, the LRMC of bulk kilowatt supply may be used as a proxy for the value of kilowatt losses in the distribution system at the time of the bulk system peak, as described below.

Losses and customer loads are indistinguishable as far as the bulk supply system is concerned. If, for example, losses do not impose burdens on bulk capacity, then the incremental costs of serving customers will also be negligible. Furthermore, in an optimally planned electricity supply system, the two conditions (discussed earlier) that must be satisfied are that: (a) the optimal price equals the marginal cost of supply; and (b) the optimal incremental cost of system improvement equals the cost of outages avoided due to improved reliability.

When losses are reduced, it is equivalent to a reduction in demand. Thus bulk system capacity additions may be deferred, yielding cost savings represented by bulk supply LRMC. Alternatively, if the generation and transmission expansion investments continue relatively unchanged (e.g., due to lumpiness), then the improved bulk supply reliability will provide cost savings (due to averted outage costs at the margin) that are equivalent to the marginal savings that could have been realized from deferred G&T investments. We note that in the first case, the cost savings accrue to the power supplying company whereas in the second case, the customers gain. Thus from a social viewpoint both cost savings are equivalent, whereas from the power company's viewpoint the former is more desirable.

Basic Price Optimizing Framework

The basic rule for maximizing net benefits of consumption is that price should be set equal to marginal cost. We briefly summarize here, the detailed framework for power price optimization presented in Chapter 3.

In Figure 3.1 (of Chapter 3) let us assume that the total benefits of electricity consumption and the costs of supply are represented by the areas under the demand curve $EFGD_0$ and supply (marginal cost) curve AHGS, respectively. Then the maximum benefits of consumption (i.e., total benefits minus costs) are realized when the price is set equal to p_0, thereby inducing the optimal level of consumption (at the market clearing point where the supply and demand curves intersect).

To the static analysis in the basic framework could be added the dynamic effect of growth of demand from year 0 to year 1, which leads to an outward shift in the demand curve from D_0 to D_1. This movement results in the iterative process (explained in Chapter 3) in which the price is periodically increased along the supply curve in order to track the movement of the intersection point of the supply and demand curves over the years.

Practical problems caused by capital indivisibilities and changes in the demand for power by time-of-use can be resolved by adopting peak load pricing. As explained in Chapter 3 and depicted in Figure 3.3, this approach (in its simplest form) leads to the result that peak period electricity users who force the utility to undertake future capacity additions should bear responsibility for capacity costs as well as variable fuel, operating and maintenance costs, while off-peak consumers ought to pay only the variable costs.

2.3 Policy Applications

The basic framework and models described in the previous section must be suitably modified and interpreted as shown below, to reflect real-world conditions and constraints, before they can be used for policy purposes.

Investment Policy

We discuss traditional least-cost system planning and operation first, and then proceed to outline how the reliability optimizing methodology developed earlier can be practically applied.

Conventional Least Cost Planning

Power system planning is a sophisticated and well developed topic with an abundant literature. Therefore, the purpose of this section is to outline the engineering-economic principles underlying conventional least-cost system planning, with appropriate references to more detailed treatment of various topics, where appropriate.

Ideally, the planning of the whole power system should be integrated. To make the practical problems of system design workable, however, it is convenient to analyze the procedure in terms of the main subsystems: generation, transmission, and distribution (Sullivan 1977, Wismer 1971, Knight 1972, Vardi and Avi-Itzhak 1981). This type of hierarchical reduction is logical from several points of view.

It is useful at this stage to briefly review the basic principles underlying power systems. An electric power system must supply electricity to consumers at the power level they want and at the time and place of their choosing, while maintaining an acceptable quality of service, i.e. voltage and frequency levels must lie within specified limits. The basic parts of a power system may be analyzed in terms of their functions: generation, transmission, and distribution. The concept of a generating source or power plant is relatively unambiguous. Both transmission and distribution facilities carry the electric power from the source to the consumer. These ideas are further clarified in Figure 2.6, a schematic diagram of a simple power system.

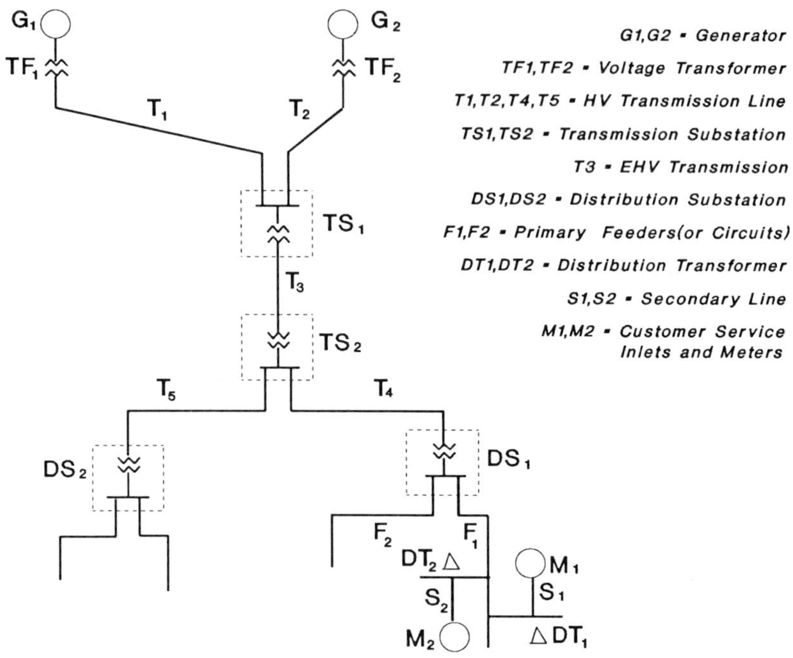

Figure 2.6: Schematic of a Simple Electric Power System

The output from two power generating plants G_1 and G_2 (hydro-electric or thermal) is increased to the high voltage (HV) level at source by the respective transformers TF_1 and TF_2 and is fed into the busbars in transmission substation TS_1 via the HV transmission links T_1 and T_2. At TS_1 the voltage is further increased to the extra high voltage (EHV) level, before the power is transmitted through the EHV line T_3 to substation TS_2. A voltage reduction occurs at TS_2, after which the electric power is fed, through HV transmission lines T_4 and T_5, into the distribution substations DS_1 and DS_2 located close to the load centres. Another voltage reduction occurs at DS_1 and DS_2, and then power flows out of the distribution substation busbars, through primary feeders such as F_1, to various distribution transformers (for example, DT_1), where the voltage is decreased further. Finally, the power is delivered to a representative consumer, through the secondary distribution line S_1, and service inlet and meter M_1.

In transporting power the general principle is that voltage level depends on distance and power flow. Transmission lines are used to carry large amounts of power over long distances, while distribution lines involve smaller power flows over shorter distances. The technical distinction between transmission and distribution facilities is usually made on the basis of their operating voltages. Although voltage standards vary greatly from country to country, broadly accepted ranges for voltage levels are as follows: EHV transmission, more than 220 kilovolts; HV transmission, 69 to 220 kilovolts; primary distribution, 6 to 35 kilovolts; and secondary distribution, 110 to 480 volts. In practice, considerable overlap is likely, and facilities operating at voltages in the range of 35 to 100 kilovolts (often called subtransmission), may be included within the distribution or transmission category, depending on their function.

An operating power system is much more complex than the one described above and in general has many generating sources linked to many load centres via an interconnected transmission network. Transmission interconnections may be used to tie the power systems of several different utilities together to form an even larger power pool. The distribution grid at a single load centre would serve thousands of consumers. Furthermore, many other components, including protective and relaying devices as well as load control and dispatching equipment, would play an important role in a large interconnected system.

As was noted in the previous section on load forecasting, various completion times are associated with projects that involve different parts of a power system, ranging from up to ten years for the design and construction of a large nuclear generating plant to one or two years for distribution schemes. Also, the problems and design philosophies associated with the different parts of the total power system are inherently dissimilar. For example, distribution system planners require detailed knowledge of localized geographic areas and must deal with many small components such as distribution transformers, switches, and reclosers. At the other extreme, generation planning is carried out on a global level and is concerned with large components such as turbines, and these generally have interaction modes that are more complex to analyze. Consequently, the models and criteria now used to design the various subsystems are also different. The following mathematical discussion is cast in the context of traditional system expansion planning, that is, choosing the design that minimizes system cost while meeting given requirements for load and reliability (Anderson 1972). The basic ideas involved, however, may be readily adapted to the new approach in which the reliability level is treated as a variable, and the total cost to society, or sum of system and shortage costs, is minimized.

Early approaches to system design and planning were rather intuitive and relied heavily on the designers' past experience. In contrast, the mathematical methods currently used rely on optimization models. The goal is to minimize an objective function, usually the present discounted value of system cost subject to various constraints, such as meeting targets for load and reliability, and maintaining technical and operating requirements (IEEE 1976, Rischl 1975).

In general, if the problem is deterministic, with known objective function and constraints -- that is, best estimates of plant costs, load, and demand in generation planning -- then some definitive nominal design may be found. With uncertainty, flexible design strategies are required. In the worst case strategy, upper and lower design limits are determined that correspond to maximum and minimum values of the objective function and constraints. Given a probabilistic distribution of values of the objective function and constraints, the statistical technique results in the average design configuration emerging from among the design variations.

The specific techniques for solving the optimization problem may be considered under three categories. Programming approaches, which include linear, nonlinear, integer; mixed-integer; and dynamic programming versions, are most useful when the objective function is simple and well defined. They rely on explicit (computerized) algorithms that are used to find the solution after a finite number of steps. Search techniques, which may be used when programming methods are difficult to apply, are generally easier to implement, but they do not guarantee convergence to the solution within a reasonable time. Special procedures are often elegant, but they are usually applicable only to particular problems (Ford and Fulkerson 1962).

Generation: The basic questions in the design of generation systems involve the size, timing, type (or mix), and location of generating plants. The economic aspects of each of the choices involved are clarified below by using a simple approach. Figure 2.7 depicts the problem of size and timing resulting from the growth of peak megawatt (MW) demand over time in a typical system with limited capacity. The installed capacity of existing generating plant MW will not be sufficient to meet the demand in two years, allowing for a suitable reserve margin. Two expansion alternatives are available that meet all the constraints. In the first, two medium-size generating plants would come on stream in the second and sixth years. In the second option, a small plant would be commissioned in the second year, followed by a large power station in the fourth year. The system planner must choose the cheaper alternative. In a real system, the expansion paths over a longer period of time would be compared. In this example, the two alternative plans have been made roughly equivalent after the sixth year to highlight the nature of the choice of size and timing of the next two-generation additions. Moreover, in predominantly hydroelectric systems, energy shortages caused by limited storage space may precede any capacity constraints.

The selection of an appropriate plant mix is best discussed in the context of the annual load duration (LDC) for a system. The LDC for a given year is shown in Figure 2.8, in which the hourly megawatt demand of the system is plotted against the number of hours of the year during which this level of demand is equaled or exceeded. The peak demand MW occurs only over a short period of time. Since the average level of megawatt demand -- that is, the total energy in megawatt-hours or area under the LDC curve divided by 8760 h -- is less than the peak demand, the load factor will be less than unity.

For simplicity, only two kinds of generating units are initially considered: gas turbines and base-load steam plants. The relevant characteristics of both these types of machines are shown by the solid lines in the upper diagram of Figure 2.8, which is a linearized graph of the total average cost per megawatt of installed capacity, that is, capital cost plus operating cost, plotted as a function of the number of hours of operation. Gas turbines have lower capital costs than steam turbines. These costs are represented by the respective intercepts a_1 and a_2, which are the investment costs of a megawatt, averaged yearly over the lifetime of the machines. Gas turbines have higher fuel costs than steam plants, however, as indicated by the slopes b_1 and b_2 respectively, of the operating cost curves.

First, consider the problem of determining the amounts of new gas turbine and base-load steam capacity that must be built to meet the given load, if the existing generating plant is ignored. From the economic point of view, gas turbines are more expensive than base-load units if they are to be used more than H hours a year. This essentially static picture is only illustrative, because the relevant stream of present discounted investment and the operating costs of the machines would have to be compared over many years. To pursue this further, the lower diagram indicates that it would be economical to serve the bottom X2 megawatts (the base load) with steam units and the upper X1 megawatts (the peak load) with gas turbines. A reserve margin may be included if necessary.

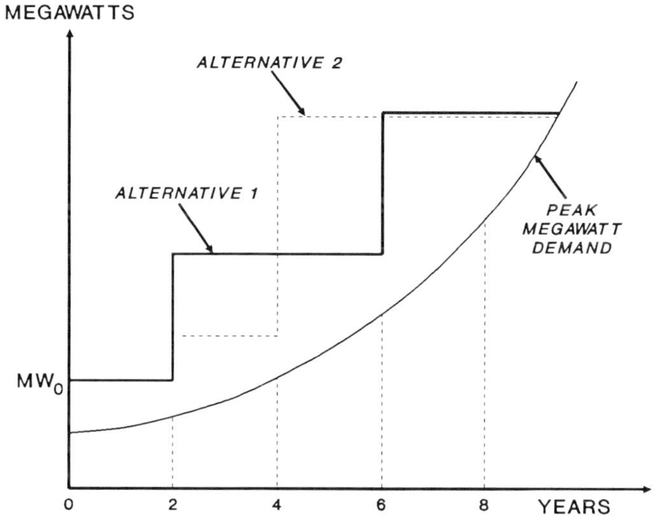

Figure 2.7: Size and Timing of Additional Generating Plants Needed to Meet the Peak Demand for Power

Next, the modifications required in generation expansion planning because of old or existing plants are illustrated using a simple example. Suppose that Y megawatts of generating capacity exist already. This plant has an operating cost given by slope B in Figure 2.8. To clarify the representation, assume $b_1 > B > b_2$. The capacity costs of the existing plant have already been incurred, and the planner must determine the optimal point of operation for these Y megawatts on the LDC. This is done geometrically by moving the line parallel to itself up or down on the diagram until the vertical distance between the appropriate points F and G on the LDC is equal to Y megawatts.

The intercept A is the shadow value of capacity of the existing plant, that is, the equivalent saving in capacity costs for the system resulting from use of the existing generating plant. The new plan requires full use of the Y megawatts of existing capacity and X1′ and X2′ megawatts, respectively, of new gas turbines and base-load steam units to be constructed to meet the LDC. Since X1′ ≠ X1 and X2′ ≠ X2, the existence of old capacity has altered the optimal mix of new plants required. In more complicated cases, if some of the existing capacity is inefficient and has very high operating costs, the optimal plant mix may include only part or none of this old plant.

In practice, many kinds of generating machines of various sizes would have to be taken into account including nuclear, coal, and oil-fired steam; gas turbine; conventional and pumped-storage hydro; and, more recently, geothermal, tidal, and wind-powered units. When generation additions are being considered to meet future load growth, however, the basic principle of selecting units with lower capital but higher operating costs for peaking purposes -- and the opposite for nonpeaking operation -- would generally hold true.

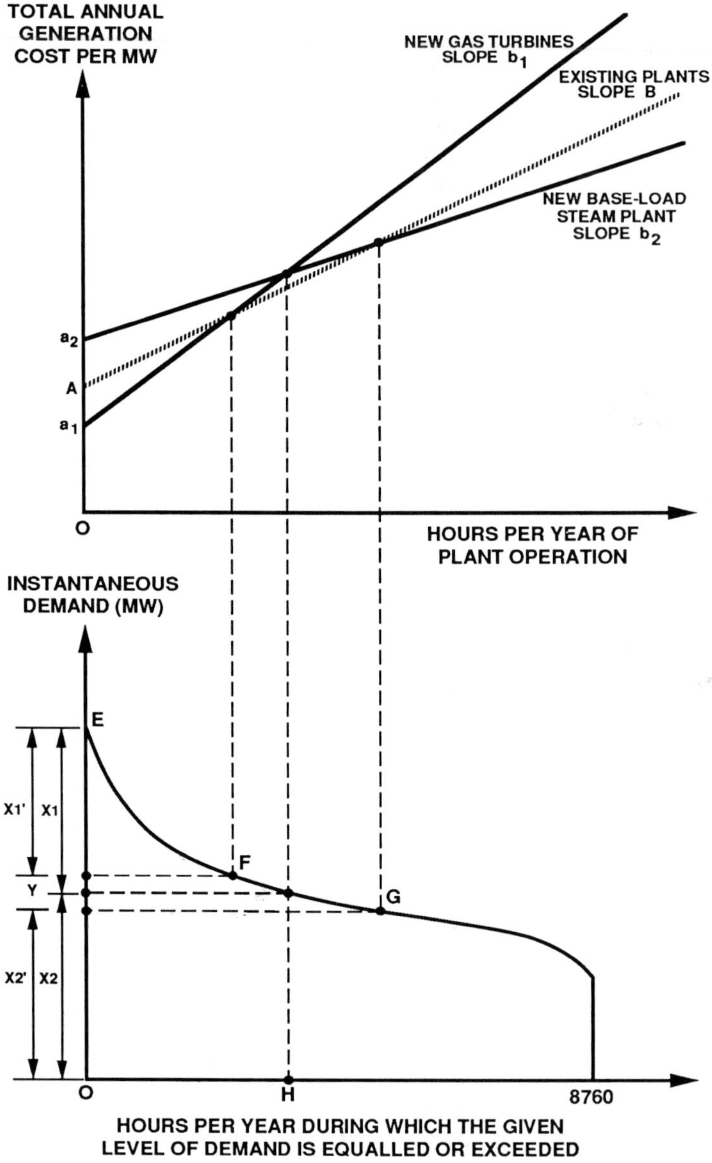

Figure 2.8: Types of Generating Plants Needed to Meet the Annual Load Duration Curve

Transmission: Usually, the problem of locating the generators is closely linked with the design of the transmission system; therefore, they should be discussed in the same context. Transmission planning typically involves solving a dynamic network-type problem. The basic objective is to select the type, timing, and location of additional lines that would connect the various generating stations to the different load centres in the least expensive way. In addition, requirements involving the load, reliability level, and other constraints must be satisfied.

A commonly used network model is the dc load flow model, which may be used in either a linear or nonlinear framework. In the single-step approaches that are essentially static, the design optimization is repeated at various points. When generation and load levels are specified -- for example, in the last year of the planning horizon -- the repetition provides instantaneous snapshots of the system's evolution. Long computing times are required in dynamic methods to handle transmission line variations in both space and time, as in the analysis of a large interconnected area during each year of the planning period. The problem of time is simplified by preselecting several expansion alternatives according to certain performance criteria. None of these techniques can guarantee a definitive solution, however, given the size of the problem and limits on calculating times (Stott 1974).

There are two general types of criteria for transmission system performance: steady-state and transient (CIDA 1977). Steady-state performance of the system consists basically of verifying that line and transformer design limits are not exceeded under normal loading and operating conditions or with single contingencies, such as loss of one line, transformer, or generator. Transient performance criteria are used to test the model for stability in various situations. One group of tests involves examining how well an interconnected system would maintain steady-state stability under both normal and contingency conditions, when excited by a small disturbance such as the switching on of a minor load. Verifying the system's dynamic stability is also important. This involves the degree of damping of high-amplitude oscillations induced by a large fault, such as failure of a major line because of a three-phase fault to ground or the loss of a block of generators.

A return to the question of generator location is appropriate at this point. The analysis of combined generation and transmission system models raises formidable problems. Therefore, although the generation and transmission subsystems are intimately related, separate models are generally used for the detailed design of each subsystem.

More specifically, a typical large computerized generation planning model would yield the appropriate plant mix, size, and timing. Sometimes a limited and often inaccurate representation of the transmission network is included. The optimal locations to supply the specified load centres would be chosen from among several potential sites within the context of a more detailed transmission planning model. In practice, it would be necessary to iterate between the models to derive the best mutually consistent solution.

Distribution: Although planning generation and transmission systems is an interrelated process, the distribution grid serving a localized geographic area or load center can be designed independently. Recently, computerized methods have been developed to determine the optimal size, timing, and location of distribution substations and primary feeders (see Chapter 11 for details). Sophisticated models and computerized algorithms, however, are not as widely used in the design of distribution systems as in generation and transmission work.

In the more conventional approach, forecast loads are imposed on the existing distribution system at various times in the future, and the network is systematically strengthened to meet these loads adequately. The location of new substations and primary feeders and the upgrading of existing ones would be determined to meet consumer demand within acceptable voltage limits under both normal and emergency conditions, as well as during peak and off-peak periods. Flexibility in the design to facilitate isolation of faulted sections of feeders and to permit customers to be switched from faulted to functioning feeders would be an important consideration.

Often, because of the large numbers of customers and components such as line sections, distribution transformers, and so on, only certain representative feeders would be analyzed in detail to establish general criteria such as maximum kilowatt-amperes for a circuit. Such criteria could then be applied to other feeders without further detailed study. In general, the larger and more complex the service area, the greater would be the advantage of using even simple computerized techniques for circuit-by- circuit analysis.

Implementing the Optimal System Reliability Model

A method for practical application of the theoretical model is summarized in Figure 2.9. First, a set of n likely future paths of expected reliability or service quality (R^{*1},...,R^{*n}) are chosen, where R^{*i} characterizes the service quality of path i during the planning time horizon as discussed earlier. This information is combined with a given price forecast and other socio-econmic and demographic data to prepare n alternative disaggregate, long-range load/demand forecasts (D^1,...,D^n) corresponding to each expected reliability path. We may normally expect that in any given future year, the demand would be greater for higher levels of expected quality of supply.

Second, power system models and data are used to design n alternative long-range system plans to meet the n demand forecasts at n alternative service quality levels (R^1,...,R^n) which conform as closely as possible to the original target set (R^{*1},...,R^{*n}). Each system plan is least-cost for the given demand and reliability path, and has an associated supply cost SC.

Third, shortage cost cost models, the demand forecasts, and the service quality information are combined to make estimates of future shortage costs QC corresponding to each system plan. The total benefits of consumption TB are also estimated for each demand forecast.

Finally, the values of TB^i, SC^i and QC^i corresponding to each alternative path i are compared and the path m which maximizes the present discounted value of net benefits: $NB^m = TB^m - SC^m - QC^m$ over the long-range time horizon is chosen as the optimal one. In practice, because the system design process involves discrete jumps in R rather than continuous variations, it is better to consider a band of optimal service quality levels around Rm rather than a unique value.

We conclude by examining several iterative feedback loops. First, if a designed system has a significantly different reliability path than the expected supply quality assumed for the demand forecast, then for consistency the latter must be revised via the reliability feedback loop. The second loop involves design feedback in which a given system plan may be fine-tuned to improve the optimum using the information concerning TB and QC. Finally, the marginal costs of supply may be derived from the optimal system plan and used to revise the original price forecast. This allows us to recouple again the joint optimal price and service quality conditions (which were uncoupled for convenience when the model was first optimized with respect to R). In summary, the feedback loops permit iteration of the model until the optimal reliability path, demand forecast and price projections are all mutually consistent.

The above cost-benefit based simulation model is sufficiently general to optimize all or part of a power system, ie., at the generation, transmission or distribution levels. For example, only the distribution system could be varied, whereas the service quality of power delivered to this distribution network is held exogenously fixed (i.e., the service quality of bulk power supplied to the distribution system is held constant). Therefore, incremental generation and bulk transmission, do not enter into the calculation, except to help value power losses in the distribution system (i.e., at the marginal cost of bulk power supplied). Under a more general optimization procedure, a given improvement in service quality may, of course, be realized more economically by strengthening generation and transmission rather than distribution, but this would be outside the scope of normal distribution system analysis.

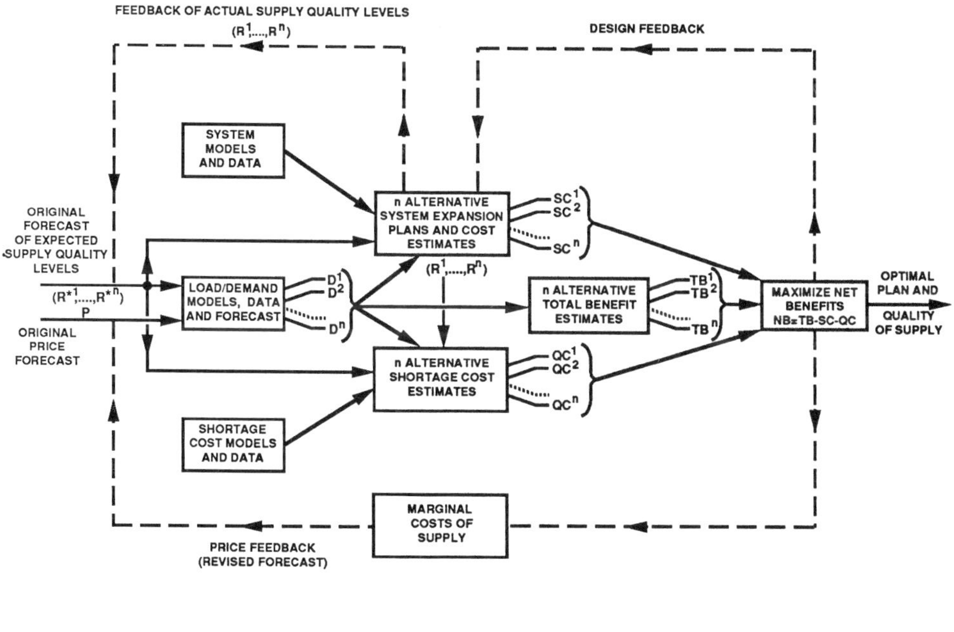

Figure 2.9: Implementation of the Reliability Optimizing Methodology
Note: p = price, R* = consumer's reliability expectation, R = reliability level, SC = supply costs, QC = outage costs, and NB = net benefits. Maximizing NB yields the optimal system plan m and the corresponding reliability Rm.

Figure 2.10 illustrates the new procedure in a different way. Suppose that in the starting year the optimal price p_0 has been set equal to LRMC(R_0), at the market clearing point A. The LRMC(R_0) curve is derived from the system plan with the optimal reliability level R_0, keeping R_0 fixed. Thus initially both price and reliability are jointly optimized.

Now if the demand curve shifts from D_0 to D_1 after some time, the optimal price is not necessarily p_1' on the same LRMC(R_0) curve; using this static LRMC corresponds to the traditional method of system planning with a fixed target reliability level R_0. The optimal reliability may have changed to R_1 and the appropriate curve is LRMC(R_1) with optimal price p_1. Thus as demand increases, the dynamic optimal long-run marginal cost curve LRMC(DYN) lies along AB.

In practice B is unlikely to be a well-defined point because although the LRMC curve is generally known from known supply-side (technological-economic) considerations, the slope and position of the demand curve are not.

Therefore, since B is a poorly defined and shifting target, a trial-and-error approach is required. As demand grows, capacity is added to expand output, while price and reliability are optimized iteratively, as described earlier. Thus when reliability is optimized, price and demand growth are assumed to be fixed, and when price is optimized, reliability and demand are assumed to be unaffected. Through successive iterations, the mutually self-consistent set of optimal price, reliability and demand levels would be reached at point B.

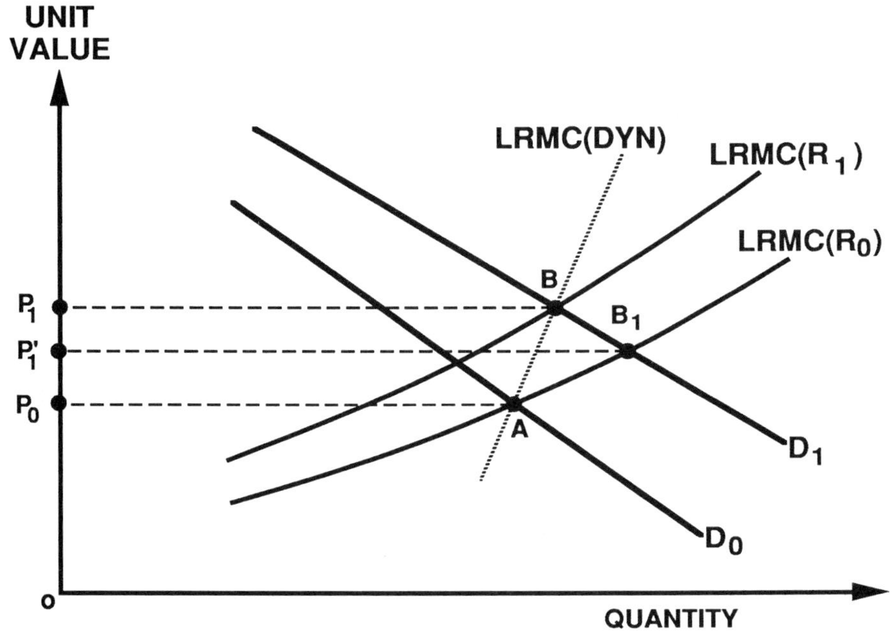

Figure 2.10: The Dynamics of Demand Growth and Optimal Reliability

The Effect of Variable Reliability on System Expansion Planning: At this stage, we summarize several important features of the reliability optimization approach presented here. The simplifying condition represented by equation (2.3) is assumed, to clarify the exposition. First, this method adds an entirely new dimension to the traditional process of system expansion planning. Usually, the power authorities examine several alternative long- range power system plans which are designed to meet a basically fixed load forecast (although some variation in the growth of demand may be considered, for sensitivity testing), at some predetermined, desired level of reliability and subject to other political, environmental and legal constraints. Then, the plan which has the lowest value of total costs is chosen. This cost minimization approach is equivalent to the maximization of net benefits criterion used in cost-benefit analysis, provided that the benefit streams of the alternatives being compared are identical.

In the new approach described in this paper, the reliability level is also a variable to be optimized. Therefore, the system planner must design a number of alternative systems to meet the future demand (which is initially assumed to be fixed) at each of several target reliability levels, but still subject to the other constraints mentioned above. Then these alternatives are compared, and the one which minimizes the total costs (defined as the sum of the outage costs and the system costs) is chosen as the optimal one. In other words, the conventional system planning criterion of minimizing only the system costs is subsumed within the new procedure, where the total social costs are minimized.

Second, another level of sophistication in system expansion planning is possible with the new model, by considering variations in the demand forecast. The main focus in this method is on the reliability level, which is optimized subject to an initially given forecast of load growth, i.e., assuming a fixed evolution of electricity prices. Ideally, from an economic point of view, both price and reliability should be optimized simultaneously, as discussed in greater detail earlier. However, electricity tariffs in the real world are most often fixed, and therefore it is more practical to assume a given evolution of prices when the first round of optimum reliability is determined. Any resulting changes in the pricing assumptions can be fed back iteratively into the model to improve the optimum.

Third, the framework for evaluating outage costs as well as system costs in this new approach, is basically economic, and more appropriate in the context of the national economy, or society, as a whole. The goods and services used as inputs to the electric power system (e.g., labour, land, physical assets, materials, etc.) are considered as scarce economic resources which could be used in alternative production, and they are valued accordingly. In particular, if markets are highly distorted, shadow prices may be used. Such an approach is particularly appropriate in the case of a publicly owned power utility, often the case in the developing countries. In contrast, the traditional system planning approach is more compatible with the financial or accounting viewpoint of a private utility. Finally, the model may be suitably disaggregated to permit optimization of an interconnected system at various levels of aggregation ranging from the global (e.g., in terms of system-wide generation reliability), to the specific (e.g., in terms of distribution reliability for small geographic areas).

Pricing Policy

As explained earlier (on the basis of an idealized model), the net economic benefits of electricity consumption would be maximized if electricity prices were set equal to the marginal cost of supply at the market clearing point (where demand equals supply). This result also satisfies the requirement that national economic resources are being efficiently allocated, since electricity is produced efficiently at least-cost, and is consumed efficiently at a level of use where the marginal benefit of consumption is equal to the marginal supply cost.

However, in the real world, many other objectives and constraints act to modify the pure economic efficiency criterion summarized above. A practical pricing policy must seek to satisfy these other requirements as well.

Objectives of Pricing Policy

There are several objectives or criteria for power pricing, not all of which are mutually consistent. First, national economic resources must be allocated efficiently, not only among different sectors of the economy, but within the electric power sector. As indicated earlier this implies that cost-reflecting prices must be used to indicate to the electricity consumers the true economic costs of supplying their specific needs, so that supply and demand can be matched efficiently. Second, certain principles relating to fairness and equity must be satisfied, including: (a) the fair allocation of costs among consumers according to the burdens they impose on the system; (b) the assurance of a reasonable degree of price stability and avoidance of large price fluctuations from year to year; and (c) the provision of a minimum level of service to customers who may not be able to afford the full cost. Third, the power prices should raise sufficient revenues to meet the financial requirements of the utility, as described earlier. Fourth, the power tariff structure must be simple enough to facilitate the metering and billing of customers. Finally, other economic and political requirements must also be considered. These might include, for example, subsidized electricity supply to certain sectors in order to enhance growth or to certain geographic areas for regional development.

Since the above criteria are often in conflict with one another, it is necessary to accept certain trade-offs between them. The LRMC approach to price setting described below has both the analytical rigor and inherent flexibility to provide a tariff structure that is responsive to these basic objectives.

Long-Run Marginal Cost (LRMC) Based Tariffs

A tariff based on LRMC is consistent with the first objective, that is, the efficient allocation of resources. The traditional accounting approach is concerned with the recovery of historical or sunk costs, while in the LRMC calculation the important consideration is the amount of future resources used or saved by consumer decisions. Since electricity prices are the amounts paid for increments of consumption, in general, they should reflect the incremental cost incurred. Supply costs increase if existing consumers increase their demand or if new consumers are connected to the system. Therefore, prices that act as a signal to consumers should be related to the economic value of future resources required to meet consumption changes. The accounting approach that uses historical assets and embedded costs implies that future economic resources will be as cheap or as expensive as in the past. This could lead to overinvestment and waste, or underinvestment and the additional costs of unnecessary scarcity.

To promote better utilization of capacity, and to avoid unnecessary investment to meet peak demands, which tend to grow very rapidly, the LRMC approach permits the structuring of prices so that they vary according to the marginal costs of serving demands: (a) by different consumer categories; (b) in different seasons; (c) at different hours of the day; (d) by different voltage levels; (e) in different geographical areas, and so on.

In particular, with an appropriate choice of the peak period, structuring the LRMC-based tariffs by time of day generally leads to the conclusion that peak consumers should pay both capacity and energy costs, whereas off-peak consumers should pay only the energy costs. Similarly, analysis of LRMC by voltage level usually indicates that the lower the service voltage, the greater the costs consumers impose on the system.

The structuring of LRMC-based tariffs also meets subcategories (a) and (b) of the second, or fairness, objective mentioned earlier. The economic resource costs of future consumption are allocated as far as possible among the consumers according to the incremental costs they impose on the power system. In the traditional approach, fairness was often defined rather narrowly and led to the allocation of arbitrary accounting costs to various rating periods and consumers, thus violating the economic efficiency criterion. Because the LRMC method deals with future costs over a long period -- for example, at least 5 to 10 years -- the resulting prices in constant terms tend to be quite stable over time. This smoothing out of costs over a long period is especially important given capital indivisibilities or "lumpiness" of power system investments.

Using economic opportunity costs (or shadow prices -- especially for capital, labor, and fuel) instead of purely financial costs, and taking external costs into consideration whenever possible, also links the LRMC method and efficient resource allocation.

The development of LRMC-based tariff structures which also meet the other objectives of pricing policy mentioned earlier, is summarized next.

Summary of Practical Tariff Setting

A two-stage process for determining a realistic tariff structure is described in detail in Chapter 3. To summarize, in the first stage of calculating LRMC, the economic (first-best) efficiency objective of tariff setting is satisfied, because the method of calculation is based on future economic resource costs rather than sunk costs, and also incorporates economic considerations such as shadow prices and external costs. The structuring of marginal costs permits an efficient and fair allocation of the tariff burden on consumers. In the second stage of developing an LRMC-based tariff, deviations from strict LRMC are considered to meet important financial, social, economic (second-best), and political criteria. This second step of adjusting strict LRMC is generally as important as the first-stage calculation.

The LRMC approach provides an explicit framework for analyzing system costs and setting tariffs. If departures from the strict LRMC are required for noneconomic reasons, then the economic efficiency cost of these deviations may be estimated roughly by comparing the impact of the modified tariff relative to (benchmark) strict LRMC. Since the cost structure may be studied in considerable detail during the LRMC calculations, this analysis also helps to pinpoint weaknesses and inefficiencies in the various parts of the power system -- for example, overinvestment, unbalanced investment, or excessive losses at the generation, transmission, and distribution levels, in different geographic areas, and so on. This aspect is particularly useful in improving system expansion planning.

Finally, any LRMC-based tariff is a compromise between many different objectives. Therefore, there is no "ideal" tariff. By using the LRMC approach, it is possible to revise and improve the tariff on a consistent and ongoing basis, and thereby approach the optimum price over a period of several years, without subjecting permanent consumers to "unfair" shocks, in the form of large abrupt price changes.

Extensions of Simple Models

The simplified models presented so far must be extended to analyze the economics of real-world power systems. First, the usual procedure adopted in marginal cost pricing studies may require some iteration. Typically, a deterministic long-range demand forecast is made assuming some given future evolution of prices. Then, using power system models and data, several plans are proposed to meet this demand at some fixed target reliability level (see below). The cheapest or least-cost system expansion plan is chosen from these alternatives. Finally, strict LRMC is computed on the basis of this least-cost plan and an adjusted LRMC tariff structure is prepared. If the new tariff that is to be imposed on consumers is significantly different from the original assumption regarding the evolution of prices, however, then the first-round tariff structure must be fed back into the model to revise the demand forecast and repeat the LRMC calculation.

In theory, this iterative procedure could be repeated until future demand, prices, and LRMC-based tariff estimates become mutually self-consistent. In practice, uncertainties in price elasticities of demand and other data may dictate a more pragmatic approach in which the LRMC results would be used after only one iteration to devise new power tariffs and to implement them. The demand behaviour is then observed over a time period; the LRMC is re-estimated and tariffs are revised to move closer to the optimum, which may itself have shifted, as described previously. An extreme form of price feedback could result in a shift of the peak outside the original peak period, especially if the latter was too narrowly defined. That is, peak load pricing may shift the demand peak, from one pricing period to another. If sufficient data on the price elasticity of demand were available, theory indicates that each potential or secondary peak should be priced to keep its magnitude just below the available capacity level. Since, in practice, the necessary information would rarely be available, a combination of techniques including use of a sufficiently wide peak period, redefining the peak period to include both the actual and potential peaks, direct switching of certain consumer loads, and so on, may be used to avoid the shifting peak problem.

Shadow Pricing

In the idealized world of perfect competition, the interaction of many small profit maximizing producers and welfare maximizing consumers gives rise to market prices that reflect the true economic costs, and scarce resources are efficiently allocated. However, conditions are likely to be far from ideal in the real world, especially in the developing countries. Distortions due to monopoly practices, external economies and diseconomies (which are not internalized in the private market), interventions in the market process through taxes, import duties, and subsidies, et cetera, all result in market (or financial) prices for goods and services, which may diverge substantially from their shadow prices or true economic opportunity costs. For example, in a country where subsidized diesel fuel is available for electricity generation, the appropriate shadow price would be the import price rather than the artificially low market price. Moreover, if there are large numbers of poor consumers, pricing based only on strict efficiency criteria may be socially and politically unacceptable. Such considerations necessitate the use of appropriate shadow prices (instead of market prices) of inputs to the electricity sector, to determine the optimal investment program as well as the LRMC.

2.4 Case Studies

In this section, the results of several case studies are presented and analyzed, to illustrate the methodologies and models for optimizing investment and pricing policy, described earlier in the paper. They are not necessarily the most complex examples possible, but are selected to provide straightforward clarification of the principles involved.

Optimizing Power System Reliability and Planning

This case study of Costa Rica deals with optimization of distribution system reliability and planning. A distribution study was selected because it is more typical of relatively recent advances in the state of the art in engineering. This supply-side consideration parallels the major demand side effort to measure economic outage costs. The engineering aspects are especially relevant, because generation and transmission system models and planning techniques tend to be much more sophisticated and better investigated than those generally used in distribution network studies, although usually distribution failures account for most consumer outages (Munasinghe 1979). Furthermore, it is possible to compute load-point outage frequency and duration indexes, which are appropriate for estimating consumer outage costs more accurately for distribution systems than for generation and transmission systems. The case study also shows how reliability expectation could affect the demand for electricity.

Before presenting the case study, however, it is useful to review some basic reliability and outage cost concepts.

Reliability

The concept of reliability in engineering may be simply defined in terms of the probability that a component, or a system, will perform its intended functions satisfactorily over some period of time, subject to actual operating conditions. The purpose of an electric power system is to supply power to consumers at some required rate, at the time and place of their choosing, while maintaining an acceptable quality of service, i.e., voltage and frequency levels must lie within specified limits.

Thus an ideal electric power system which unfailingly supplies power to consumers, whenever required, is by definition a perfectly reliable one, and conversely, a system which is never able to deliver electricity to users could be termed totally unreliable. All real-world power systems lie between these two extremes, and furthermore, as the system reliability level is improved, there is an inherent trade-off between the increased costs of supply, and the reduction in the inconvenience and costs imposed on consumers due to power shortages. Therefore, it is important to develop criteria and methods of assessing and ranking systems according to reliability levels.

As in the case of any other product or service, electric power shortages occur when the demand exceeds the supply, due to the failure of planners to correctly predict all the uncertainties in supply and demand. The stochastic nature of demand manifests itself through unforeseen increases in the load level, e.g., the sudden build-up in air-conditioning load due to unusually warm weather. Similarly, the randomness of supply is characterized by the unexpected failure or outage of the various components which make up a power system, the unavailability of water for hydroelectric generation, and so on. For a detailed definition of a component outage, see Endrenyi (1978) and *IEEE* (1973).

From the consumers' viewpoint, power shortages manifest themselves in a number of ways, including complete interruptions of supply (or blackouts), frequency and voltage reductions (or brownouts), and instability effects such as erratic frequency fluctuations and power surges. While all these phenomena are likely to inconvenience and impose costs on electricity users, the consequences of supply interruptions are most severe, and probably also the easiest to define. For example, a blackout will disrupt productive activity, and prevent the enjoyment of leisure, whereas a voltage reduction may have lesser effects such as the inconvenience caused by the dimming of lights, and so on. Mains frequency variations would affect a power system's own automatic control equipment, as well as other synchronous devices such as electric clocks, while power surges could also cause some damage to equipment, but these consequences are difficult to isolate. Therefore, although the term outage cost (which is used throughout) is meant to encompass the expenses incurred by consumers due to all types of shortages, in practice, the principle emphasis will be on the impact of supply interruptions. Moreover, the effects of random or unexpected interruptions will be stressed, since they are likely to impose much greater hardships on users than known or planned power cuts.

The main safeguard against the incidence of shortages is the provision of sufficient redundancy and slack capacity throughout the system to meet unexpected contingencies. Once a failure has occurred, in the short run, voluntary decreases in demand, voltage, and frequency reductions, and either operator-controlled or automatic load shedding may occur. The connecting and disconnecting of generators and loads to the system, also give rise to technical problems involving sequencing, matching, and synchronization, but these are usually handled routinely. If the load is not curtailed in an orderly manner, an uncontrolled system failure is likely to occur, accompanied by effects such as instability, e.g., frequency fluctuations and power surges, the tripping of and damage to generators and lines, and so on, finally culminating in widespread and unpredictable blackouts.

Bulk transmission reserve capability is ensured by following basic design principles, such as having adequate thermal margins (i.e., to prevent overheating), steady-state, dynamic, and transient stability against disturbances (e.g., the build-up of large voltage oscillations following a lightning stroke), avoiding bottlenecks in power flows, and providing alternative pathways to feed load centers through the use of ring systems, etc. In an interconnected system, it is important to avoid prolonged and far-reaching blackouts caused by the progressive overloading and cascading failure of many lines, following the initial outage of a single link (e.g., due to a three-phase fault to ground). The design of distribution networks also provides a certain measure of redundancy by the adoption of techniques such as allowing extra safety margins for substation capacity and power flow along feeders, as well as the use of switches to isolate faulted sections of line, and to temporarily connect customers to circuits which are not affected by the outage.

There are two different (but interrelated) approaches for maintaining the quality of supply to customers. The first viewpoint which deals with the assessment of system security is essentially short-term, with the objective of guiding system operations on an hourly or daily basis. The second approach is concerned with reliability evaluation for use in the long-range planning and design of power systems and in the long-run potential tradeoff between system costs and outage costs; it is the more relevant one here.

Reliability concepts are required to establish target reliability levels for purposes of system design, as well as to analyze and compare the future levels of system reliability with target values. Historically, the early use of reliability criteria in system planning relied on simple rule-of-thumb, deterministic approaches, such as the concept of ensuring an adequate reserve margin or use of a single- or double-worst fault design criterion (i.e., the capability of the system to meet the load, after the loss of the largest or the two largest generating units, respectively).

However, in light of the basically probabilistic character of system reliability and component outages, the next stage of sophistication in reliability analysis led to the development of combinatorial (probabilistic) methods. With these combinatorial concepts and models, it was possible to compute the long-term expectation that a large and complex system would be in a given state or condition (e.g., unable to meet the load), by using relevant load data in conjunction with information regarding the probable availability of the individual system components (e.g., a generator), appropriately "combined" according to the axioms of probability. Familiar reliability indices such as the loss-of-load probability (LOLP), and loss-of-energy probability (LOEP) were estimated using this approach (Knight 1972, Munasinghe 1979, Sullivan 1977, Endrenyi 1978, Vardi and Avi-Itshak 1981, IAEA 1984).

The latest generation of probabilistic tools for power system reliability analysis are based on the idea of overlaying the combinatorial technique with the concept of the system viewed as a stochastic process which evolves over time. With this latter approach, the system is commonly modelled as a discrete-state, continuous-transition Markov process (Endrenyi 1978). At any given moment the system may change from one state to another, due to events such as component outages, or actions designed to restore normal operation. Once all the possible states are identified, and the rates of transition between these states (which are determined by the corresponding contingencies or restorative control actions) are known, the system may be analyzed. The transition rate matrix, which specifies the transition rates between all possible pairs of states, provides a useful mathematical summary of the Markov process. The latter may also be visually depicted by means of a state-space diagram, with a number of closed boxes representing the system states, and a series of lines joining the boxes, representing the possible transitions. Among the advantages of this type of model is the fact that it facilitates the estimation of the expected frequencies and durations of outages, the modelling of the impact of exogenous shocks, e.g., sudden load, and the treatment of nonindependent events, e.g., failure and repair of different system components. We discuss the probabilistic measures of system reliability below.

The loss-of-load probability or LOLP criterion (also called the loss-of-load expectations), is the reliability index which is commonly used, especially in generation planning. System failure is most often defined in terms of inability to meet the daily peak load and, therefore, in the context of a generation study, the LOLP may be interpreted as the average number of days over some long period of time, during which the peak demand is expected to exceed the available generating capacity. For example, the target LOLP frequently used in the United States is 1 day in 10 years, while in European countries, the corresponding standard used varies from 1 day in 15 years to 1 day in 2.5 years. In the case of transmission and distribution planning, time is measured continuously, rather than in terms of the daily load cycle, and therefore the LOLP is given in terms of the long-term average fraction of total time the system is expected to be in a state of failure, e.g., defined as an interruption of supply, or sometimes in terms of a voltage reduction or component overload.

The LOLP is often supplemented by the expected-loss-of-load (XLOL) index, which indicates the expected magnitude of the unsupplied load, given that a failure has occurred. Another index which measures the magnitude of system failure is the loss-of-energy expectation or LOEE (also called the expected unserviced energy). The LOEE is the expected amount of energy not supplied due to outages in the long run, but expressed as a fraction of the total energy demanded. The expected frequency or mean recurrence time between outages, and the expected duration of such events when they occur, are a more recently discovered pair of reliability indices (FAD), which are widely used in transmission and distribution system planning, but rather infrequently in generation studies.

Each of the indices has its strengths and weaknesses, and does not give a complete description of outages. For example, although it is a measure of the probability of system failure, the LOLP does not indicate the expected magnitude of the load lost during such a contingency, and furthermore, since it is a long-run concept, the impact of operating procedures and criteria which could modify the risk of failure are effectively ignored. On the other hand, the XLOL provides the expected magnitude but neither the probability nor the average duration of an outage. Like the LOLP, the FAD measure also does not give the mean size of the failure, but since it is commonly specified at the load-point or particular point of consumption, the number of affected customers and the magnitude of the interrupted load may be deduced.

Another problem with these commonly used reliability measures is that the underlying models, the methods of computation, and their interpretation may all be different and the various indices are generally neither comparable, nor consistent (in the sense that they could yield different rankings of systems, by reliability level). Often, even the same index may be calculated in more than one way. For example, it is possible to derive the LOLP using a conventional model in which each generator is represented in two or more discrete states, or alternatively, by modelling the whole generating system as a normal distribution (Allan and Takieddin 1977). Similarly, the FAD measure of outages may be interpreted differently, depending on whether it is defined at the load-point as mentioned earlier, or specified as a system-wide generation/transmission index (Allan and Takieddin 1978, Marko 1978). Finally, if the LOLP and LOEP criteria are compared, the former index will rank a system which has a long-lasting failure on a single day above a system with two very short failures on separate days, whereas the latter measure is likely to reverse the order of ranking.

Two broad categories of mathematical techniques are used to calculate the values of power system reliability indices. The class of analytical methods rely on the direct (or analytical) manipulation of the basic probability axioms (IAEA 1984). In contrast, the Monte Carlo methods use the computer to first stimulate various random events in a manner analogous to throwing dice (e.g., mean time to failure, or repair time of components), and then estimate system reliability levels (Noferi et al. 1975).

Outage Costs

Outage costs may be defined as direct or indirect, respectively, according to whether they are incurred because an outage actually takes place, or simply because an outage is expected to occur. For example, during an outage, direct outage costs are likely to be incurred since normal productive activity is disrupted. In contrast, indirect outage costs are incurred in the absence of an outage itself, because consumers may adapt their behaviour patterns in ways that are less efficient, or more costly, but less susceptible to outage disruptions; or, they may purchase alternative (standby) sources of energy. Although indirect outage costs cannot be attributed to any particular outage, they depend on the general level of reliability, and represent real resource costs which should be considered, when attempting to estimate the economic costs associated with alternative reliability levels. Generally, direct outage costs are related more to the short-term impact of unexpected outages whereas indirect outage costs arise from long-term considerations of outage expectation, including the effects of planned power cuts. In practice, it would often be difficult to estimate outage costs associated with a particular level of reliability because such costs may consist of a combination of direct and indirect components. A number of recent studies have sought to analyze and measure outage costs (see Chapter 7).

Production is a process in which capital and labour are combined with other inputs such as raw materials and intermediate products to produce a time-stream of outputs. Under conditions approximating those of perfect competition, the net social benefit of a marginal unit of output in a given time period equals the value of the output minus the value of inputs. For infra-marginal units of output, producers' and consumers' surplus must be included when measuring the net benefits of production. When market distortions are present in the economy, appropriate shadow prices have to be used to value inputs and outputs. The net social benefit resulting from a time-stream of marginal outputs equals the present value of this stream of net social benefits.

When an outage disrupts production, the net benefits derived from such activities are reduced, i.e., direct outage costs are incurred, since the costs of inputs are increased, and/or the value of outputs is reduced. Specifically, an outage can cause raw materials, intermediate products, or final outputs to spoil, and it can also result in productive factors being made idle. The spoilage effect leads to an opportunity cost equal to the value of the final product not being made available as a result of the outage, minus the value of additional inputs not used because the final product was not produced. However, if the value of the output is not easily determined, as in the case of household and public sector outputs which are not directly sold on the market, then it is necessary to use the cost of producing the spoiled product or output as a minimum estimate of the resulting cost of the outage.

Distribution System Reliability

A case study was carried out to test empirically the new methodology described earlier. We begin by adapting, for practical application, the analytical approach summarized in Figures 2.2 and 2.9.

Let the annual stream of future power system supply costs (such as investment, maintenance, etc.), net of marginal generation costs of kWh not supplied due to outages, and associated with some given system expansion program i, be denoted by SC^i_t, for t = 0,..,T. Also, let the corresponding costs of poor supply quality be given by QC^i_t. Next we define the present discounted values:

supply costs:

$$SC^i = \sum_{0}^{T} SC^i_t/(1+r)^t, \text{ and}$$

shortage costs:

$$QC^{i} = \sum_{0}^{T} QC^{i}_{t}/(1+r)^{t},$$

where r is the appropriate economic discount rate, and T is the planning period.

Ideally, the reliability R should be represented by a composite and disaggregate vector of many attributes of supply quality such as the frequency and duration of outages, or voltage variations experienced by individual consumers. In practice, we work with an aggregate index that approximates R for part or all of the power system. One such energy based index may be defined:

$$R^{i} = 1\text{-}\{[\sum_{0}^{T} \frac{OE^{i}_{t}}{(1+r)^{t}}] / [\sum_{0}^{T} \frac{TE^{i}_{t}}{(1+r)^{t}}]\}$$

where OE^{i}_{t} = electric energy not supplied because of outages in year t, and TE^{i}_{t} = total electric energy that would have been supplied in year t if there were no outages. An alternative index based on the effects of poor supply voltage, is described later.

It is also convenient to define:

$$TB^{i} = \sum_{0}^{T} TB^{i}_{t}/(1+r)^{t},$$

where TB^{i}_{t} are the total benefits of electricity consumption associated with system design i, in the absence of costs due to poor supply quality. Then choosing the system plan m which maximizes net benefits also yields the optimal reliability level Rm. Note that we are interested only in relative changes, and therefore TB need not be estimated completely. Instead only the differential benefits or change in total benefits (due to induced demand) between a given system design i and some baseline system plan b needs to be evaluated:

$$dTB^{ib} = TB^{i} - TB^{b}$$

The method for practical application of the theoretical model follows the procedure summarized earlier in Figure 2.9.

Electricity Load and System Costs in Guanacaste, Costa Rica

The optimization case study was carried out in the service area of the Rural Electric Co-operative of Guanacaste, Costa Rica (Munasinghe, Scott et al. 1982). In 1978, the

Co-operative served 7867 consumers and provided 22,057 megawatt-hours of energy. Of this amount, about 50 percent was consumed by residential consumers and 25 percent each by commercial and industrial consumers. The primary industrial activities in the area were sugar milling, grain processing, ice making and mechanized ranching. Hotels and restaurants provided the major share of commercial electricity demand. Residential consumers in the Guanacaste area own a variety of electricity-using appliances and consume 1.5 megawatt-hours of electricity per household per year.

Voltage Drop and Supply Quality

The quality of power supply in Guanacaste encompasses both voltage variations and outages. Consumers close to substations usually have good supply voltage (that is, service at the nominal or standard voltage) while those further away usually have supply below the nominal voltage. Thus, on average, longer lines usually have poorer voltage. Outages occur for a variety of reasons, mainly due to natural occurrences like a tree branch falling across a distribution line or a truck colliding with a utility pole. The frequency of such outages for a particular line are likely to depend on the length of the line. The duration of outages will also tend to vary with the length of a particular feeder as the time taken to locate a fault will be greater if the line is longer. More generally, poorly maintained or overloaded systems will exhibit both low voltage and high outage rates. For our purposes, we will focus on the voltage drop and define three broad quality of supply categories: medium, low, and very low. The high quality of supply category, which is more appropriate to an outage sensitive urban industry, is not relevant for this study.

Consider the difference between the voltage drop level for a single consumer and that of the whole network. All consumers being served by a particular rural electric system do not suffer the same voltage drop. Since we seek only a broad categorization of the quality of electricity supply, let us define the quality of supply (in a given year) based on the "average" voltage drop for the whole network, as specified by an aggregate voltage drop index V,

$$V = \sum_{1}^{n} V_i L_i \; / \; \sum_{1}^{n} L_i$$

where L_i is the total load experiencing a voltage drop V_i percent from the nominal level (and there are n such levels V_i).

The three quality of supply levels were defined according to the voltage variation parameter V, such that:

medium supply quality - voltage variation less than or equal to 5 percent of nominal voltage.

low supply quality - voltage variation between 5 percent and 10 percent of nominal voltage.

very low supply quality - voltage variation greater than or equal to 10 percent of nominal voltage.

Residential Electricity Consumption

An explicit model that relates planned electricity consumption of households to price, income, and supply quality, is presented in Chapter 5. Based on this work, we may write:

$$\ln X_{er} = C + a_L D_L + a_M D_M + t_1 \ln p_s + k_0 \ln p_x + k_1 (\ln p_x)^2 + t_3 \ln Y + u \qquad (2.4)$$

where

X_{er} = electricity demand per household
p_x, p_s = prices of electricity and substitute energy
Y = household income
$D_L = 0$, $D_M = 1$ for medium supply quality
$D_L = 1$, $D_M = 0$ for low supply quality
$D_L = 0$, $D_M = 0$ for very low supply quality

and C, a_L, a_M, t_1, k_0, k_1 and t_3 are constants to be estimated.

Since proportional changes in all price and income variables do not affect demand, we also impose a linear restriction:

$$t_1 + k_0 + k_1 \ln p_{xav} + t_3 = 0 \qquad (2.5)$$

where p_{xav} is the mean price of electricity in the sample. The data sample consisted of 86 observations from 9 different rural electric networks in Costa Rica, each having approximately 10 years of observations. The networks were selected such that there were three of them in each supply quality category: medium, low and very low. The determination of the supply quality for each network for each year was made, based on the criteria described previously. The ex-post residential price charged for electricity and the per capita residential electricity demand were obtained from official statistics. Data on per capita incomes for persons in each network were available by county. Costa Rica which is a relatively small country (under 20,000 square miles) is divided into approximately 80 counties, and the networks considered in the study typically covered one or two counties. For the sample considered p_{xav} = Colones 0.256 (US\$1 = Colones 8.5 in 1978).

Table 2.1 shows the OLS coefficient estimates and the t-values for the residential demand function, Eq.(2.4), subject to the linear restriction, Eq.(2.5). The equation's R squared is 0.939 and all coefficients except t_1 are significant at the 95 percent level.

Commercial and Industrial Electricity Consumption

A demand equation similar to Eq.(2.4) may be derived for commercial electricity consumers in Costa Rica.

$$\ln X_{ec} = C + b_L D_L + b_M D_M + r_1 \ln p_s + n_0 \ln p_x + n_1 (\ln p_x)^2 + r_3 \ln V_c + u \qquad (2.6)$$

where

X_{ec} = commercial electricity demand per customer
V_c = value added per commercial establishment
D_L = 0, D_M = 1 for medium supply quality
D_L = 1, D_M = 0 for low supply quality
D_L = 0, D_M = 0 for very low supply quality

and C, b_L, b_M, r_1, n_0, n_1 and r_3 are constants to be estimated. Note that now there is no homogeneity condition to be satisfied as in the case of residential consumers.

Table 2.1

Coefficients for the Residential Demand Function

Symbol	Value	t-Statistic
C	-0.492	-
a_L	0.223	3.19
a_M	0.385	5.58
t_1	0.055	0.39
k_0	-0.943	-7.92
k_1	-0.097	-7.92
t_3^*	0.802	-

* t_3 was calculated using the linear restriction
$t_1 + k_0 + k_1 \ln(0.256) + t_3 = 0$

The sample of 86 observations that was used in the case of residential consumers was also used for commercial consumers. Table 2.2 shows the OLS regression results, when the demand function, Eq.(2.6), was estimated. The equation had an R squared of 0.742 and all coefficients except b_L and r_1 were significant at the 95 percent level. The theoretical demand equation for industrial electricity consumers is quite similar to the function for commercial users, but it was not possible to estimate an industrial demand function, because of data constraints. However, as shown later, approximate results were derived in this case.

Demand Forecasts For Different Qualities of Supply

The rural electric network presently existing in Guanacaste provides electricity of medium supply quality, and has done so for many years in the past. Two separate estimates of future demand were made for this network. The first was based on time trend analysis of the historical electricity consumption data for the various consumer categories. The second was based directly on the specific demand functions estimated earlier (except for industrial consumers and public lighting, accounting for about a quarter of the total load, where only time trend forecasts were available). The two approaches yielded very similar results. An "average" demand forecast lying between these two estimates was used as the base case demand corresponding to a medium quality of supply expected by consumers.

Table 2.2

Coefficients for the Commercial Demand Function

Symbol	Coefficient Value	t-Statistic
C	3.000	-
b_L	-0.141	-0.73
b_M	0.586	3.46
r_1	-0.828	-1.87
n_0	-1.487	-5.08
n_1	-0.287	-3.59
r_3	0.430	2.43

The demand forecasts corresponding to rural electric systems providing electricity of low as well as very low qualities of supply were then derived using the base case demand forecast and the appropriate coefficients from the demand functions relating demand and quality of supply (assuming no changes in other explanatory variables). If E^M_{com}, E^L_{com} and E^{VL}_{com} respectively represent the total commercial demand for systems of medium, low, and very low supply qualities, and E^M_{res}, E^L_{res} and E^{VL}_{res} are the corresponding values for households, it is possible to show that:

$$E^L_{res} = 0.85\ E^M_{res},$$

$$E^{VL}_{res} = 0.68\ E^M_{res},$$

$$E^L_{com} = 0.56\ E^M_{com} \text{ and}$$

$$E^{VL}_{com} = 0.56\ E^{M}_{com}$$

The effect on residential consumers (i.e., the smaller one), was used to calculate the changes in industrial demand when the quality of supply changes. Since industrial consumers constituted less than 25 percent of total demand throughout the planning period, any errors due to misstating the effect on industrial consumers are likely to be small.

Since the existing rural electric network in Guanacaste provides electricity of medium quality, even if no improvements are made in the starting year (1978), it would have taken some years before the supply quality deteriorated -- due to the growth of demand and lack of maintenance. Therefore, for the poorer service quality systems, the decline in reliability and lowered load growth was assumed to occur over a six year transition period. This reflects both the gradual change in system quality as well as the gradual change in consumer expectations. The computed load forecasts for the medium, low, and very low quality networks are summarized in Table 2.3.

Table 2.3

Load Forecasts for Different Systems

	Medium		Low		Very Low	
Year	MWh Total	Peak Load (kW)	MWh Total	Peak Load (kW)	MWh Total	Peak Load(kW)
1978	22,057	5,036	22,057	5,036	22,057	5,036
1979	25,826	5,896	24,638	5,625	23,913	5,460
1980	29,831	6,811	27,289	6,231	25,614	5,848
1981	34,051	7,774	30,046	6,860	27,177	6,205
1982	38,483	8,786	32,705	7,467	28,920	6,603
1983	43,011	9,820	35,259	8,050	30,727	7,016
1984	47,639	10,876	37,942	8,663	32,592	7,441
1985	52,436	11,972	40,811	9,318	34,560	7,891
1986	57,404	13,106	44,650	10,194	37,808	8,632
1987	62,532	14,277	48,609	11,098	41,159	9,397
1988	67,795	15,478	52,669	12,025	44,597	10,182
1989	73,268	16,728	56,887	12,988	48,167	10,997
1990	78,948	18,025	61,265	13,988	51,874	11,844
1991	84,837	19,369	65,798	15,023	55,712	12,720
1992	90,977	20,771	70,520	16,101	59,711	13,633
1993	97,341	22,224	75,396	17,214	63,851	14,578
1994	103,958	23,735	80,483	18,378	68,154	15,561
1995	110,772	25,290	85,702	19,567	72,580	16,571
1996	117,766	26,887	91,051	20,783	77,119	17,608
1997	124,917	28,520	96,506	22,034	81,750	18,665
1998	132,202	30,183	102,060	23,302	86,471	19,743
1999	139,595	31,871	107,680	24,585	91,254	20,835
2000	147,132	33,592	113,403	25,892	96,125	21,947

System Supply Costs (SC)

A typical rural electric distribution system can be divided into several distinct sub-systems: the distribution substation, primary system, distribution transformers, and secondary system. Voltage drops occur in both the primary and the secondary systems and the final voltage at the consumer end depends on both voltage drops. Several primary and secondary systems were designed and coupled to yield four separate plans which could provide electricity at four different qualities of supply, measured by an aggregate voltage variation index (VVI) defined as follows:

$$\mathrm{VVI} = \frac{\sum_{0}^{T} \sum_{1}^{p} L^{k}_{i} . V^{k}_{i}/(1+r)^{k}}{\sum_{0}^{T} \sum_{1}^{p} L^{k}_{i}/(1+r)^{k}}$$

where

L^{k}_{i} = the total load of all customers suffering a voltage drop of V^{k}_{i} percent during year k
r = an appropriate discount rate
T = number of years in the plan

and there are p voltage drop levels V_1, $V_2 ... V_p$.

Note that the two indices VVI and V (defined earlier) are similar, except that the former has an added time dimension. The value of VVI is used to correlate the system plan with the corresponding demand forecast. Table 2.4 shows the four networks and their corresponding values of VVI. The system design process provided the annual investment and O&M costs for each of the system plans, and the corresponding losses. The marginal economic costs of power supply at the transmission level were also estimated (using Colones 1572 per kW and 0.128 per kWh). In order to account for the finite horizon of the system designs, salvage values for all plants were calculated, based on the useful life remaining at the end of the planning period. The results of the system costs calculations are summarized in Table 2.5.

Table 2.4

Characteristics of the Overall Plans Used in the Optimization

Plan	Primary* System	Secondary System	VVI	Overall* Plan Quality	Demand Forecast Used
A	1/IM	Good	4.50	IM	Medium
B	2/M	Good	5.00	M	Medium
C	3/M	Intermediate	8.56	L	Low
D	4/L	Poor	11.62	VL	Very Low

* IM = Improved Medium; M = Medium; L = Low; VL = Very Low

Table 2.5

System Costs

Present Discounted Value (in Thousands of Colones)

Discount Rate	Plan Number	Plan Quality	Total Investment and O&M Costs	Total Cost of Electricity Transmission	Total Cost of Losses	Total Salvage Value	Total Costs of System[a]	Relative Costs of System
0.08	Plan A	IM	53,548	324,913	21,027	9,562	389,926	104,453
	Plan B	M	41,191	324,913	22,134	8,339	379,899	94,426
	Plan C	L	33,045	261,443	25,169	5,814	306,843	28,370
	Plan D	VL	20,978	227,010	42,407	4,922	285,473	0
0.10	Plan A	IM	45,882	266,271	17,482	6,373	323,262	84,913
	Plan B	M	35,845	266,271	18,403	5,695	314,824	76,475
	Plan C	L	29,022	215,696	20,910	3,970	2,658	23,309
	Plan D	VL	18,411	188,105	35,182	3,349	238,349	0
0.12	Plan A	IM	40,277	221,935	14,785	4,378	272,610	70,156
	Plan B	M	31,714	221,935	15,564	3,885	265,328	62,874
	Plan C	L	25,908	181,050	17,672	2,712	221,918	19,464
	Plan D	VL	16,448	158,602	29,682	2,278	202,454	0

[a] Total Cost of System = Total Investment and O&M Costs + Total Cost of Electricity at Transmission Level + Total Cost of Losses - Total Salvage Value.

Quality of Supply Costs and System Optimization

As discussed earlier, the quality of supply costs consist of outage or interruption costs, and costs due to variations in voltage. Furthermore, the functional forms of these costs vary for different consumer categories. Table 2.6 shows the main categories of electricity consumers surveyed in Guanacaste, and the quality of supply costs they experienced. The survey and estimation methodology is outlined below (see Chapters 8 and 9 for details).

Table 2.6

Consumer Types and Outage and Voltage Variation Costs

Consumer Category and Type	Outage Cost	Voltage Variation Cost
Residential	yes	yes
Industrial		
Sugar Mills	yes	no
Grain Processors	yes	no
Ice Factories	no[a]	yes
Farms/Ranches	no[a]	yes
Commercial		
Hotels	yes	no
Restaurants	yes	yes

[a] Relatively small unless outage duration is long (e.g., over 6 hours).

Residential Outage Costs (QC Component)

A comprehensive survey of residential consumers confirmed that power supply was especially important in the evening hours between 7:00 p.m. and 10:00 p.m., while outages during other times did not impose much costs or inconvenience on them. Furthermore, in the evening hours, consumers felt that electricity was most crucial between 8:00 p.m. and 9:00 p.m. when many of them watched television or ate their evening meal. They indicated that they were willing to pay to avoid unplanned outages of different durations between the hours of 7:00 p.m. and 10:00 p.m., but not for outages outside this period. They were willing to pay the most between the hours of 8:00 p.m. and 9:00 p.m. The direct amounts that consumers were willing to pay to avoid outages between 8:00 p.m. and 9:00 p.m. were found to be proportional to the duration of the outage. Alternative estimates of outage cost per unit of time were also obtained, to serve as a check, by asking consumers what they would consider as adequate payment to compensate for a doubling of their present outage rate during the hours of 8:00 p.m. to 9:00 p.m. and also what they would be willing to pay to have their present outage rate halved between 8:00 p.m. and 9:00 p.m. The answers to these questions yielded estimates of outage costs comparable to the estimate based on direct willingness to pay. These three questions on the cost of outages, though related, were not asked consecutively so as to minimize the chances that a respondent would extrapolate the answer to a previous question to answer a subsequent one. Based on these responses a household outage cost function of the form shown in Figure 2.11, was assumed.

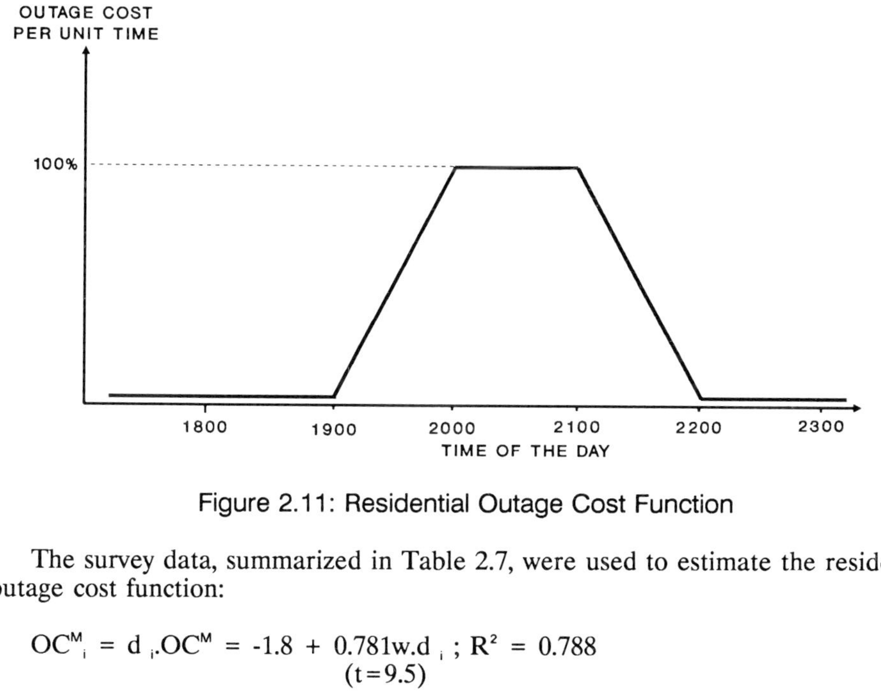

Figure 2.11: Residential Outage Cost Function

The survey data, summarized in Table 2.7, were used to estimate the residential outage cost function:

$$OC^M_i = d_i.OC^M = -1.8 + \underset{(t=9.5)}{0.781w.d_i} \; ; \; R^2 = 0.788$$

where OC^M_i = the outage cost of residential consumers for an outage of duration d_i hours, occurring between 8 p.m. and 9 p.m., and w = wage rate of residential consumers in colones per hour.

Total household outage costs may be estimated for the plan period, based on the above function, future household incomes, and the engineering estimates of frequency and duration of outages associated with each system plan.

Industrial and Commercial Outage Costs (QC Component)

Detailed surveys of industrial and commercial firms were conducted to determine the effects of electricity outages and variations in voltage on these firms. General questions concerned subjects such as wage rates, and input and output costs. Specific questions dealt with the effects of electricity outages and variations in supply voltage. For example, how would outages disrupt production? Would outages result in spoiled products or idle factors? Would actions such as purchasing backup energy sources be taken to reduce costs during outages.

On the basis of this information, both linear and piecewise linear outage cost functions were estimated. Linear functions had the form:

$$OC = a + b.d;$$

where OC is the outage cost per kilowatt-hour of energy used per year, d is the outage duration in hours, and a and b are constants to be estimated.

Table 2.7

Outage Costs and Wage Rates for Residential Consumers

OC/Hour (Colones)	Wage Rate/Hour (Colones)
40.0	45.50
2.0	2.84
23.0	17.50
4.0	6.82
4.0	4.55
4.0	7.39
4.5	5.68
80.0	90.90
3.0	7.39
3.0	14.77
10.0	28.41
25.0	34.09
5.0	8.52
4.0	5.11
4.0	11.36
30.0	56.80
50.0	42.85
3.0	4.55
20.0	56.80
50.0	56.80
30.0	22.70
10.0	14.20
20.0	4.55
25.0	34.09
10.0	28.41
10.0	14.20
3.0	25.57

Piecewise linear functions had the form:

$$OC = a_1 + b_1.d \qquad \text{for } d \leq d^*$$

$$OC = a_2 + b_2.d \qquad \text{for } d > d^*$$

where a_1, a_2, b_1 and b_2 are constants and d^* is some critical length of outage (determined by the nature of the industrial process).

The outage cost functions estimated for different industrial and commercial electricity consumers are shown in Table 2.8. These functions were used with future electricity demand forecasts, and outage frequency and duration estimates, to compute total industrial and commercial outage costs for the different network designs.

Table 2.8

Outage Cost Functions for Commerce and Industry

Industry or Commercial Type	OC Function Type	Estimated Coefficients a	b	Remarks
Grain Processors	piecewise linear	0	.02799	d ≤
		.05226	.00187	d > 2
Sugar Mills	piecewise linear	0	.00324	d ≤ 1.5
		-.000162	.00432	d > 1.5
Restaurants	linear	0	.0004359	all d
Hotels	linear	0	.0003615	all d

Voltage Variation Costs (QC Component)

The cost of voltage variations for the various consumer types was estimated using the expression:

$$x = 0.0023V^{1.659} \qquad V \leq 39.0$$

and

$$x = 1.0 \qquad V > 39.0$$

where x is the fractional reduction in lifetime of windings when an appliance is operated at a voltage V percent below the nameplate rating. The values were based on manufacturers' test data. This functional form was also verified with Costa Rica field information, as described below.

As one example, a mechanized ranch which used relatively large motors for pumping water reported that damage due to low voltage operation had reduced the lifetime of motor windings from the normal 10 years to 8 years. This consumer's service voltage was approximately 15 percent below the nameplate rating. When V = 15, the formula indicates that x = 0.20, yielding a motor lifetime of 8 years -- which is consistent with the observed result.

The future voltage variation costs associated with each system design was estimated using the above formula, the index VVI, the growth of relevant consumer appliances (g), the nominal appliance lifetimes (L), and the costs of replacement (B), as given in Table 2.9. Finally, the total costs incurred by consumers, due to poor supply quality, are summarized in Table 2.10.

Table 2.9

Voltage Variation Cost Parameters

Consumer Type	L (in Years)	B (in Colones)	r_n
Residential	10	192.5	0.08
Industrial			
Ice Factories	12	5,500	0.10
Farms/Ranches	10	16,000	0.10
Commercial			
Restaurants	10	5,000	0.10

Differential Consumption Benefits Due to Demand Curve Shifts

As discussed earlier, differential benefits arise because the quantity of electricity demand changes when the expected quality of supply changes. The total benefits, TB, were defined based on a perfect quality of supply RP, and therefore, the relevant demand curve to use in estimating these benefits is the one corresponding to perfect supply quality. Figure 2.12 shows the curve DP corresponding to perfect supply quality. Suppose X_e^{VL}, X_e^{L}, and X_e^{M} represent the quantities of electricity demanded when the qualities of supply are very low, low, and medium, respectively. Then the areas BCIH and ABHG represent the differential benefits between medium and low quality systems, and between the low and very low quality systems, respectively. In practice, none of the empirically estimated curves corresponded exactly to perfect quality. However, since the estimated curve for medium supply quality DM was valid for the entire range of voltage variation 0 percent to 5 percent, it is a close enough approximation to DP (i.e., zero voltage variation). This is essentially the same as assuming that the demand is saturated when the quality of supply is medium, or that further increases in supply quality will not lead to significant demand increases.

Table 2.10

Cost of Poor Supply Quality

(Present Discounted Value in Thousands of Colones)

			Outage Cost				Voltage Variation Cost					
Discount Rate	Plan Number	Plan Quality	Resi-dential	Indus-trial	Comm-ercial	Total	Resi-dential	Indus-trial	Comm-ercial	Total	Total Quality of Supply Costs	Relative Quality of Supply Costs
0.08	Plan A	IM	2,081	2,991	962	6,034	0	0	0	0	6,034	-3,244
	Plan B	M	2,339	3,221	1,037	6,597	14	21	1	36	6,633	-2,645
	Plan C	L	2,326	2,432	783	5,541	752	739	22	1,513	7,054	-2,224
	Plan D	VL	3,170	2,923	943	7,036	1,186	1,027	29	2,242	9,278	0
0.10	Plan A	IM	1,826	2,603	837	5,266	0	0	0	0	5,266	-2,548
	Plan B	M	2,049	2,799	902	5,750	11	16	0	27	5,777	-2,037
	Plan C	L	2,028	2,128	685	4,841	580	563	17	1,160	6,001	-1,813
	Plan D	VL	2,726	2,534	817	6,077	922	792	23	1,735	7,814	0
0.12	Plan A	IM	1,626	2,299	739	4,166	0	0	0	0	4,666	-2,014
	Plan B	M	1,820	2,469	795	5,084	9	12	0	21	5,105	-1,575
	Plan C	L	1,794	1,849	609	4,332	417	400	13	830	5,162	-1,518
	Plan D	VL	2,376	2,226	718	5,320	725	617	18	1,360	6,680	0

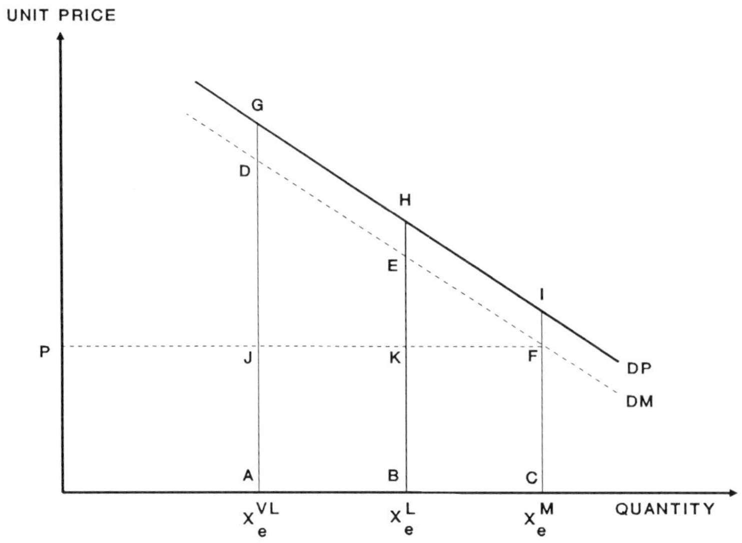

Figure 2.12: Measuring the Differential Benefits of Induced Demand

Referring to Figure 2.12, consider the differential benefits between the low and very low quality systems (measured relative to DM). This would be represented by area ABED and can be divided into two parts. Area ABKJ, representing the revenue change or the increase in consumers expenditure in moving from a system of very low quality to one of low quality, and the area JKED, representing the change in consumers' surplus. The electricity demand for residential and commercial consumers (estimated earlier), may be written in the form:

$$X_e = Ap_x^{(m_0 + m_1 \ln p_x)}$$

where A is a known constant for a given year, consumer type and supply quality; p_x is the electricity price; and m_0 and m_1 are estimated parameters. The area under the demand curve may be numerically integrated to calculate the differential benefits for these two consumer categories. A benefit to revenue ratio (BRR) was defined as the ratio between differential benefits and sales revenues (for example, the ratio of areas ABED to ABKJ). The differential benefits of industrial consumers were estimated by using the known industrial electricity revenues and assuming that the industrial BRR value was the average of the residential and commercial BRR values.

Table 2.11 shows the differential benefits for the different plans. These results are presented as total benefits (and revenues), by arbitrarily setting total benefits (and revenues) for the very low quality system to zero, for all consumer categories. This normalization simplifies the presentation and does not affect final result, because the goal is to pick the best plan and not to estimate the absolute value of net benefits. The numbers presented in the table are identical for Plans A and B (having improved medium and medium qualities of supply respectively), because the total demand is the same in each case.

System Optimization

As discussed earlier, the optimization procedure is based on selecting the system plan with the highest value of net benefits. Table 2.12 summarizes selected information from the previous tables and also shows the value of net benefits for each plan. Net benefits are highest for Plan B, for the entire range of reasonable discount rates 0.08 to 0.12. The fact that this number is negative has no significance because the net benefits for the very low quality plans were arbitrarily normalized to zero, and only relative net benefits are compared for purposes of optimization. The net benefits from Table 2.12 are plotted against supply quality in Figure 2.13. The smooth curves joining the four points (for each value of discount rate) are merely intended to be indicative. However, these curves suggest that the optimal value of VVI that maximizes NB lies between 5 and 6 -- closest to the discrete medium (M) quality plan B modelled in this study.

The values presented in the table indicate that if the demand function DP (corresponding to perfect quality) had been known and used as the benchmark instead of DM (the demand function corresponding to medium quality), the results of our analysis would not have changed. Using DP instead of DM would have increased the differential benefits between the very low and low plans as well as between the low and medium plans. The benefit difference between the medium and improved medium plans would still have been zero. These benefit increases would only reinforce the result -- that the movement from very low through low to medium supply quality was justified. A shift from the medium to the improved medium plan would still lead to a reduction in net benefits because the increase in system costs would be larger than the decrease in quality of supply costs.

In conclusion, the study shows that induced demand effects are much larger than the effects of outages and voltage variations, up to the point where demand begins to saturate when the quality of supply is improved beyond the medium level. Therefore, it appears that for power systems whose supply quality is generally poor to begin with (especially rural electric networks), the induced demand effects of variations in quality of supply could be a critical factor in determining the optimal quality of supply.

Optimizing Power System Losses

In this section, a case study is presented which illustrates the practical application of the loss optimizing framework described earlier. The most important conclusion is that, within practical limitations, loss reductions at distribution levels provide capacity and energy at far less cost than obtaining capacity and energy through the construction of new generation and transmission facilities. The case studies indicate that utilities can obtain savings up to $15 for every $1 of cost associated with reducing distribution system losses. Therefore, losses should be recognized as a dominant factor in the engineering criteria used in system planning, design, and operation.

Table 2.11

Revenues and Benefits of Alternate Plans
(Present Discounted Value in Thousand of Colones)

			Total Revenues in 1000s of Colones				Total Benefits in 1000s of Colones[a]			
Discount Rate	Plan Number	Plan Quality	Residential	Commercial	Industrial	Total	Residential	Commercial	Industrial	Total
0.08	Plan A	IM	31,668	24,560	15,783	72,011	43,883	45,667	25,569	115,119
	Plan B	M	31,668	24,560	15,783	72,011	43,883	45,667	25,569	115,119
	Plan C	L	16,891	0	8,419	25,310	26,812	0	14,524	41,336
	Plan D	VL	0	0	0	0	0	0	0	0
0.10	Plan A	IM	25,407	19,496	12,610	57,513	35,106	36,084	20,428	91,618
	Plan B	M	25,407	19,496	12,610	57,513	35,106	36,084	20,428	91,618
	Plan C	L	13,561	0	6,732	20,293	21,440	0	11,609	33,049
	Plan D	VL	0	0	0	0	0	0	0	0
0.12	Plan A	IM	20,686	15,701	10,225	46,612	28,495	28,912	16,564	73,971
	Plan B	M	20,686	15,701	10,225	46,612	28,495	28,912	16,564	73,921
	Plan C	L	11,049	0	5,460	16,511	17,393	0	9,421	26,814
	Plan D	VL	0	0	0	0	0	0	0	0

[a] Incremental benefits relative to the very low (VL) quality system.

Table 2.12

Global Characteristics of the Four Alternate Plans

				Present Discounted Value in Thousands[b] of Colones			
Disc. Rate	Plan Number	Plan Quality	Voltage Variation Index	System Costs	Quality of Supply Costs	Total Benefits	Net[a] Benefits
0.08	Plan A	IM	4.50	104,453	-3,244	115,119	13,910
	Plan B	M	5.00	94,426	-2,645	115,119	23,338
	Plan C	L	8.56	28,370	-2,224	41,336	15,190
	Plan D	VL	11.62	0	0	0	0
0.10	Plan A	IM	4.50	84,913	-2,548	91,618	9,253
	Plan B	M	5.00	76,475	-2,037	91,618	17,180
	Plan C	L	8.56	23,309	-1,813	33,049	11,553
	Plan D	VL	11.62	0	0	0	0
0.12	Plan A	IM	4.50	70,156	-2,014	73,971	5,829
	Plan B	M	5.00	62,874	-1,575	73,971	12,672
	Plan C	L	8.56	19,464	-1,518	26,814	8,868
	Plan D	VL	11.62	0	0	0	0

[a] Net Benefits = Total Benefits - System Costs - Supply Costs.
[b] Net benefits are relative since all values are measured with respect to an arbitrary zero value for the very low quality Plan D. This does not affect the choice of a plan which maximizes net benefits (Plan B).

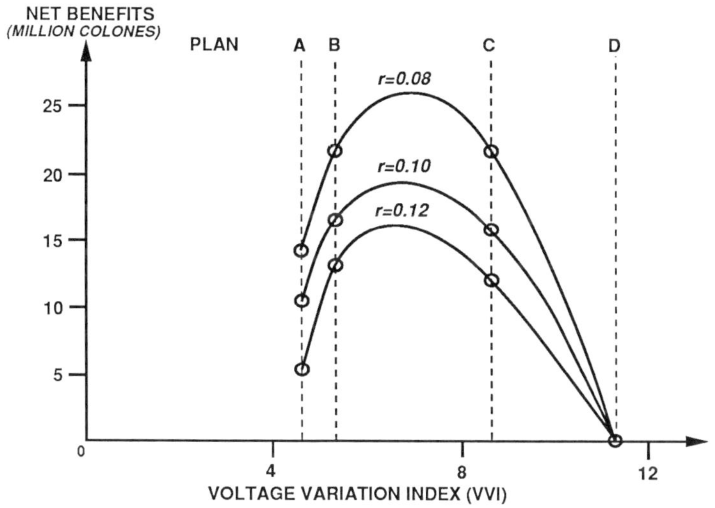

Figure 2.13: Net Benefits versus Supply Quality (Voltage Variation Index) for Different Values of Discount Rate (r)

Primary Conductor Loading

A portion of the distribution system of Punjab State, India, was used in evaluating losses on a primary system for a variety of conductor loading levels. The objectives of this case study were to determine:

1. the investment costs for various primary systems serving an area over a 20-year time frame. The major variation between systems was the maximum conductor loading permitted;

2. the cost of losses (demand and energy) associated with each system;

3. the ratio of savings through loss reduction to the cost of reducing those losses;

4. probable economic conductor loading limits.

Specifically, case studies were made for the following conditions:

1. Primary models were studied over a 20-year time frame with maximum conductor loadings at 100, 75, 50, and 25 percent.

2. Losses were evaluated for two cost levels: (a) $130/kW/year and $0.011/kWh (about 4 UScents per average kWh); and (b) $250/kW/year and $0.037/kWh (about 9 UScents per average kWh).The range of values 4 to 9 UScents per average kWh lost is representative of a wide variety of all-hydro, hydro-thermal, and all-thermal systems found in developing countries (LDC's). Some systems may have higher values of LRMC per kWh but hardly any would be lower.

3. The present worth of investment and O&M costs was derived using a discount rate of 12 percent. This long-term average discount rate (or opportunity cost of capital) is on the conservative side, since it is at the upper end of the scale appropriate to most LDC's. A lower value would further enhance the arguments for and urgency of distribution loss reduction.

The results of this case study are tabulated in Table 2.13 and shown graphically in Figure 2.14. Table 2.13 shows that for a system with maximum conductor loading of 100 percent and the higher value of losses, the present worth (P.W.) of losses will be $881,600 and the present worth of future investment and O&M costs is $4900. If maximum conductor loading is reduced to 75 percent, the P.W. of investment rises to $15,500 and P.W. of losses reduces to $717,300. The incremental investment of $10,600 produces a reduction in losses of $164,300 for a benefit to cost ratio (BCR) of 15.5 to 1 ($15.5 saved per $1 of cost).

Figure 2.14 indicates that the breakeven point (BCR = 1) for maximum conductor loading, which corresponds to the minimum value of NSC, is probably in the general neighbourhood of 15 to 25 percent -- a substantially lower level than loading limits based on existing criteria.

Table 2.13

Value of Loss Savings by Reducing Maximum Conductor Loading (No Power Factor Correction)

Maximum Conductor Loading (%)	P.W. Investment (1000$)		P.W. Cost of Losses (1000$)		Incremental Benefit-to Cost Ratio[2]
	Total	Incremental[1]	Total	Incremental[1]	
Losses Priced at $250/kW/year and $0.037/kWh					
100	4.9	-	881.6	-	-
75	15.5	10.6	717.3	164.3	15.5
50	30.8	15.3	554.0	163.3	10.7
25	104.0	73.2	332.9	221.1	3.0
Losses Priced at $130/kW/year and $0.011/kWh					
100	4.9	-	445.4	-	-
75	15.5	10.6	362.6	82.8	7.8
50	30.8	15.3	280.0	82.6	5.4
25	104.0	73.2	168.2	111.8	1.5

[1] Incremental with respect to next higher level of maximum conductor loading.
[2] Incremental loss savings divided by incremental investment.

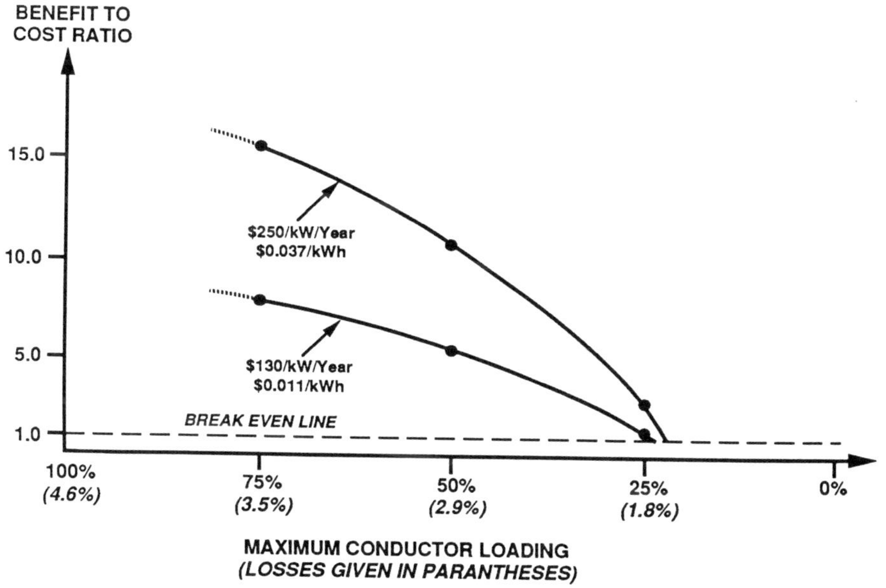

Figure 2.14: Incremental Benefit to Cost Ratio for Successive Reductions in Conductor Loading

Power Factor Control

Power factor is controlled by most utilities with the aid of capacitors on the primary system. Many devices, especially motors, require lagging kVAR's to operate. The lagging kVAR's may be supplied from the generators which requires the kVAR's to be transported through the entire system (transmission, subtransmission, stations, and primaries), or they may be supplied from capacitors located near the load centres.

Basically, power factor control reduces demand which results in reduced losses, better voltage regulation, and releases generation, transmission, and distribution capacity to serve other existing loads, or saves on investment by delaying the need to add new facilities to serve load growth.

In this case study, benefits have been evaluated for reducing losses only. This was accomplished by correcting the power factor from 80 to 95 percent for the Primary Conductor Loading systems. The system additions (without capacitors) to maintain conductor loading at 100, 75, 50, and 25 percent were left unchanged. Capacitors were added to those systems over the 20 years as required to maintain a 95 percent power factor. Consequently, capacitor additions are the same for each conductor loading condition. The benefits between systems vary because the load levels and losses vary with conductor loading. The maximum benefits from power factor control are related to the system with 100 percent maximum loading because there are more losses to eliminate by reducing kVAR flow.

A summary of the study is tabulated in Table 2.14 and graphed in Figure 2.15. For the higher costs of losses and 100 percent conductor loading, the P.W. of capacitor costs of $19,700 result in a P.W. saving in losses of $294,400. This situation provides $14.9 of savings for every $1 of cost -- an excellent benefit-to-cost ratio. Even the lowest benefit condition (lowest cost of losses and 25 percent conductor loading) provides $2.9 in savings for every $1 of cost. It is clear that power factor control via capacitors reduces losses with a high benefit-to-cost ratio.

Optimizing Power Prices

This case study demonstrates how the pricing methodology described earlier, can be actually implemented to develop a practical electricity tariff structure.

The first step in setting electricity prices is to calculate strict long-run marginal costs of supplying electricity (LRMC). Data constraints and the objective of simplifying metering and billing procedures suggest that these costs be calculated for (and hence tariffs be differentiated by): (1) various voltage levels (EHV/HV, MV, and LV); and (2) different time periods (peak or off-peak).

Next, tariffs based on strict LRMC should be adjusted to ensure that financial requirements are met. Since average costs of supplying electricity are typically rising, marginal cost-based tariffs are apt to produce a financial surplus. Such surpluses, however, can be disposed of in a manner which does not unduly distort consumer behaviour. For example, connection charges can be subsidized; or low-income consumers can be provided with a subsidized block of electricity to meet their basic needs.

Table 2.14

Value of Loss Savings by Correcting Power Factor from 80 to 95 Percent at Various Maximum Conductor Loading Levels

Maximum Conductor Loading (%)	P.W. of Capacitor Costs (1000$)	P.W. Cost of Losses (1000$) No Capacitor	P.W. Cost of Losses (1000$) With Capacitor	Loss Savings	Benefit to Cost Ratio[1]
Losses Priced at $250/kW/year and $0.037/kWh					
100	19.7	881.6	587.2	294.4	14.9
75	19.7	717.3	477.7	239.6	12.2
50	19.7	554.0	369.0	185.0	9.4
25	19.7	332.9	221.7	111.2	5.6
Losses Priced at $130/kW/year and $0.011/kWh					
100	19.7	445.4	296.7	148.7	7.5
75	19.7	362.6	241.5	121.1	6.1
50	19.7	279.8	186.4	93.1	4.7
25	19.7	168.2	112.0	56.2	2.9

[1] Loss savings divided by capacitor costs.

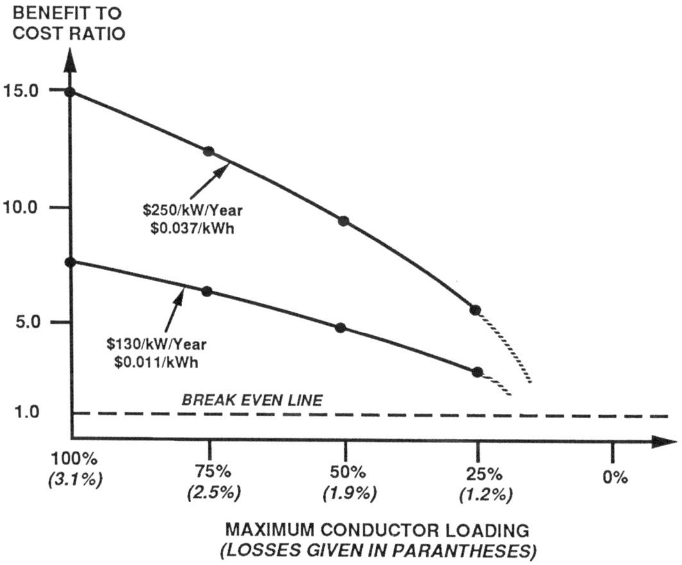

Figure 2.15: Benefit to Cost Ratios versus Maximum Conductor Loading for Power Factor Correction from 80 to 95 percent

Price distortions elsewhere in the economy may also justify the deviation of tariffs from strict LRMC. For example, if alternative energy sources are priced below marginal cost, pricing electricity below LRMC may be justified in order to prevent excessive use of alternative forms of energy.

In summary, the LRMC pricing approach provides an explicit framework for analyzing system costs and setting tariffs. Departures of tariffs from strict LRMC can be analyzed within this framework; and, such departures can be designed so as to reduce any economic efficiency costs, while still satisfying other socio-political constraints.

Below, strict LRMC's are estimated at various voltage levels in the Pakistan (WAPDA) power system. These costs are separated into: (1) marginal capacity costs, (2) marginal energy costs, and (3) customer costs; and, they are determined for periods of peak and off-peak demand. The period of peak demand is assumed to occur between 4 p.m. and 8 p.m. The principal sources of data for these estimates are WAPDA's expansion programme for the Fifth Plan (1978/79-1982/83) and their load forecast for the same period. Then, a set of tariffs is proposed which is based on these cost estimates after adjusting for financial and distributional requirements.

Estimating Strict LRMC

Sector Background: The Water and Power Authority of Pakistan (WAPDA) is the major partner in the power sector, accounting for about 80 percent of the public power supply in Pakistan. The Karachi Electric Supply Corporation (KESC) serves the Karachi metropolitan area and accounts for the remaining 20 percent of the public power supply in Pakistan. Finally, several other companies such as the Multan Electric Supply Company (MESCO) and the Rawalpindi Electric Power Company (REPCO) distribute electricity which they purchase in bulk from WAPDA.

Between 1970/71 and 1976/77, the average annual growth rates of generation, sales, and maximum demand for electricity in Pakistan were 6.9, 5.1 and 10 percent, respectively. By the end of 1976/77, the combined generation and maximum demand of WAPDA and KESC totalled 10,742 GWh and 2,003 MW, respectively, while sales to 2,325,000 consumers equalled 7,150 GWh. Thus, the benefits of electricity were available to about 22 percent of Pakistan's total population.

Capacity Costs: Marginal capacity costs include the future costs of generation, transmission and distribution. The relevant generation capacity costs are those of the plants which will be operated to meet peak demand. For the WAPDA system, these are gas turbine and gas-fired thermal units. Given the relative importance of each of these types of plants in meeting peak demand, and their costs per kilowatt, generation capacity costs at the bus-bars in border prices are calculated to be Rs.8875/kW installed. These calculations are summarized in Table 2.15.

Transmission and distribution capacity costs are estimated for EHV/HV, MV, and LV voltage levels. These voltage levels correspond to 66/132 kV, 1/33 kV, and 220/400 V, respectively. The planned investment programs for each of these voltage categories during 1980-1990 are summarized in Table 2.16, while projected peak demands at each voltage level over 1980-1990 are presented in Tables 2.17 and 2.18. Using these projections and the average incremental cost approach (see below), incremental EHV/HV, MV, and LV capacity costs are estimated to be Rs.5997/kW, Rs.1478/kW, and Rs.3203/kW, respectively.

Table 2.15

Generation Capacity Costs at the Busbar; Rs./kW Installed (1979 Prices)

Unit	Capacity Cost in Market Prices[1]	Capacity Cost in Border Prices[2]	Weighted Cap. Cost in Border Prices[3]
Gas-fired steam	16100	12558	6905
Gas turbine	5610	4376	1970
Weighted total			8875

[1] Capacity costs are adjusted to account for a 30 percent reserve margin and 2.1 percent station use of energy.

[2] This assumes a 50 percent foreign exchange component and a 50 percent local cost component of capacity costs. Conversion factors for the foreign exchange component (which includes 40 percent import duty), and the local component (which consists of local materials and skilled labor) are 0.71 and 0.85, respectively.

[3] The relative weights are 0.55 for steam and 0.45 for gas turbine.

Assume that in the year i, ΔMW_i and I_i are the increase in demand served (relative to the previous year), and the investment cost, respectively. Then the average incremental cost of capacity (AIC) is

$$AIC = \left[\sum_{0}^{T} I_i/(1+r)^i \right] / \left[\sum_{L}^{T+L} \Delta MW_i/(1+r)^i \right]$$

where r is the discount rate, T is the planning horizon, and L is the average time delay between the investments and commissioning dates of new facilities.

Energy Costs: Peak and off-peak energy costs at the generators equal the running costs of the units used during each of these periods to supply incremental amounts of energy. This is usually the running cost of the most inefficient plant used to meet demand during each period. As discussed above, increments in peak demand for the WAPDA system are met by using a mix of gas-fired thermal units (at Lyallpur, Multan, and Guddu) and gas turbines; while increments in off-peak demand are met by using one of the gas-fired thermal units. Using information on heat rates of these various plants, the BTUs/m^3 of gas, and the economic cost of producing gas, energy costs at the generators can be calculated. They are estimated to be Rs.266/kWh and Rs.225/kWh for peak and off-peak periods, respectively.

Table 2.16

Summary of WAPDA Investment Program 1980-1990
(1979 Rs. x 106)

Year	EHV/HV Transmission			MV Distribution			LV Distribution		
	Cost in Domestic Prices (1)	Cost in Border[1] Prices (2)	Discounted Cost in Border[2] Prices (3)	Cost in Domestic Prices (4)	Cost in Border[3] Prices (5)	Discounted Cost in Border Prices (6)	Cost in Domestic Prices (7)	Cost in Border[3] Prices (8)	Discounted Cost in Border Price (9)
1980	1224	967	863	215	182	163	322	272	243
1981	1226	970	774	274	231	184	412	348	277
1982	1587	1279	854	286	242	172	428	362	258
1983	1700	1344	254	323	273	173	485	410	261
1984	1632	1289	731	330	278	158	493	416	236
1985	1672	1322	670	335	284	144	503	425	215
1986	1698	1343	607	341	288	130	511	431	195
1987	1729	1367	552	348	294	119	522	442	179
1988	1847	1461	527	454	299	108	531	449	162
1989	1740	1375	442	366	310	100	549	464	149
1990	1969	1319	379	368	311	89	552	467	134
Total 1980-1988			6471						
Total 1980-1989						1451			2175

1 This assumes a 40 percent foreign exchange component, a 55 percent share for local materials and skilled labor, and a 5 percent share for unskilled labor. Conversion factors for the foreign exchange component (which includes a 40 percent import duty); local materials, and unskilled labor ar Rs 0.71, Rs 0.85 and Rs 0.75, respectively.

2 The assumed discount rate is 12 percent.

3 The same as 1 except that we assume no foreign exchange component of distribution investment.

Table 2.17

System Peak Demand at EHV, MV, and LV, 1979-1990 (MW)[1]

Year	Gen. Peak (1)	Station Use (2)	EHV/HV Trans. Loss (3)	HV Peak (4)	HV Cons. (5)	HV Peak Cons. (6)	MV Loss (7)	MV Peak (8)	MV Cons. (9)	MV Peak Cons. (10)	LV Loss (11)	LV Cons. (12)
1979	1972	39	316	1617	221	1396	145	1251	177	1074	317	757
1980	2149	54	301	1794	246	1548	161	1387	197	1190	351	839
1981	2342	54	304	1984	276	1708	162	1546	221	1325	370	955
1982	2553	56	306	2191	310	1881	163	1708	248	1460	381	1073
1983	2783	56	278	2449	347	2102	183	1919	278	1641	415	1226
1984	3033	61	243	2729	392	2337	203	2134	313	1821	441	1380
1985	3306	66	264	2976	413	2563	223	2340	331	2009	486	1523
1986	3538	71	283	3184	442	2742	239	2503	354	2149	420	1629
1987	3821	76	306	3439	478	2961	258	2703	382	2321	561	1760
1988	4127	83	330	3714	516	3198	278	2920	413	2507	607	1900
1989	4457	89	357	4011	557	3454	300	3154	446	2708	655	2053
1990	4813	96	385	4332	602	3730	325	3405	481	2924	707	2217

[1] The loss factors utilized in this table are those which measure losses as a percentage of incoming during periods of peak demand. These are summarized in Table 2.19.

Table 2.18

Summary of Projected Peak Demand at EHV/HV, MV, and LV, 1979-1990
(MW)

Year	EHV/HV Peak Demand[1] (1)	δEHV/HV Peak Demand (2)	Discounted δEHV Peak Demand[2] (3)	MV Peak Demand[3] (4)	δMV Peak Demand (5)	Discounted δMV Peak Demand (6)	LV Peak Demand[4] (7)	δLV Peak Demand (8)	Discounted δLV Peak Demand (9)
1979	1617	-	-	1251	-	-	757	-	-
1980	1794	177	158	1387	136	121	839	82	73
1981	1984	190	151	1546	159	127	955	116	92
1982	2191	207	147	1708	162	115	1073	118	84
1983	2441	258	164	1919	211	134	1226	153	97
1984	2729	280	159	2134	215	122	1380	154	87
1985	2976	247	125	2340	206	104	1523	143	72
1986	3184	208	94	2503	163	74	1629	106	48
1987	3439	255	103	2703	200	81	1760	131	53
1988	3714	275	99	2920	217	78	1900	140	50
1989	4011	297	96	3154	234	75	2053	153	49
1990	4332	321	92	3405	251	72	2217	164	47
Total 1981-1990						982		679	
Total 1982-1990			1079						

1 Column (1) is taken from column (4) in table 2.17.
2 The assumed discount rate is 12 percent.
3 Column (4) is taken from column (8) in Table 2.17.
4 Column (7) is taken from column (12) in Table 2.17.

Lower and upper bounds for the marginal opportunity cost of natural gas are the marginal extraction and transportation cost of gas at the power station and the value of imported liquid fuels which the gas displaces. Estimates of these costs are Rs.0.107 - Rs.0.345/m^3 for natural gas and Rs.2.483/l for high-speed diesel fuel; and, the heat values are 32,500 - 34,430 BTUs/m^3 of natural gas and 35,500 BTUs/l of diesel fuel. The relative weights of Guddu, Lyallpur/Multan, and the gas turbines during peak periods are 0.345, 0.531, and 0.115, respectively; during off-peak periods, the relative weight of Guddu is 1.0. Station use of 2.1 percent is allowed for in the estimate.

Summary of Strict LRMC: In order to determine capacity and energy costs at various voltage levels, line losses must be considered. Projected peak and off-peak loss factors for EHV/HV transmission, and MV and LV distribution for 1979-1990 are summarized in Table 2.19. The capacity costs determined above are converted to domestic prices and averaged yearly in Table 2.20. Then in Table 2.21, these yearly averaged costs are combined with expenditure on O&M and A&G and adjusted for peak line losses to determine strict long-term marginal capacity costs at various voltage levels. In the same table, long-term marginal energy costs are calculated at each voltage level by adjusting for average line losses.

Practical Tariff Setting:

Comparison with Existing Tariffs: The existing WAPDA tariffs and LRMC's of energy and capacity are summarized in Table 2.22. The current tariff structure is complex, with a variety of energy and minimum charges which are generally not directly related to the costs of supplying electricity to consumers. Specifically, the kilowatt charges are significantly below marginal capacity costs; also, existing tariffs are lowest for those categories of consumers for which supply costs are highest. This suggests that significant efficiency gains might be achieved by revising tariffs so as to make them more consistent with supply costs.

Proposed New Tariff Structures: A number of considerations underlie the proposed new tariff structure. First, the target average price is Rs.0.482/kW sold. This price is based on WAPDA's financial requirements as agreed under the World Bank's Third Power Loan. Second, the principal changes in tariff structure should be directed at increasing existing capacity charges closer to estimated long-term marginal capacity costs, while not raising the average price much above Rs.0.482. Third, the tariff for low-income residential and private tube-well consumers is a sensitive political and distributional issue and thus, any increase in this tariff should be kept to a minimum.

The new tariff is summarized in Table 2.23. The principal changes in the tariff categories are:(1) An additional block of consumption (greater than 200 kWh) is added to the domestic category, which should be effective for higher income households.(2) In the bulk supply categories C.1 and C.2, the number of subcategories is reduced from three to two by consolidating (b) and (c), which results in a simpler tariff structure without reducing the consistency between tariffs and LRMC (Munasinghe 1981). Similarly, in the bulk supply category C.3, the two subcategories are consolidated into one. The rationale for this is essentially the same as for (2) above.

Table 2.19

WAPDA System Loss Factors, 1979–1990

Voltage Level	1979	1980	1981	1982	1983	1984	1985	1986	1987	1988	1989	1990	Average 1979–1990
Generation													
% Generation	2.0	2.5	2.3	2.2	2.0	2.0	2.0	2.0	2.0	2.0	2.0	2.0	
% Incoming Av.	2.0	2.5	2.3	2.2	2.0	2.0	2.0	2.0	2.0	2.0	2.0	2.0	2.1
% Incoming Peak	2.0	2.5	2.3	2.2	2.0	2.0	2.0	2.0	2.0	2.0	2.0	2.0	2.1
EHV/HV													
% Generation	12.00	10.5	10.0	9.0	8.0	6.5	6.5	6.5	6.5	6.5	6.5	6.5	
% Incoming Av.	12.00	10.5	10.0	9.0	8.0	6.5	6.5	6.5	6.5	6.5	6.5	6.7	7.0
% Incoming Peak	16.0	14.0	13.0	12.0	10.0	8.0	8.0	8.0	8.0	8.0	8.0	8.0	9.0
MV													
% Generation	6.0	6.0	5.5	5.3	5.0	5.0	5.0	5.0	5.0	5.0	5.0	5.0	
% Incoming Av.	8.0	8.0	7.3	7.1	6.7	6.7	6.7	6.7	6.7	6.7	6.7	6.7	7.0
% Incoming Peak	10.4	10.4	9.5	8.7	8.7	8.7	8.7	8.7	8.7	8.7	8.7	8.7	9.0
LV													
% Generation	14.0	14.0	13.2	12.5	12.0	11.5	11.5	11.5	11.5	11.5	11.5	11.5	
% Incoming Av.	25.4	25.4	24.0	22.7	25.4	25.4	24.0	22.7	21.8	20.9	20.9	20.9	21.0
% Incoming Peak	29.5	29.5	27.9	26.5	25.3	24.2	24.2	24.2	24.2	24.2	24.2	24.2	25.0

Table 2.20

Capacity Costs (1979 Rs.)

Voltage Level	Capacity Costs Rs./kW (Border Prices)	Capacity Costs Rs./kW (Domestic Prices)[1]	Average Yearly Capacity Costs, Rs./kW/Year (Domestic Prices)2
Generation	8875	10440	1330
HV Transmission	5997	7055	856
MV Distribution	1478	1739	222
LV Distribution	3203	3768	480

1 The conversion factor equals 0.85.

2 The assumed lifetimes for generation, HV, MV, and LV investments are 25, 40, 25, and 25 years, respectively. With an assumed opportunity cost of capital of 12 percent, the resulting yearly averaged factors are 0.1275, 0.1213, 0.1275, and 0.1275 for the respective capacity investments.

Table 2.21

Summary of Strict Long-Run Marginal Capacity and Energy Costs (Domestic Prices)

	Capacity Cost (Rs./kW/Year)												Energy Cost (Rs./kWh)	
	Generation			EHV/HV			MV			LV				
Voltage Level[2]	Annual Capacity	O&M + A&G[1]	Total	Ann. Cap.	O&M + A&G	Cum Total	Ann. Cap.	O&M + A&G	Cum Total	Ann. Cap.	O&M + A&G	Cum Total	Peak	Off-Peak
Generation	1330	156	1486										0.266	0.255
EHV/HV			1633	856	212	2701							0.292	0.275
MV						2968	222	122	3312				0.321	0.296
LV									4416	480	264	5160	0.428	0.374

1 O&M plus A&G costs equal 1.5, 3, 7 and 7 percent of capacity costs/kW for generation, EHV/HV, MV, and LV capacities, respectively.

2 The losses of incoming peak, as presented in Table 2.19, equal 9 percent for EHV/HV and MV, and 25 percent for LV. Station use equals 2.1 percent. Average losses of incoming equal 7 percent for EHV/HV and MV, and 21 percent for LV.

Table 2.22

Summary of Main Features of the Existing WAPDA Tariff (as of July 1, 1979)

Category of Consumer	Existing Tariff: Energy Charge (Paise/kWh)[1]	Existing Tariff: Cap.Charge (Rs./kW/Month)	LRMC[5]: Energy[4] (Paise/kWh)	LRMC[5]: Capacity (Rs./kW/Month
A1. **Domestic**	31 first 40 kWh 30 balance (Rs. 5/month)	-	42.8(p) & 37.5(op)	430
A2. **Commercial**	80 first 100 kWh 95 balance (Rs. 15/month minimum)	-	42.8(p) & 37.5(op)	430
Industrial				
B1. 70-kW Connected Load	52	20.00-25.00 (minimum charge)	42.8(p) & 37.5(op)	430
B2. Up to 500 kW Connected Load	36	50.00	42.8(p) & 37.5(op)	430
B3. Supply of 11kV	35	45.00	32.1 (p) & 29.6 (op)	276 276
B4. Supply at 33, 66, or 132 kV	32	42.00	29.2 (p) & 27.5(op) (66/132 kV)	225 225
Bulk Supply				
C1. Supply at 400V				
(a)[2]	32	45.00	42.8 (p)	430
(b)[2]	40	45.00	&	430
(c)[3]	35	45.00	37.5 (op)	430
C2. Supply at 11kV				
(a)[2]	28	41.00	32.1 (p)	276
(b)[2]	35	41.00	&	276
(c)[3]	34	41.00	27.5 (op)	276
C3. Supply at 33, 66, and 132 kV			29.2 (p) &	
(a)	34	39.50	27.5 (op)	225
(b)	33	39.50	(66/132 kV)	225
Tube Wells				
D1. for Reclamation	31	-	42.8(p) &	430
D2. Drainage for Agriculture and Irrigation			37.5(op)	430
(i) in N.W. Frontier	13	9	42.8(p) & 37.5(op)	430 430
(ii) Other Areas	19	11	42.8(p) & 37.5(op)	430 430

1 A fuel adjustment surchage is added to the energy charges listed in this table; 100 paise = Rs.1.
2 Includes Licensees and Non-Licensees.
3 Includes railways, cantonment boards, and other institutions with their own distribution facilities.
4 (p) = peak; (op) = off-peak.
5 No adjustment made for diversity factor.

Table 2.23

LRMC and Proposed New WAPDA Power Tariff
(for 1980-1981)

Category	LRMC[5] Cap. (Rs/ kW/m)	LRMC[5] Energy[1] (Rs./ kWh)	LRMC[5] Av Eff. Cost[2] (Rs/ kWh)	New Tariff Cap[6] (Rs/ kW/m)	New Tariff Energy (Rs/ kWh)	New Tariff Av. Eff. Price (Rs/ kWh)	Load Factor x Div. Factor (%)	Application of New Tariff to 1978-1979 Consumption: No. of Customers (1000)	Av. Cons/ Cust. (kWh)	Total Revenue (1000 Rs)	Rev/ Cust. (Rs)
A1. Domestic											
a. 30 kWh	430	0.43/0.38	1.58	-	0.31	0.31	-	1183.77	17.9	6568.7	5.55
b. 30-200 kWh	430	0.43/0.38	to	-	0.45	0.399	-	599.91	82.64	19781.0	32.97
c. 200 kWh	430	0.43/0.38	3.35[3]	-	0.75	0.587	-	82.87	394.24	19177.7	231.41
A2. Commercial			1.58								
a. 100 kWh	430	0.43/0.38	to	-	0.85	0.85	-	462.95	60.41	24233.9	52.35
b. 100 kWh	430	0.43/0.38	3.35[3]	-	0.95	0.90	-				
B. Industrial											
B1. 70 kWh	430	0.43/0.38	1.58to	-	0.52	0.52	-	98.21	946.94	48357.7	492.39
B2. 70-500 kWh	430	0.43/0.38	3.35[3]	90	0.29	0.60	40	2.33	20285.7	28359.4	12171
B3. 11/33kV	276	0.32/0.30	1.07	80	0.27	0.49	50	0.403	194289	38366.3	95201
B4. 66/132kV	225	0.29/0.28	0.77	70	0.26	0.42	60	0.009	1353417	5115.9	568433
C. Bulk Supply											
C1. 400V a.	430	0.43/0.38	2.76	-	0.50	0.50	-	0.529	54875	15967	30183
b.	430	0.43/0.38	2.76	65	0.25	0.60	25				
C2. 11/33kV a.	276	0.32/0.30	1.26	-	0.43	0.43	-	0.236	174314	17895	75826
b.	276	0.32/0.30	1.26	60	0.23	0.44	40				
C3. 66/132kV	225	0.29/0.28	0.88	55	0.21	0.36	50	0.014	85643	432	30857
D. Tube Wells											
D1. Govt.	430	0.43/0.38	2.76	-	0.48	0.48	-			23272.8	-
D2. a. NWFP (Priv)	430	0.43/0.38	2.76	31	0.15	0.32	25	11.14	4352.7	25163.6	4748.8
b. Other	430	0.43/0.38	2.76	33	0.20	0.38	25			4465.1	
E. Public Lighting	430	0.43/0.38	1.82[4]	-	0.70	0.70	-	1.315	4417	4065.8	3091.9
F. Traction				-	0.30	0.30	-	0.008	444896	1067.8	133475

Av. Price (New Tariff) = 48.5 paise/kWh.

1 LRMC of energy; peak/off-peak.
2 LRMC of cap./(730 x L.f x div.fac.) + LRMC of Energy.
3 Assuming (load factor x div. factor) = 20 to 50 percent.
4 10-h lighting per day.
5 No adjustment made for diversity factor.
6 Using time-of-day metering for capacity charge. Peak demand metering less desirable but adequate.

The major changes in the tariff rates are:(1) The energy charge for residential consumers in the second block (31-200 kWh) is increased about 30 percent to Rs.045/kWh, while for the new third block (above 200 kWh) the charge is Rs.075/kWh or 115 percent higher than the existing level.(2) For industrial consumers, capacity charges are raised closer to marginal capacity costs, while energy charges are lowered. (3) For bulk supplies, tariffs are set to encourage a shift to higher bulk supply voltages, i.e., average price per kilowatt-hour falls as voltage rises. (4) Tariffs for private tube wells are raised as much as distributional and political considerations are judged to allow.

Tariff Study Summary

The tariffs-setting procedure described above involves two principal steps: (1) Estimation of strict long-term marginal costs of supply, and (2) adjustment of strict LRMC to yield a practical tariff structure which meets other requirements such as financial needs, distributional considerations, and price stability.

The results of the first step in this procedure demonstrate that existing WAPDA tariffs are badly distorted, with capacity or kilowatt charges significantly below marginal capacity cost and energy charges quite close to marginal energy cost. As a result, a revised set of tariffs is suggested. However, tariff setting is a continuous process; and therefore, this proposed new tariff structure should be thought of as reflecting one step in a transitional process toward tariffs which reflect LRMC even more accurately. Several issues remain to be resolved in future tariff revisions. First, the relation between WAPDA and KESC tariffs need to be explored. KESC tariffs are currently lower than WAPDA's due to its high load density and relatively small needs for new investment. This may lead to resentment among WAPDA's MV and HV consumers, who, in effect, subsidize rural and tube-well consumers and pay a large share of investment costs which result -- in part -- from load growth in response to subsidized tariffs. Second, the marginal opportunity cost, or economic value, of natural gas used in the power sector needs to be more accurately determined. Third, load forecasting techniques need to be improved. Finally, one should address the issue of whether or not a conscious policy of having electricity tariffs deviate from LRMC is the most effective means of accomplishing income distributional objectives within the economy.

2.5 Summary and Conclusions

This paper has presented advances and developments that have occurred, especially in the recent past, with respect to the engineering- economic analysis of electric power systems. The two chief areas of interest relate to the formulation of optimal investment and pricing policies. The relevant theory and a comprehensive analytical framework are presented first.

The basic economic principle underlying the methodology is that decision makers should seek to provide electricity services in a manner which maximizes the net benefits of consumption of society as a whole. In the area of investment policy, this principle may be used to optimize overall planning, reliability, and losses. Often the maximization of net benefits criterion may be reduced to the simpler rule of minimizing total costs, such as for least-cost investment planning.

In pricing policy, the basic economic principle requires that prices reflect the marginal costs of supply, thus ensuring that economic efficiency criteria are satisfied. However, these marginal costs must be systematically adjusted to yield a practical tariff structure that meets other national policy objectives such as supplying the basic electricity needs of poor consumers, ensuring financial viability, price stability, and so on.

Several case studies are provided which illustrate the practical application of the methodological framework to optimize investment planning and reliability, system losses, and tariffs.

Appendix A2.1: Reliability Optimizing Model

The task of the utility is to ensure that the supply of electricity services (S) exceeds the demand (D) at all times in the future. Even if planned supply exceeds planned demand in expected value terms during a given time period t, i.e., $S_t > D_t$, there is a stochastic element in both supply and demand. Therefore, actual demand D_t, may exceed actual supply St, so that consumers will suffer shortage costs.

We might represent this most simply as:

$$\tilde{S}_t - \tilde{D}_t = S_t - D_t + U_t$$

where U_t is a random variable (e.g., with a zero mean and standard deviation σ^2) which represents the stochastic nature of both supply and demand. This is shown in Figure A2.1, where a shortage condition (or excess demand) exists in the shaded region.

The system would have a higher level of reliability (R_t), the smaller the probability of shortages. Clearly, the reliability would be higher (lower) when the planned safety margin, $S_t - D_t$, was larger (smaller), and when uncertainty, σ^2, was smaller (greater). To clarify the presentation and keep the optimization model as operational as possible, it is developed in terms of expected values. The random nature of supply and demand is captured by the reliability parameter R and its effects on shortage (or outage) costs QC and the system supply costs SC. As explained more fully below, the choice of an appropriate (multidimensional) reliability index R which can be easily related to both QC and SC is an important requirement for application of the model.

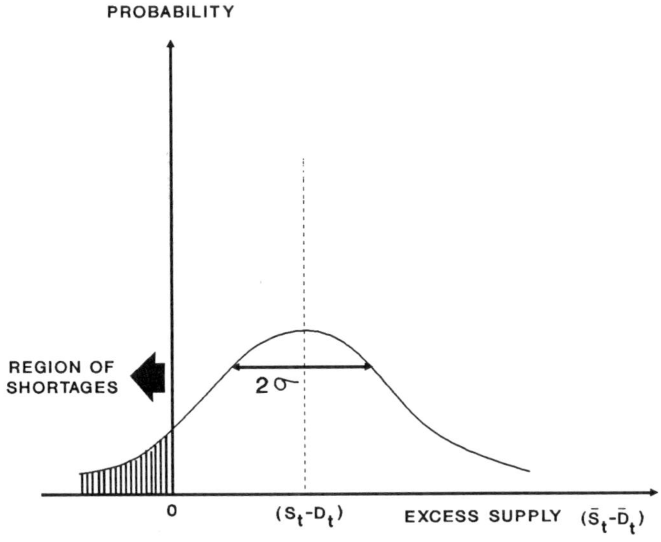

Figure A2.1: The Condition for Shortages

Consider a power system in which the expected value of electricity demand over a long period of τ years is represented by D_t where t varies from 0 to τ. D is a function of other variables and may be written as

$$D_t = D_t(P_t, Y_t, R^*_t, Z_t) \qquad (A2.1)$$

where P is the electricity price, Y is the income variable which represents the level of economic activity, R* is the value of reliability expected by consumers, and Z is a vector of other independent variables affecting demand. Initially, it is assumed that the values of arguments P_t, Y_t, and Z_t on the right-hand side of (A2.1) are exogenously fixed. Prespecifying P_t allows us to uncouple the joint optimal pricing and reliability conditions at this stage. This constraint may be subsequently relaxed, as described later.

The economic costs suffered by electricity consumers due to electric power shortages associated with each system plan may be represented as a function of the generalized reliability level R_t, forecast demand, and reliability expectation

$$QC_t = QC_t(R_t, D_t, R^*_t) \qquad (A2.2)$$

Corresponding power system supply costs, incurred as a consequence of providing electricity services may be written as

$$SC_t = SC_t(R_1,...,R_\tau, D_1,... D_\tau) \qquad (A2.3)$$

The supply costs described in (A2.3) correspond to least cost system designs. Note that SC should include the capital and operating expenses as well as the kilowatt and kilowatt-hour losses in the system, appropriately valued (at the marginal cost of supply), but net of the marginal supply costs of kilowatt-hours not delivered due to outages. Finally, since reliability expectation depends on past reliability, we may write

$$R^*_t = R^*_t(R_{t-1},...R_{t-s}) \qquad (A2.4)$$

A simple expression for the net benefit (NB) of electricity consumption to society (in present discounted value terms) is

$$NB = TB - QC - SC \qquad (A2.5)$$

where

$$TB = \sum_{t=0}^{T} TB_t/(1 + r)^t$$

and SC and QC are similarly defined present discounted value summations, while $TB_t = TB_t(D_t)$ is the total benefit of electricity consumption in the absence of outages which depends only on the demand, and r is the appropriate discount rate. Because of the long lifetime of some investments, the discounting time horizon T may be chosen to be greater than the planning time horizon τ to eliminate end effects.

In order to identify the marginal conditions which maximize NB, we formulate the Lagrangian

$$L = \sum_{t=0}^{T} \{(TB_t - QC_t - SC_t)/(1 + r)^t$$

$$-\lambda[\overset{*}{R}_t - \overset{*}{R}_t(R_{t-1},...,R_{t-s})]$$
$$-\mu[D_t - D_t(\overset{*}{R}_t)]\}$$

The three first-order conditions of relevance are

$$(\partial L/\partial R_j) = -[(\partial QC_j/\partial R_j)/(1+r)^j + \sum_{t=0}^{T}(\partial SC_t/\partial R_j)/(1 + r)^t]$$

$$+ \sum_{t=0}^{T}\lambda_t(\partial\overset{*}{R}_t/\partial R_j) = 0$$

$$(\partial L/\partial\overset{*}{R}_j) = -(\partial QC_j/\partial\overset{*}{R}_j)/(1+r)^j$$

$$+ \mu_j (\partial D_j/\partial\overset{*}{R}_j) - \lambda_j = 0$$

$$(\partial L/\partial D_j) = [(\partial TB_j/\partial D_j) - (\partial QC_j/\partial D_j]/(1 + r)^j$$
$$+ \sum_{t=0}^{T}(\partial SC_t/\partial D_j)/(1 + r)^t - \mu_j = 0$$

for $j = 0, 1, ..., T$

Combining these three equations yields

$$\{-(\partial QC_j/\partial R_j)/(1 + r)^j - \sum_{t=0}^{T} (\partial SC_t/\partial R_j)/(1 + r)^t\} \quad \text{(A2.6)}$$

$$+ \sum_{t=0}^{T}(\partial\overset{*}{R}_t/\partial R_j)\{[- (\partial QC_t/\overset{*}{R}_t) + (\partial D_t/\overset{*}{R}_t)$$

$$\times \{(\partial TB_t/\partial D_t) - (\partial QC_t/\partial D_t)\}]/(1 + r)^t$$
$$- \sum_{u=0}^{T}(\partial SC_u/\partial D_t)/(1 + r)^u\} = 0$$

The first term in (A2.6) captures the direct impact of reliability changes on QC and SC, while the rest of the equation represents the corresponding indirect effects via the reliability expectation. To simplify interpretation of the above condition, we might assume to first order that $(\partial R^*_t/\partial R_j).(\partial QC_t/\partial R^*_t)$ and $(\partial R^*_t/\partial R_j).(\partial D_t/\partial R^*_t)$ are negligible. In other words, the effect of changes in reliability level R on both QC and

D, via the reliability expectation R*, are small. This assumption would be especially valid if the range of variation of R were not very wide. Therefore

$$-(\partial QC_j/\partial R_j)/(1 + r)^j - \sum_{t=0}^{T} (\partial SC_t/\partial R_j)/(1 + r)^t = 0 \qquad \text{(A2.7)}$$

Equation (A2.7) indicates that the net benefits to society will be maximized at the point where the marginal outage costs associated with a change in reliability are exactly offset by the corresponding change in supply costs. Although this equation relates to reliability in time period j, in practice, changes in reliability are likely to be implemented over several time periods. Therefore, a more useful form of (A2.7) would be

$$\sum_{j=0}^{T} [-(\partial QC_j/\partial R_j).\Delta R_j/(1 + r)^j$$

$$- \sum_{t=0}^{T} \sum_{j=0}^{T} [-(\partial SC_t/\partial R_j).\Delta R_j/(1 + r)^t] = 0 \qquad \text{(A2.8)}$$

where ΔR_j is a small change in reliability during period j.

For practical application of this rule, let us consider n alternative long-range system expansion programs, where the reliability level of the ith program in year t is R^i_t. Suppose that expansion plan (i+1) is derived from plan i, by a small change in reliability given by

$$\Delta R^i_t = R^{i+1}_t - R^i_t$$

for i = 0, 1, ..., n. Examining (A2.7) and (A2.8), the corresponding change in net benefits may be written:

$$\Delta NB^i = NB^{i+1} - NB^i = -\Delta QC^i - \Delta SC^i \qquad \text{(A2.9)}$$

where

$$\Delta QC^i = \sum_{j=0}^{T} (\partial QC^i_j/\partial R^i_j).\Delta R^i_j/(1 + r)^j$$

and

$$\Delta SC^i = \sum_{j=0}^{T} \sum_{t=0}^{T} (\partial SC^i_t/\partial R^i_j).\ \Delta R^i_j/(1 + r)^t$$

To interpret (A2.9), it is assumed that the change from system expansion plan i to plan (i+1) involves an overall unambiguous improvement in reliability, i.e., that each component ΔR^i_t is nonnegative. In general, this implies correspondingly that $\Delta QC^i \leq 0$ and $\Delta SC^i \geq 0$. In this case, (A2.9) yields

$$|\Delta NB^i| = |\Delta QC^i| - |\Delta SC^i|$$

Therefore, in order to maximize the net economic benefits of supplying electricity to society, the reliability level should be increased in successive system plans as long as the corresponding decrease in incremental outage costs exceeds the increase in incremental supply costs, and vice versa.

References

Anderson, D., "Models for Determining Least-Cost Investments in Electricity Supply," *Bell Journal of Economics*, vol. 3, Spring 1972, pp. 267-301.

Allan, R.N. and F.N. Takieddin, "Generation Modelling in Power System Reliability Evaluation," in *Reliability of Power Supply Systems* (IEEE Conf. Publ. 148) London, England, Feb. 1977, pp. 47-50.

Allan, R.N. and F.N. Takieddin, "Network Limitations on Generating Systems Reliability Evaluation Techniques", Paper A77-8070-5, in Proc. IEEE Power Eng. Soc. Winter Meet., New York, Jan.-Feb. 1978.

Baldwin, C.J., C.H. Hoffman, and P.H. Jeynes, "A Further Look at Cost of Losses," *AIEE Trans. (Power App. Syst.)*, vol. 80, 1961.

Banco Central de Costa Rica, *Cifras Sobre Produccion Agropecuaria*, San Jose, Costa Rica, July 1978.

Banco Central de Costa Rica, *Cifras de Cuentas Nacionales de Costa Rica*, Serie 1957-77, San Jose, Costa Rica, July 1978.

Boiteux, M., "La Tarification des Demandes en Pointe," *Rev. Generale de l'Electricite*, vol. 58, 1949, pp. 321-340.

Boiteux, M. and P. Stasi, "The Determination of Costs of Expansion of an Interconnected System of Production and Distribution of Electricity," in J. Nelson (ed.), *Marginal Cost Pricing in Practice*, Prentice-Hall, Englewood Cliffs, NJ , 1974.

Brown, G., Jr., and M.B. Johnson, "Public Utility Pricing and Output Under Risk," *Amer. Economic Rev.*, Mar. 1969, pp. 119-128.

Canadian International Development Agency, Proc. Regional Power systems Engineering Seminar, Kathmandu, Nepal, Feb.-Mar. 1977.

Crew, M.A. and P.R. Kleindorfer, *Public Utility Economics*, St. Martins Press, New York, 1979.

Crew, M.A. and P.R. Kleindorfer, "Reliability and Public Utility Pricing," *Amer. Economic Rev.*, vol. 68, Mar. 1978.

Dias-Bandaranaike, R. and M. Munasinghe, "The Demand for Electricity Services and the Quality of Supply," *The Energy Journal*, vol. 4, Apr. 1983.

Dupuit, P., "De l'Utilite et de sa Mesuer," *La Reforma Soziale*, Turin, Italy, 1932.

Endrenyi, J., *Reliability Modeling in Electric Power Systems*, Wiley, New York, 1978.

Ford, L.R. and D.R. Fulkerson, *Flows in Networks*, Princeton Univ. Press, Princeton, NJ, 1962.

Green, H.A.J., *Aggregation in Economic Analysis: An Introductory Survey*, Princeton Univ. Press, Princeton, NJ, 1964.

Holt, S.B., Jr. and D.M. Crawford, "Distribution Substation Planning Using Optimization Methods," in *Application of Optimization Methods in Power System Engineering* (IEEE Tutorial Course), pp. 69-76.

Hotelling, H., "The General welfare in Relation to Problems of Railway and Utility Rates," *Econometrica*, vol. 6, July 1938, pp. 242-269.

IAEA, *Expansion Planning for Electrical Generating Systems: A Guidebook*, International Atomic Energy Agency, Vienna, 1984.

IEEE Tutorial Course, no. 76CH1107-2-PWR, Application of Optimization Methods in Power System Engineering, IEEE, New York, 1976.

IEEE Standard 346-1973, "IEEE Standard Definitions in Power Operations Terminology Including Terms for Reporting and Analyzing Outages of Electrical Transmission and Distribution Facilities and Interruptions to Customer Service," New York, IEEE, 1973.

Jaramillo,P. and E. Skoknic, "Costo Social de las Restricciones Energia Electrica," Oficina de Planificacion, Santiago, Chile, Aug. 1973.

Johnson, B.F., "Evaluation of the Costs of Losses in Power Systems," Consumers Power Co., Tech. Rep., Apr. 1976.

Knight, U.G., *Power System Engineering and Mathematics*, Pergamon, New York, 1972.

Lolander, D., "Memorandum Concerning Estimation of the Costs of the National Economy in the Event of a Loss of Supply", Vattenfall, Sweden, 1948.

Marginal Costing and Pricing of Electrical Energy (Proc. of the State of the Art Conf.), Canadian Electrical Association, Montreal, Que., May 1978.

Marko, G.E., "A Method of Combining High Speed Contingency Load Flow Analysis with Stochastic Probability Methods to Calculate a Quantitative Measure of Overall Power System Reliability," paper No. A780-53-1 in Proc. IEEE Power Eng. Soc. Summer Meet, Los Angeles, July 1978.

Ministerio de Economia, Industria y Comercio, Censos Nacionales de 1973 Agropecuaria, San Jose, Costa Rica, July 1974.

Munasinghe, M. *The Economics of Power System Reliability and Planning*, Johns Hopkins Univ. Press, Baltimore, MD, 1979.

Munasinghe, M., "The Costs Incurred by Residential Electricity Consumers Due to Power Failures," *J. Consumer Res.*, vol. 6, Mar. 1980, pp. 361-369.

Munasinghe, M., "System Planning to Achieve Optimal Reliability Levels", Proceedings of the Third International Conference on Analysis, Forecasting and Planning of Public Utilities, INSEAD, Paris, June 1980, pp. 303-24.

Munasinghe, M., "An Integrated Framework for Energy Pricing in Developing Countries", *The Energy Journal*, vol.1, July 1980, pp.1-30.

Munasinghe, M. "Integrated National Energy Planning in Developing Countries," *Natural Resources Forum*, Oct. 1980, pp. 359-374.

Munasinghe, M., "Principles of Modern Electricity Pricing," *Proc. IEEE*, vol. 69, Mar. 1981, pp. 332-348.

Munasinghe, M. "Optimal Electricity Supply: Reliability, Pricing and System Planning," *Energy Economics*, vol. 3, July 1981, pp. 140-152.

Munasinghe, M., "Practical Application of Integrated National Energy Planning Using Microcomputers", *Natural Resources Forum*, Feb. 1986, pp. 17-38.

Munasinghe, M. and M. Gellerson, "Economic Criteria for Optimizing Power System Reliability Levels", *Bell Journal of Economics*, Vol.10, Spring 1979, pp. 353-65.

Munasinghe, M. and S. Rungta, Eds., *Costing and Pricing Electricity in Developing Countries*, Asian Development Bank, Manila, Philippines, 1984.

Munasinghe, M. and G. Schramm, *Energy Economics, Demand Management and Conservation Policy*, Van Nostrand Reinhold, New York, 1983.

Munasinghe, M. and W. Scott, "Long Range Distribution System Planning Based on Optimum Economic Reliability Levels," paper A78576-1 in Proc. IEEE Power Eng. Soc. Summer Meet., Los Angeles, July 1978.

Munasinghe, M. and W. Scott, *Energy Efficiency: Optimization of Electric Power Distribution System Losses*, Energy Department Paper No. 6, The World Bank, Washington, DC, July 1982.

Munasinghe, M., W.G. Scott, and R. Dias-Bandaranaike, *Optimizing Rural Electricity Supply*, Energy Dept. Report, The World Bank, Washington DC, 1982.

Munasinghe, M. and J.J. Warford, *Electricity Pricing*, Johns Hopkins Univ. Press, Baltimore, MD, 1982.

Nickel, D.L., "Distribution Transformer Loss Evaluation: Proposed Techniques," presented at the IEEE, PES Winter Meet., New York, Feb. 1980.

Noferi, P.L., L. Paris, and L. Salvaderi, "Monte Carlo Methods for Power System Reliability Evaluation in Transmission and Generation Planning," in Proc. Annual Reliability and Maintainability Symp., Washington, DC, Jan. 1975, pp. 460-469.

Ontario-Hydro, "Survey on Power System Reliability: Viewpoint of Large Users," Rep. PMA 76--5, Toronto, Ont., Canada, 1976.

Rischl, R., "Optimal System Expansion Planning. A Critical Review", *Systems Engineering for Power: Status and Prospects*, Proc. USERDA Conf. (Henniker, NH, Aug. 1975) NTIS CONF-750867, pp. 233-260.

Ruggles, N.,"The Welfare Basis of the Marginal Cost Pricing Principle," *Rev. Economic Studies*, vol. 17, 1949-1950, pp. 29-46.

Ruggles, N.,"Recent Developments in the Theory of Marginal Cost Pricing," *Rev. Economic Studies*, vol. 17, 1949-1950, pp. 107-126.

Sanghvi, A., "Optimal Electric Supply Reliability Using Customer Shortage Costs", *Energy Economics*, vol.5, April 1983.

Sherman, R. and M. Visscher, "Second Best Pricing With Stochastic Demand," *Amer. Economic Review*, Mar. 1978, pp. 41-53.

Steiner, P., "Peak Loads and Efficient Pricing," *Quart. J. Economics*, Nov. 1957, pp. 585-610.

Stott, B., "Review of Load-Flow Calculation Methods," *Proc IEEE*, vol. 62, July 1974, pp. 916-929.

Sullivan, R.L., *Power System Planning*, McGraw-Hill, New York, 1977.

Swedish Joint Committee for the Electricity Council, "Cost of Interruptions in Electricity Supply", Stockholm, Sweden, Sept. 1969.

Symposium on Peak Load Pricing, *Bell J. Economics*, vol. 7, Spring 1976, pp. 197-250.

System Control, Inc., "Analysis of Electric Power System Reliability", prepared for the California Energy Resources Conservation and Development Commission, Palo Alto, CA, 1978.

Telson, M.L., "The Economics of Alternative Levels of Reliability for Electric Generating Systems," *Bell J. Economics*, vol. 6, Fall 1975, pp. 679-694.

Turvey, R., *Optimal Pricing and Investment in Electricity Supply*, MIT Press, Cambridge, MA, 1968.

Turvey, R. and D. Anderson, *Electricity Economics*, Johns Hopkins Univ. Press, Baltimore, MD, 1977.

Vardi, J. and B. Avi-Itzhak, *Electric Energy Generation: Economics, Reliability and Rates*, MIT Press, Cambridge, MA, 1981.

Williamson, O.E., "Peak Load Pricing and Optimal Capacity under Indivisibility Constraints," *Amer. Economic Rev.*, vol. 56, no. 4, Sept. 1966, pp. 810-827.

Wismer, D.A., *Optimization Methods for Large Scale Systems*, McGraw-Hill, New York, 1971.

Chapter 3

PRINCIPLES OF MODERN ELECTRICITY PRICING[1]

3.1 Introduction and Overview

Modern societies have become increasingly dependent on various types of energy sources, among which electric power has occupied a dominant position. Traditionally, electric power pricing policy in most countries has been determined mainly on the basis of financial or accounting criteria, e.g., raising sufficient sales revenues to meet operating expenses and debt service requirements while providing a reasonable contribution towards the capital required for future power system expansion.

However, in recent times several new factors have arisen, including the rapid growth of demand, the increase in fuel oil prices and prices of fossil fuel and nuclear plant, the dwindling availability of cheaply exploitable hydroelectric resources, and the expansion of power systems into areas of lower consumer density at relatively high unit costs. These developments have led to increasing emphasis being laid on the use of economic principles in order to produce and consume electric power efficiently, while conserving scarce resources, and meeting various national objectives. In particular, a great deal of attention has been paid to the use of marginal cost pricing policies in the electric power sector. We note that price is an effective "soft" technique of demand management especially in the long run. The effects of pricing policy are also greatly enhanced by coordinating it properly with other "soft" demand management tools such as financial/tax incentives, and "hard" demand management techniques including load control, etc., that are more useful in the short run.

The objectives of power tariff policy in the national context, and a pricing framework based on long-run marginal costs (LRMC) which meets these requirements, are summarized in this section. In Section 3.2, the economic principles underlying the LRMC approach are described, and in Section 3.3, a framework for calculating strict LRMC is presented. The process of adjusting LRMC to devise a practical tariff structure which meets other national constraints is discussed in Section 3.4. Section 3.5 contains a review of recent results of modern pricing structures in several countries and a discussion of the implications of the Public Utilities Regulatory Policies Act (PURPA) of 1978 in the U.S., and this is followed by several technical appendices.

Requirements of a Power Tariff

The modern approach to power pricing recognizes the existence of several objectives or criteria, not all of which are mutually consistent. First, national economic resources must be allocated efficiently, not only among different sectors of the economy, but within the electric power sector. This implies that cost-reflecting prices must be used

[1] Edited version of a paper published in *Proceedings of the IEEE*, vol. 69, no.3, March 1981, pp. 332-48.

to indicate to the electricity consumers the true economic costs of supplying their specific needs, so that supply and demand can be matched efficiently.

Second, certain principles relating to fairness and equity must be satisfied, including: (a) the fair allocation of costs among consumers according to the burdens they impose on the system; (b) the assurance of a reasonable degree of price stability and avoidance of large price fluctuations from year to year; and (c) the provision of a minimum level of service to persons who may not be able to afford the full cost.

Third, the power prices should raise sufficient revenues to meet the financial requirements of the utility, as described earlier. Fourth, the power tariff structure must be simple enough to facilitate the metering and billing of customers. Fifth, and finally, other economic and political objectives must also be considered. These might include, for example, subsidized electricity supply to certain sectors in order to enhance growth or to certain geographic areas for regional development.

Since the above criteria are often in conflict with one another, it is necessary to accept certain tradeoffs between them. The LRMC approach to price setting described below has both the analytical rigor and inherent flexibility to provide a tariff structure that is responsive to these basis objectives.

LRMC-Based Tariffs

A tariff based on LRMC is consistent with the first objective, that is, the efficient allocation of resources. The traditional accounting approach is concerned with the recovery of historical or sunk costs, while in the LRMC calculation the important consideration is the amount of future resources used or saved by consumer decisions. Since electricity prices are the amounts paid for increments of consumption, in general they should reflect the incremental cost incurred. Supply costs increase if existing consumers increase their demand or if new consumers are connected to the system. Therefore, prices that act as a signal to consumers should be related to the economic value of future resources required to meet consumption changes. The accounting approach that uses historical assets and embedded costs implies that future economic resources will be cheap or as expensive as in the past. This could lead to overinvestment and waste, or underinvestment and the additional costs of unnecessary scarcity.

To promote better utilization of capacity, and to avoid unnecessary investments to meet peak demands, which tend to grow very rapidly, the LRMC approach permits the structuring of prices so that they vary according to the marginal costs of serving demands: (a) by different consumer categories; (b) in different seasons; (c) at different hours of the day; (d) by different voltage levels; (e) in different geographical areas; and so on.

In particular, with an appropriate choice of the peak period, structuring the LRMC-based tariffs by time of day generally leads to the conclusion that peak consumers should pay both capacity and energy costs, whereas offpeak consumers should pay only the energy costs. Similarly, analysis of LRMC by voltage level usually indicates that the lower the service voltage, the greater the costs consumers impose on the system.

The structuring of LRMC-based tariffs also meets subcategories (a) and (b) of the second, or fairness, objective mentioned earlier. The economic resource costs of future consumption are allocated as far as possible among the customers according to the incremental costs they impose on the power system. In the traditional approach, fairness was often defined rather narrowly and led to the allocation of arbitrary accounting costs to various rating periods and consumers, thus violating the economic efficiency criterion. Because the LRMC method deals with future costs over a long period -- for example, at least 5 to 10 years -- the resulting prices in constant terms tend to be quite stable over time. This smoothing out of costs over a long period is especially important given capital indivisibilities or "lumpiness" of power system investments.

Using economic opportunity costs (or shadow prices -- especially for capital, labor, and fuel) instead of purely financial costs, and taking externalities into consideration whenever possible, also link the LRMC method and efficient resource allocation.

The development of LRMC-based tariff structures which also meet the other objectives of pricing policy mentioned earlier, are discussed next.

Practical Tariff Setting

The first stage of the LRMC approach is the calculation of pure or strict LRMC that reflect the economic efficiency criterion. If price was set strictly equal to LRMC, consumers could indicate their willingness to pay for more consumption, thus signaling the justification of further investment to expand capacity.

In the second stage of tariff setting, ways are sought in which the strict LRMC may be adjusted to meet the other objectives, among which the financial requirement is most important. If prices were set equal to strict LRMC, it is likely that there will be a financial surplus. This is because marginal costs tend to be higher than average costs when the unit costs of supply are increasing. In principle, financial surpluses of the utility may be taxed away by the state, but in practice the use of power pricing as a tool for raising central government revenues is usually politically unpopular and rarely applied. Such surplus revenues can also be utilized in a way that is consistent with the other objectives. For example, the connection charges can be subsidized without violating the LRMC price, or low-income consumers could be provided with a subsidized block of electricity to meet their basic requirement, thus satisfying sociopolitical objectives. Conversely, if marginal costs are below average costs -- typically as a result of economies of scale -- then pricing at the strict LRMC will lead to a financial deficit. This will have to be made up, for example, by higher lump sum connection charges, flat rate charges, or even government subsidies.

Another reason for deviating from the strict LRMC arises because of second-best considerations. When prices elsewhere in the economy do not reflect marginal costs, especially for electric power substitutes and complements, then departures from the strict marginal costs pricing rule for electricity services would be justified. For example, in rural areas, inexpensive alternative energy may be available in the form of subsidized kerosene and/or gas. In this case, pricing electricity below the LRMC may be justified, to prevent excessive use of the alternative forms of energy. Similarly, if incentives are provided to import private generators and their fuel is also subsidized, then charging the full marginal cost to industrial consumers may encourage them to purchase their own or captive power plant. This is economically less efficient from a national perspective. Since the computation of strict LRMC is based on the power utilities' least cost expansion program, LRMC may also need to be modified by short-term considerations if previously unforeseen events make the long-run system plan suboptimal in the short run. Typical examples include a sudden reduction in demand growth and a large excess of installed capacity that may justify somewhat reduced capacity charges, or a rapid increase in fuel prices, which could warrant a short-term fuel surcharge.

As discussed earlier, the LRMC approach permits a high degree of tariff structuring. However, data constraints and the objective of simplifying metering and billing procedures usually requires that there should be practical limit to differentiation of tariffs by : (a) major customer categories -- residential, industrial, commercial, special, rural, and so on; (b) voltage levels -- high, medium, and low; (c) time of day -- peak and off-peak; and (d) geographic region. Finally, various other constraints also may be incorporated into the LRMC based tariffs, such as the political requirement of having a uniform national tariff, subsidizing rural electrification, and so on. In each case, however, such deviations from LRMC will impose an efficiency cost on the economy.

Summary

In the first stage of calculating LRMC, the economic (first best) efficiency objectives of tariff setting are satisfied, because the method of calculation is based on future economic resource costs rather than sunk costs, and also incorporates economic considerations such as shadow prices and externalities. The structuring of marginal costs permits an efficient and fair allocation of the tariff burden on consumers. In the second stage of developing an LRMC-based tariff, deviations from strict LRMC are considered to meet important financial, social, economic (second best), and political criteria. This second step of adjusting strict LRMC is generally as important as the first stage calculation.

The LRMC approach provides an explicit framework for analyzing system costs and setting tariffs. If departures from the strict LRMC are required for noneconomic reasons, then the economic efficiency cost of these deviations may be estimated roughly by comparing the impact of the modified tariff relative to (benchmark) strict LRMC. Since the cost structure may be studied in considerable detail during the LRMC calculations, this analysis also helps to pinpoint weaknesses and inefficiencies in the various parts of the power system -- for example, overinvestment, unbalanced investment, or excessive losses at the generation, transmission, and distribution levels, in different geographic areas, and so on. This aspect is particularly useful in improving system expansion planning.

Finally, any LRMC-based tariff is a compromise between many different objectives. Therefore, there is no "ideal" tariff. By using the LRMC approach, it is possible to revise and improve the tariff on a consistent and ongoing basis, and thereby approach the optimal price over a period of several years, without subjecting long-standing consumers to "unfair" shocks, in the form of large abrupt price changes.

3.2 Economics of Marginal Cost Pricing

Marginal cost pricing theory dates back to the pathbreaking efforts of Dupuit (1932) and Hotelling (1938). Ruggles (1947, 1948) provides a useful review of the early literature. The development of the theory, especially for application in the electric power sector, received a strong impetus from the 1950's (Boiteux 1949, Boiteux and Stasi 1964, Steiner 1957, Turvey 1968, Williamson 1966). Recent work has led to developments in peak load pricing, incorporation of the effects of uncertainty and the costs of power shortages, etc. (see Chapters 2 and 8). This section briefly reviews the basic principles of marginal cost pricing and some recent developments.

Basic Marginal Cost Theory

The rationale for setting price equal to marginal cost may be clarified using Figure 3.1. Let $EFGD_0$ be the demand curve (which determines the kilowatt-hours of electricity demanded per year, at any given average price level), while AGS is the supply curve (represented by the marginal cost, MC, of supplying additional units of output).

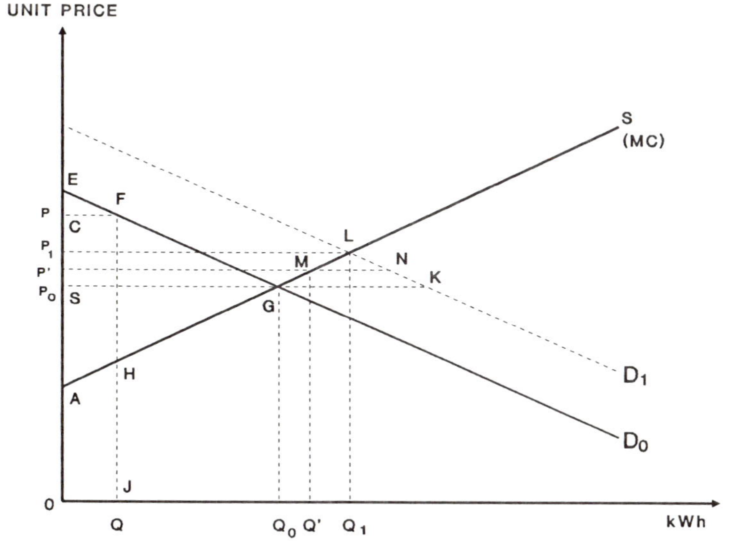

Figure 3.1: Supply and Demand Diagram for Electricity Consumption

At the price p, and demand Q, the total benefit of consumption is represented by the consumers willingness to pay, i.e., the area under the demand curve OEFJ, and the cost of supply is the area under supply curve OAHJ. Therefore, the net benefit, or total benefit minus supply costs, is given by the area AEFH. Clearly, the maximum net benefit AEG is achieved when price is set equal to marginal cost at the optimum market clearing point G, i.e., (p_0, Q_0). In mathematical terms, the net benefit (NB) is given by

$$NB = \int_0^Q p(q)dq - \int_0^Q MC(q)\, dq$$

where p(q) and MC(q) are the equations of the demand and supply curves, respectively. Maximizing NB yields:

$$d(NB)/dQ = p(Q) - MC(Q) = 0$$

which is the point of intersection of the demand and marginal cost curves (p_o, Q_o). A more rigorous dynamic mathematical exposition of this result is presented in Appendix A3.1.

Next, we add to this static analysis, the dynamic effect of growth of demand from year 0 to year 1, which leads to an outward shift in the demand curve from D_o to D_1. Assuming that the correct market clearing price p_o exists in year 0, excess demand GK will occur in year 1. Ideally, the supply should be increased to Q_1 and the new optimal market clearing price established at p_1. But data concerning the demand curve D_1 may be incomplete, making it difficult to locate the point L.

Fortunately, system data permit the marginal cost curve to be determined more accurately. Therefore, as a first step, the supply may be increased to an intermediate level Q', at the price p'. Observation of the excess demand MN indicates that both the supply and the marginal cost price should be further increased. Conversely, if we overshoot L and end up in a situation of excess supply, then it may be necessary to wait until the growth of demand catches up with the overcapacity. In this iterative manner, it is possible to move along the marginal cost curve towards the optimal market clearing point. Note that, as we approach the optimum, it is also shifting with demand growth, and therefore we may never hit this moving target. However, the basic principle of setting price equal to the marginal cost and expanding supply until the market clears, is still a valid and useful operational rule.

Capital Indivisibilities and Peak Load Pricing

Owing to economies of scale, capacity additions to power systems (especially generation) tend to be large and long-lived, resulting in lumpy investments. Suppose that in year 0, the maximum supply capacity is QM_1, as shown in Figure 3.2, while the optimal price and output combination (p_0,Q_0) prevails, corresponding to the demand curve D_0 and the short-run marginal system cost curve SRMSC (e.g., fuel, operating and maintenance costs). As demand grows from D_0 to D_1 over time, with capacity fixed, the price must be increased to p_1 to clear the market. When the demand curve has shifted to D_2 and the price is p_2, new plant is added on. Once the capacity increases to QM_2, p_3 becomes the optimal price corresponding to demand D_3 and the SRMSC line. Generally, the resulting large price fluctuations over time will be unacceptable to consumers. This practical problem may be avoided by adopting a long-run marginal cost (LRMC) approach and peak load pricing.

The basic static peak load pricing model shown in Figure 3.3 has two demand curves; for example, D_{pk} could represent the peak demand during the x daylight and evening hours of the day, while D_{op} is the off-peak demand during the remaining (24 - x) hours when electric loads are light. For simplicity, a single type of plant is assumed with the SRMSC of fuel, operating and maintenance costs given by the constant a, and the LRMC of adding to capacity (e.g., investment costs suitably averaged over years and distributed over the lifetime output of the plant) given by the constant b. The diagram indicates that the pressure on capacity arises due to peak demand D_{pk}, while the off-peak demand D_{op} does not infringe on the capacity QM. The optimal pricing rule now has two parts corresponding to two distinct rating periods (i.e., differentiated by the time of day): (i) peak period price of p = a + b; and (ii) off-peak period price of p = a. The logic of this simple result is that peak period users, who are the cause of capacity additions, should bear responsibility for the capacity costs as well as fuel, operating and maintenance costs, while off-peak consumers only pay the latter costs.

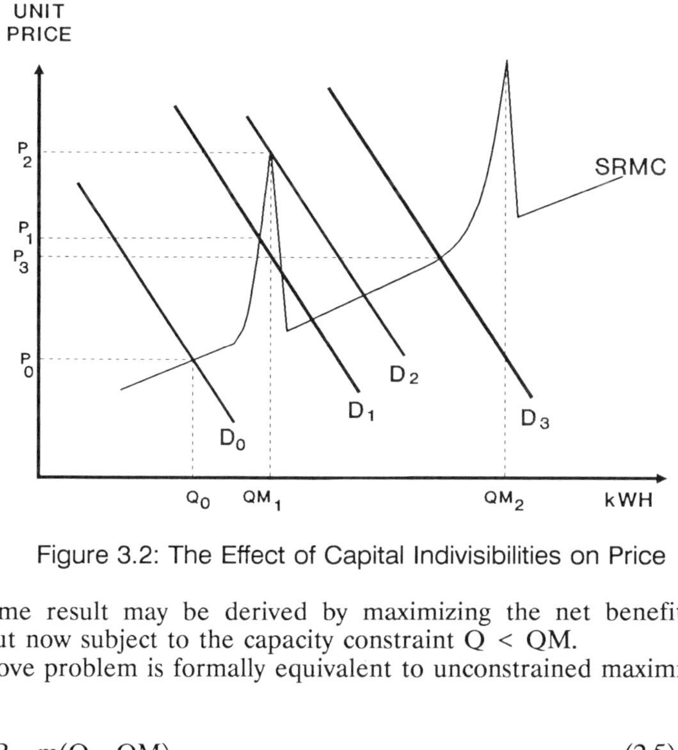

Figure 3.2: The Effect of Capital Indivisibilities on Price

The same result may be derived by maximizing the net benefits defined in Eq.(2.4), but now subject to the capacity constraint Q < QM.

The above problem is formally equivalent to unconstrained maximization of the Lagrangian

$$L = NB - m(Q - QM) \qquad (2.5)$$

The first order (necessary) condition is

$$\partial L/\partial Q = p - MC - m = 0 \qquad (2.6)$$

In Eq.(2.6), MC may be interpreted as the short-run marginal system cost (SRMSC) of supplying more output with capacity fixed, and m has the usual meaning of the marginal cost of adding to capacity. Thus MC = a and m = b with reference to Figure 3.3. During off-peak periods, Q < QM and m = 0, yielding the result p = MC = a. During the peak period, Q = QM and m > 0, indicating that p = MC + m = a + b (see also Appendix A3.1).

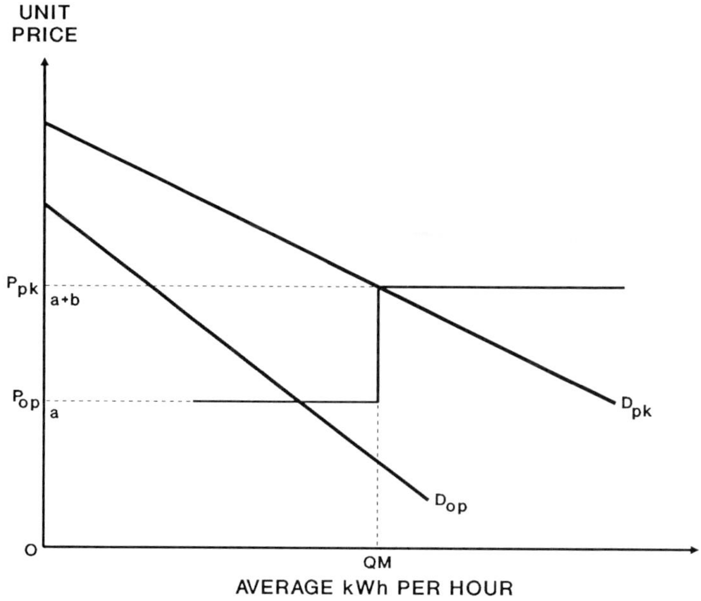

Figure 3.3: Basic Peak Load Pricing Model

Extensions of Simple Models

The simplified models presented so far must be extended to analyze the economics of real-world power systems. First, the usual procedure adopted in marginal cost pricing studies may require some iteration as shown in Figure 3.4. Typically, a deterministic long-range demand forecast is made assuming some given future evolution of prices. Then, using power system models and data, several plans are proposed to meet this demand at some fixed target reliability level (see below). The cheapest or least cost system expansion plan is chosen from these alternatives. Finally strict LRMC is computed on the basis of this least cost plan and an adjusted LRMC tariff structure is prepared. If the new tariff that is to be imposed on consumers is significantly different from the original assumption regarding the evolution of prices, however, then this first-round tariff structure must be fed back into the model to revise the demand forecast and repeat the LRMC calculation.

In theory, this iterative procedure could be repeated until future demand, prices, and LRMC-based tariff estimates become mutually self-consistent. In practice, uncertainties in price elasticities of demand and other data may dictate a more pragmatic approach in which the LRMC results would be used after only one iteration to devise new power tariffs and to implement them. The demand behavior is then observed over some time period; the LRMC is re-estimated and tariffs are revised to move closer to the optimum, which may itself have shifted, as described previously. An extreme form of price feedback could result in a shift of the peak outside the original peak period, especially if the latter was too narrowly defined; that is, peak load pricing may shift the demand peak, from one pricing period to another. If sufficient data on the price elasticity of demand were available, theory indicates that each potential or secondary peak should be priced to keep its magnitude just below the available capacity level. Since the necessary information would rarely be available in practice, a combination of techniques including use of a sufficiently wide peak period, redefining the peak period to include both the actual and potential peaks, direct switching of certain consumer loads, and so on, may be used to avoid the shifting peak problem.

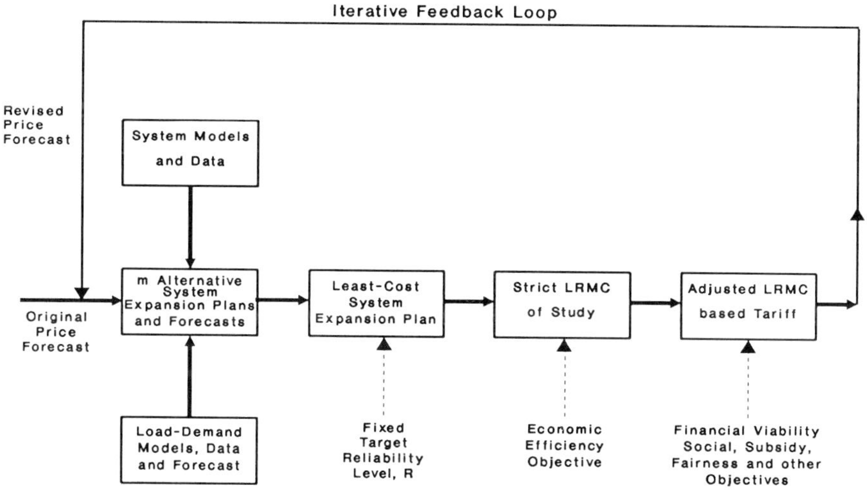

Figure 3.4: The Use of Price Feedback in Estimating LRMC Based Tariffs

Second, the interrelated issues of supply and demand uncertainty, reserve margins, and costs of shortages raise certain problems. Since the least cost system expansion plan to meet the demand forecast is generally determined assuming some (arbitrary) target level of system reliability, e.g., loss-of-load probability (LOLP), reserve margin, etc., the marginal costs depend on the target reliability level (Munasinghe 1979). However, economic theory suggests that reliability should also be treated as a variable to be optimized, and both price and capacity (or equivalently, reliability) levels should be optimized simultaneously. The optimal price is the marginal cost price, while the optimal reliability level is achieved when the marginal cost of capacity additions are equal to the expected value of economic cost savings to consumers due to electricity supply shortages averted by those capacity increments. These considerations lead to a more generalized approach to system expansion planning, described in Chapter 2 and briefly summarized below.

Consider a simple expression for the net benefits NB of electricity consumption, which is to be maximized:

$$NB(D,R) = TB(R) - SC(D,R) - OC(D,R)$$

where TB is total benefits of consumption if there were no outages; SC is supply costs (i.e., system costs); OC is outage costs (i.e., costs to consumers of supply shortages); D is demand; and R is reliability.

In the traditional approach to system planning both D and R are exogenously fixed, and therefore NB is maximized, when SC is minimized, i.e., least cost system expansion planning. However, if R is treated as a variable:

$$\frac{d(NB)}{dR} = -\frac{\partial}{\partial R}(SC + OC) + \frac{\partial}{\partial D}(TB - SC - OC).\frac{\partial D}{\partial R} = 0$$

is the necessary first-order maximization condition. Assuming $\partial D/\partial R = 0$, yields: $\partial(SC)/\partial R = -\partial(OC)/\partial R$.

Therefore, as described earlier, reliability should be increased by adding to capacity until the above condition is satisfied. An alternative way of expressing this result is that since TB is independent of R, NB is maximized when total costs: TC = (SC + OC) are minimized. The above criterion effectively subsumes the traditional system planning rule of minimizing only system costs, but it raises new problems stemming from the need to accurately estimate outage costs (see Chapters 7, 8 and 9 for details).

Third, consider again the choice between SRMC and LRMC for pricing. The SRMC may be defined as the cost of meeting additional electricity consumption, (including the costs of shortages) with capacity fixed. The LRMC is the cost of providing an increase in consumption (sustained indefinitely into the future) in a situation where optimal capacity adjustments are possible. When the system is optimally planned and operated (i.e., capacity and reliability are optimal), SRMC and LRMC coincide. However, if the system plan is temporarily suboptimal, significant deviations between SRMC and LRMC will have to be carefully resolved. For example, in the post 1973 period many utilities replaced oil fired plant with coal fired units to realize fuel cost savings. This resulted in significant excess capacity, and low marginal capacity costs in the short run, thus justifying a reduction in demand charges below the LRMC level. However, as peak demand grows and the system approaches optimality again, the capacity charges should rise smoothly towards LRMC. This transition could become undesirably abrupt if the initial reduction in demand charges was too large and demand growth was overstimulated.

Finally, if there are substantial outage costs outside the peak period, then the optimal marginal capacity costs may be allocated among the different rating periods in proportion to the corresponding marginal outage costs. It has been suggested that capacity costs should be allocated to different ratings periods in inverse proportion to LOLP, but this would be only an approximation because aggregate reliability indices such as LOLP are poor proxies for prorating outage costs.

Shadow Pricing

In the idealized world of perfect competition the interaction of many small profit maximizing producers and welfare maximizing consumers gives rise to market prices that reflect the true economic costs, and scarce resources are efficiently allocated. However, conditions are likely to be far from ideal in the real world. Distortions due to monopoly practices, external economies, and diseconomies (which are not internalized in the private market), interventions in the market process through taxes, import duties and subsidies, etc., all result in market (or financial) prices for goods and services, which may diverge substantially from their shadow prices or future economic opportunity costs. For example, in a country where subsidized diesel fuel is available for electricity generation, the appropriate shadow price would be the import price rather than the artificially low market price. Moreover, if there are large numbers of poor consumers, pricing based only on strict efficiency criteria may be socially and politically unacceptable. Such considerations necessitate the use of appropriate shadow prices (instead of market prices) of inputs to the electricity sector, to determine the optimal investment program as well as LRMC (Munasinghe and Warford 1978).

3.3 Calculating Strict LRMC

Strict LRMC may be defined practically as the incremental cost of optimal adjustments in the system expansion plan and system operations attributable to a small increment of demand which is sustained into the future. The term long-run incremental cost may also be used interchangeably with LRMC, because the changes refer to small but finite variations. LRMC must be structured within a disaggregated framework, based chiefly on technical grounds. This structuring may include: differentiation of marginal costs by time of day, voltage level, geographic area, season of the year, and so on. The degree of structuring and sophistication of the LRMC calculation depends on data constraints and the usefulness of the results, given the practical problems of computing and applying a complex tariff; e.g., in theory, the LRMC of each individual consumer at each moment of time, may be estimated. The basic concepts for calculating strict LRMC are summarized below while details of theory and illustrative case studies may be found in Munasinghe and Warford (1982), and Turvey and Anderson (1977).

Cost Categories and Rating Periods

The three broad categories of marginal costs are: capacity costs, energy costs and consumer costs. Marginal capacity costs are basically the investment costs of generation, transmission and distribution facilities associated with supplying additional kilowatts. Marginal energy costs are the fuel and operating coasts of providing additional kilowatt-hours. Marginal customer costs are the incremental costs directly attributable to consumers including costs of hook-up, metering, and billing. Relevant operation and maintenance costs (O&M), as well as administrative and general costs (A&G) must also be allocated to these basic cost categories. Furthermore, where appropriate, these elements of LRMC must be structured by time of use, voltage level, and so on.

The first step in structuring is the selection of appropriate rating periods. The system load duration curves and generation schedules, should be examined to determine periods during which demand presses on capacity and supply costs are highest. These cyclic critical periods may be due to daily demand variations (e.g., evening lighting load), or seasonal variations in both demand (e.g., summer airconditioning peak load), and supply (e.g., dry season for hydro systems). To illustrate the principles of structuring and calculating strict LRMC, we begin with an all thermal system that does not exhibit marked seasonability of demand, choosing only two rating periods by time of day, i.e., peak and off-peak. Seasonal variations in LRMC and the analysis of hydroelectric systems are discussed later.

Marginal Capacity Costs

Consider in Figure 3.5(a), the typical system annual load duration curve (LDC) ABEF for the starting year 0, divided into two rating periods: peak and off-peak. As demand grows over time, the LDC increases in magnitude, and the resultant forecast of peak demand is given by the curve D in Figure 3.5(b), starting from the initial value MW. The LRMC of capacity may be determined by asking the following question: what is the change in system capacity costs ΔC associated with a sustained increment ΔD in the long run peak demand (as shown by the shaded area of Figure 3.5(a) and the broken line $D+\Delta D$ in Figure 3.5(b)). Consequently, the LRMC of generation would be $(\Delta C/\Delta D)$, where the increment of demand ΔD is marginal both in time, and in terms of MW. In theory, ΔD can be either positive or negative, and generally the ratio $(\Delta C/\Delta D)$ will vary with the sign as well as the magnitude of ΔD. If many such values of $(\Delta C/\Delta D)$ are computed, it is possible to average them to obtain LRMC.

In an optimally planned system, the new incremental load would normally be met by advancing future plant or inserting new units such as gas turbines or peaking hydro plant (see Figure 3.5(b)). Using a computerized generation planning model, it is easy to determine the change in capacity costs ΔC by simulating the expansion path and system operation, with and without the demand increment ΔD. If a more sophisticated tariff structure having many rating periods is used, then the LRMC in any rating period may be estimated by running the computerized system expansion model with a sustained load increment added to the LDC during that particular period. This method that simulates the optimal system planning process is based on the dynamic LRMC concept.

When constraints due to time, data and resources preclude this ideal approach, more approximate methods may be used. Several practical methods of estimating LRMC are analyzed in NARUC Reports 66 and 67 (1980). Simple considerations based on a more static interpretation of LRMC often yield very good results. Suppose that gas turbines are used for peaking; then the required LRMC of generating capacity ($LRMC_{Gen.Cap.}$) may be approximated by the cost per KW installed, annuitized over the expected lifetime. This figure must be adjusted for the reserve margin (RM%) and losses due to station use (L_{su}%). Thus a typical expression would be:

$$LRMC_{Gen.Cap.} = (\text{Annuitized Cost per kW}).(1 + RM\%)/(1 - L_{su}\%).$$

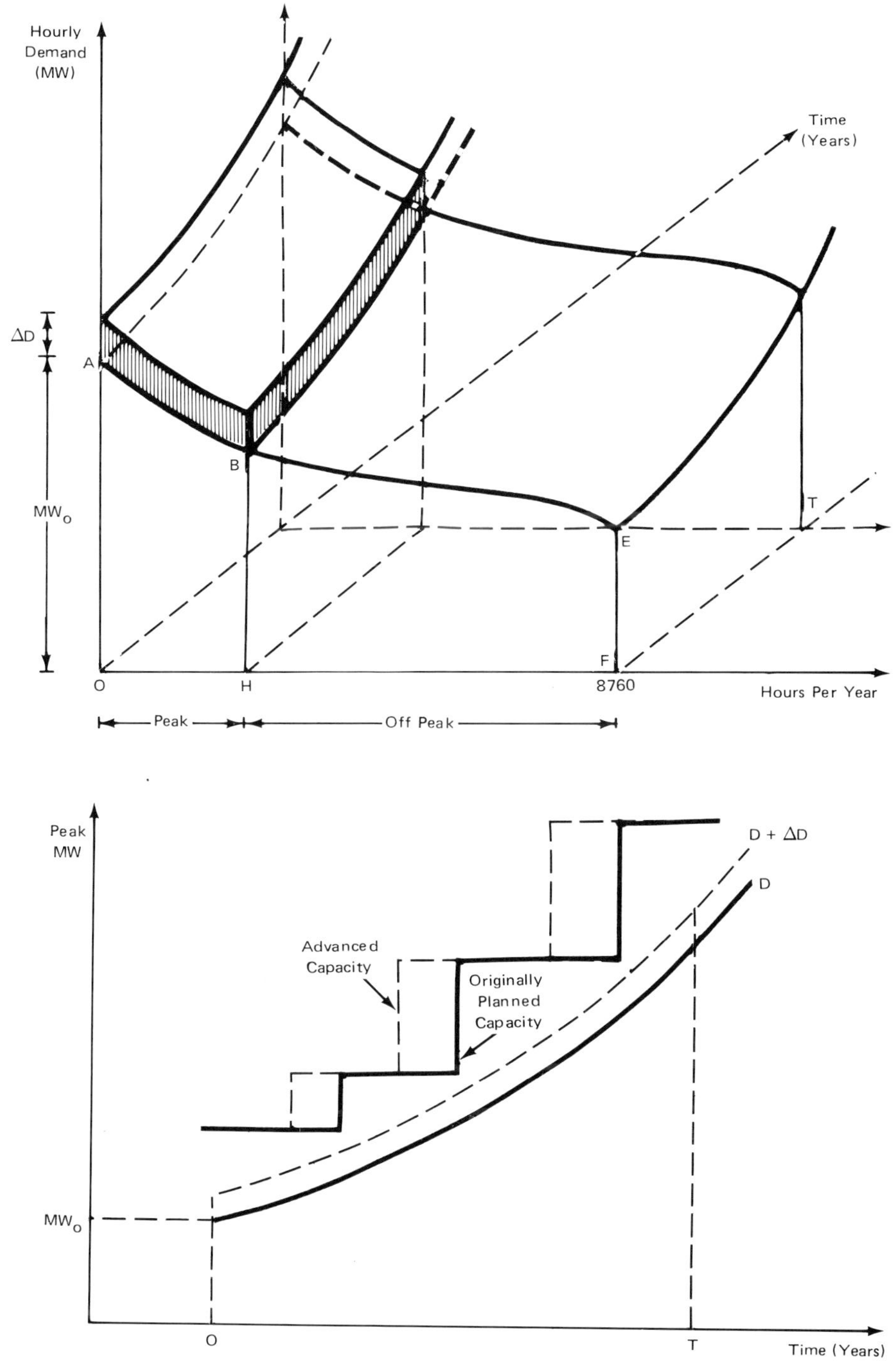

Figure 3.5: (a) Typical Annual LDC (b) Forecast of Peak Power Demand

In our basic model, all capacity costs are to be charged to peak period consumers. Therefore, if the capacity costs of base load generating units are included in the calculations, it is very important to net out potential fuel savings due to displacement of less efficient plant by these new base load units (see Appendix A3.1 for details). Even intuitively, it would not be sensible to incorrectly charge peak consumers the high-capacity costs of expensive base load units (e.g., nuclear), thus encouraging them, for example to install their own captive gas turbine plant.

Next, the LRMC of transmission and distribution (T&D) are calculated. Generally, all T&D investment costs (except customer costs -- discussed later) are allocated to incremental capacity, because the designs of these facilities are determined principally by the peak kilowatts that they carry rather than the kWh. However, particularly at the distribution level, the size of a given feeder may depend on local peak demand which may not occur within the system peak period and this could complicate the problem of allocating distribution capacity costs among the various rating periods (Boiteux and Stasi 1964). The concept of structuring by voltage level may be introduced at this stage. Consider three supply voltage categories: high, medium, and low (HV, MV, LV). Since consumers at each voltage level are charged only upstream costs, capacity costs at each voltage level must be identified.

The simplest approach is to use the average incremental cost (AIC) method to estimate the LRMC of T&D. Suppose that in year i, ΔMW_i, and I_i are the increase in demand served (relative to the previous year), and the investment cost respectively. Then, the AIC of capacity is given by:

$$AIC = \left[\sum_{0}^{T} I_i/(1+r)^i \right] \Big/ \left[\sum_{L}^{T+L} \Delta MW_i/(1+r)^i \right]$$

where r is the discount rate (e.g., the opportunity cost of capital), T is the planning horizon (e.g., 10 years), and L is the average time delay between the investment and commissioning dates for new facilities. We note that in the AIC method, the actual additional increments of demand are considered as they occur, rather than the hypothetical fixed demand increment ΔD used (more rigorously) in calculating generation LRMC. However, because there is no problem of plant mix with T&D investments, AIC and the hypothetical increment method will yield similar results, while AIC is also usually much easier to calculate using readily available planning data. An alternative method of determining marginal T&D costs at several different voltage levels would be to use historical data to fit regression equations such as:

(Transmission Costs) = a + b.(Peak Demand).

However there is no guarantee that such past relationships would hold true in the future, as the system expands.

Assume that the AIC of EHV and HV transmission has been computed and annuitized over the lifetime of the plant (e.g., 30 years) to yield the marginal costs $\Delta LRMC_{HV}$. Then, the total LRMC of capacity during the peak period, at the HV level would be:

$$LRMC_{HVCap.} = LRMC_{Gen.Cap.}/(1 - L_{HV}\%) + \Delta LRMC_{HV}$$

where $L_{HV}\%$ is the percentage of incoming peak power that is lost in EHV and HV network. This procedure may be repeated at the MV and LV levels. The LRMC of T&D calculated in this way is based on actual growth of future demand, and averaged over many consumers. However, facilities associated with given generating sites or loads should be specifically allocated to these uses rather than averaged out, e.g., transmission spur line, exceptionally low or high distribution costs for one or more given customers.

Marginal Energy Costs and Treatment of Losses

The system lamda concept is useful in calculating marginal energy costs. The LRMC of peak period energy will be the running costs of the machines to be used last in the merit order, to meet the incremental peak kilowatt-hour represented by ΔD. In our model, this would be the fuel and operating costs of gas turbines, adjusted by the appropriate peak loss factors at each voltage level. Similarly, the LRMC of off-peak energy would usually be the running costs of the least efficient base load or cycling plant used during this period. Exceptions occur when the marginal plant used during a rating period was not necessarily the least efficient machine that could have been used. For example, less efficient plants which have long start-up times and are required in the next rating period, may be operated earlier in the loading order than more efficient plant. This would correspond to minimization of operating costs over several rating periods rather than on an hourly basis. Again since the heat rate of the plants could vary with output level, the simple linear relationship usually assumed between generation costs and kilowatt-hours may need to be replaced by a more realistic nonlinear model (NARUC Report 67 1980). We note that the loss factors for adjusting off-peak costs will be smaller than the peak period loss factors when current flows are greatest (Cichetti et al. 1977).

The treatment of losses raises several important issues. While total normal technical losses (including station use) vary from system to system, if these are significantly greater than about 15 percent of gross generation, then loss reduction should have a high priority. When engineering losses in excess of acceptable levels are routinely passed onto the customer, this may act as a disincentive to improvements in technical or administrative efficiency. Losses due to theft and unpaid bills are also often loaded onto paying customers. Here again, the issue is whether these nontechnical losses could be reduced by appropriate measures, or if incremental consumption always has an avoidable component of such losses associated with it. Theft in U.S. systems has been estimated to average about 2 percent of gross generation, but norms in developing countries may have to be set somewhat higher (Donziger 1979, Munasinghe and Warford 1982). The LRMC analysis at the generation, transmission and distribution levels helps to establish whether these incremental costs are excessive because of overinvestment, high losses, or both.

Consumer Costs

It has proved difficult to allocate part of the distribution system investment costs to customer costs, on the basis of a skeleton system required to serve a hypothetical minimum load. Similarly a regression analysis of past data to fit equations such as:

$$\text{(Distribution Costs)} = a + b.\text{(Peak Demand)} + c.\text{(Number of Customers)}$$

has not been too successful because peak demand and the number of customers are usually highly correlated. Therefore, general distribution network costs may be considered as capacity costs, while customer costs are defined as those which can be readily allocated to users. Initial customer costs consist of noncurrent expenses attributable to items such as service drop lines, meters and labor for installation. These costs may be charged to the customer as a lump sum or distributed payments over several years.

Recurrent customer costs that occur due to meter reading, billing, administrative and other expenses, could be imposed as a recurring flat charge, in addition to kilowatt and kilowatt-hour charges. In general, the allocation of incremental (non-fuel operation, maintenance and administrative costs among the categories: capacity, energy and customer costs, varies from system to system and requires specific analysis. However, these costs are usually small and their allocation will not greatly affect the results.

Analysis of Hydroelectric Generation

Next, we briefly mention several specific issues which arise in the analysis of hydro systems, and when seasonal variations in LRMC are important (further details are provided in Appendix A3.2). Generally, in an all hydro system the LRMC of generating capacity would be based on the cost of increasing peaking capability (i.e., additional turbines, penstocks, expansion of powerhouse, etc.), while incremental energy costs would be the costs of expanding reservoir storage. When there is significant spilling of water (e.g., during the wet season), incremental energy costs would be very small (e.g., O&M costs only), and at times when demand does not press on capacity, incremental capacity costs may be ignored. However if the system is likely to be energy constrained and all incremental capacity is needed primarily to generate more energy because the energy shortage precedes the capacity constraint for many years in the future, then the distinction between peak and off-peak costs, and between capacity and energy costs, tends to blur. In an extreme case, because hydro energy consumed during any period (except when spilling) usually leads to an equivalent drawdown of the reservoirs, it may be sufficient only to levy a simple kilowatt-hour charge at all times, e.g., by applying the AIC method to total incremental system costs.

In a mixed hydro-thermal system, an important general guideline is that if the hydro is used to displace thermal plant during a rating period, then the running cost of the latter is the relevant incremental energy costs. If pumped storage is involved, the marginal energy costs or value of water used would be the cost of pumping net of appropriate losses. Also, if the pattern of operation is likely to change rapidly in the future (e.g., shift from gas turbines to peaking hydro as the marginal peaking plant, or vice versa), then the value of the LRMC would have to be calculated as a weighted average, with the weights depending on the share of future generation by the different types of plant used.

3.4 Adjusting Strict LRMC

Once strict LRMC has been calculated, the first stage of tariff setting is complete. In the second stage, the actual tariff structure which meets economic second best, social, financial, political and other constraints must be derived by modifying strict LRMC, and this topic is dealt with below. This process of adjusting LRMC will, in general, result in deviations in both the magnitude and structure of strict LRMC. Changes in tariff structure at this stage will be based mainly on sociopolitical factors, e.g., differentiation by type of consumer (residential, commercial, industrial and so on), or by income level (low-, middle-, and high-income residential). Practical considerations such as the difficulties of metering and billing will also affect the final tariff structure.

The constraints which necessitate deviations in the final tariffs relative to strict LRMC fall into two categories (Munasinghe and Warford 1978). The first group consists of distortions which may be analyzed basically within an economic framework, i.e., second best considerations and subsidized (or lifeline) tariffs for low income consumers. In these cases, it is possible to quantify the extent of the deviation from strict LRMC by using an appropriate pricing model and explicit system of shadow prices instead of market prices. Strict (shadow-priced) LRMC also deviates from the market-priced LRMC, and this is done to correct for distortions in the economy. Therefore, the constraints which force further departures from strict LRMC (in the second stage of the tariff setting procedure) may also be considered consequently as distortions which impose their own shadow values on the calculation. The second group includes all other considerations such as financial viability, sociopolitical constraints and problems of metering and billing where strict economic analysis is difficult to apply. These two groups of constraints may be interrelated, e.g., subsidized tariffs can simultaneously have economic welfare, financial and sociopolitical implications.

Second-Best Considerations

Where prices elsewhere in the economy, especially of substitutes and complements for electric power, do not reflect marginal costs, a "second best" departure from a strict marginal cost pricing policy for electricity services may be required. More generally, price distortions affecting inputs into the production of electric power and outputs of other sectors which are electricity intensive (e.g., aluminum) should also be considered. The former type of distortion may be dealt with by direct shadow pricing of inputs as discussed earlier, but the latter case (although quite rare) requires more detailed analysis of the market for the output. As an example of price distortion for an energy substitute, consider the subsidies for imported generators and/or diesel fuel, which exist in some countries. This may make it advantageous for users to set up their own captive plant, even though to the economy as a whole this is not the least cost way of meeting the demand. The appropriate solution in this case might be for the government to revoke such subsidies or restrict imports of private generating plant. However, if transportation policy dictates the need to maintain subsidies for diesel fuel, or if strong pressure groups make such changes politically unfeasible, the low cost of (subsidized) private generation may require the setting of an optimal grid supplied electricity price which is below strict LRMC. The extent of the deviation from strict LRMC is determined by the magnitude of the subsidy and degree of substitutability of the alternative energy source (Munasinghe and Warford 1982).

A related question concerns the availability of subsidized kerosene which in many countries is aimed mainly at providing basic energy requirements for low income consumers at prices they can afford. The subsidy may also prevent low income household especially in developing countries from shifting to use of noncommercial fuels, e.g., wood, the overuse of which leads to deforestation and associated environmental consequences, or animal dung, which has a high opportunity cost as a fertilizer. However, undesirable leakages may occur if the cheap "poor man's" fuel is mixed with more expensive gasoline or diesel and used by relatively wealthy automobile owners or industrialists. If we assume the kerosene subsidy as given, then once again the price of electricity must be reduced proportionately.

Subsidized or Lifeline Rates

In addition to the second best economic arguments (e.g., associated with subsidized kerosene), sociopolitical or equity arguments are often advanced in favor of "lifeline" rates for electric power, especially where the costs of electricity consumption are high in comparison to the relevant income levels. While the ability of electric power utilities to act as discriminating monopolists permits such tariff structuring, the appropriateness of the "lifeline" rate policy and the size of the rate blocks requires detailed analysis.

The concept of a subsidized "social" block, or "lifeline" rate, for low income consumers has another important economic rationale, based on the income redistribution argument. We clarify this point with the aid of Figure 3.6 which shows that respective demand curves AB and GH of low (I_1) and average (I_2) income domestic users, the social tariff P_s over the minimum consumption block O to Q_{min}, and marginal cost based price level P_e. If the actual tariff $P = P_e$, then the average household will be consuming at the "optimal" level Q_2, but the poor household will not be able to afford the service.

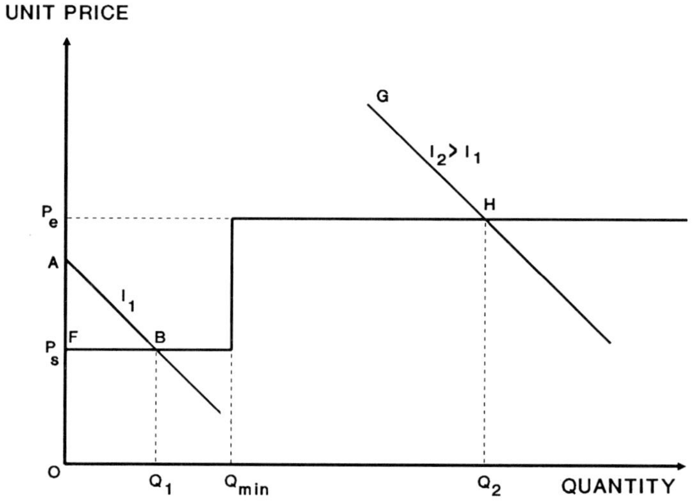

Figure 3.6: Welfare Economic Basis for the Social and Lifeline Rate

If increased benefits accruing to the poor have a high social weight or value, the consumer surplus portion ABF should be multiplied by the appropriate social weight (greater than unity). Then, although in nominal market prices the point A lies below P_e, the weighted distance OA could be greater than the marginal cost of supply. The adoption of the increasing block tariff shown in Figure 3.6, consisting of the lifeline rate P_s, followed by the full tariff P_e, helps to capture this "weighted" consumer surplus of the poor user, but does not affect the optimum consumption pattern of the average consumer, if we ignore the income effect due to reduced expenditure of the average consumer for the first block of consumption, i.e., up to Q_{min}. In practice, the magnitude Q_{min} should be based on acceptable criteria for identifying "low income" groups, and reasonable estimates of their minimum consumption levels (e..g, sufficient to supply basic requirements for lighting, heating, appliances). In the U.S. where the average household consumes over 700 kWh/month, the minimum consumption level may typically be of the order of 100 to 200 kWh/month, while in the developing countries where average electricity use is much lower, Q_{min} is usually around 50 kWh/month. For the price P_s, one simple welfare model yields:

$$P_s = \text{strict LRMC x (poor persons income/critical income)}$$

where the critical income is like a nationally established poverty line (Munasinghe and Warford 1978). The utility revenue constraints and the ability to pay of the poor consumer would also be considered in determining P_s and Q_{min} (Francfort and Woo 1977). This approach may be reinforced by an appropriate connections policy (e.g., subsidized house connections, et cetera). In the U.S., the rights of poor and disadvantaged customers are recognized in the Public Utilities Regulatory Policies Act (PURPA) of 1978, with respect to lifeline rates (Section 114), and protection against unfair terminations of service (Section 115g).

Financial Viability

The financial constraints most often encountered relate to the revenue requirements of the sector, and are often embodied in criteria such as some target financial rate of return on assets, or an acceptable rate of contribution towards the future investment program. In principle, for state-owned power utilities, the most efficient solution would be to set price equal to marginal cost and rely on government subsidies (or taxes) to meet the utilities' financial needs. In practice, some measure of financial autonomy and self-sufficiency is an important goal for the sector. Because of the premium that is placed on public funds, a marginal cost pricing policy which results in failure to achieve minimum financial targets for continued operation of the power sector, would rarely be acceptable. The converse and more typical case, where marginal cost pricing would result in financial surpluses well in excess of traditional revenue targets, often leads to consumer resistance. Therefore in either case, changes in revenues have to be achieved by adjusting the strict marginal cost based tariffs.

A widely used criterion of financial viability is the utility potential to earn an acceptable rate of return on assets, for example, the net operating income after taxes given as a fraction of net fixed assets in operation plus, in some cases, adequate working capital. In the case of private utilities -- for example, in the U.S. -- the regulatory authorities have traditionally imposed a fair rate of return as an upper limit on earnings and therefore, on average price per unit sold (Garfield and Lovejoy 1964). Where utilities are government owned, as in most developing countries, the target rate of return is usually considered a minimum requirement to help resist sociopolitical pressures that tend to keep prices too low. If the asset base is defined in revalued terms, then this requirement is more consistent with the forward looking approach of LRMC. Another future oriented financial criterion that is especially useful when the system expands rapidly, requires the utility to make a reasonable contribution to its future investment program from its own revenues. This self-financing ratio is often expressed by the amount of internally generated funds available after operating expenses and debt service, as a fraction of capital expenditures.

The application of the financial criteria often raises serious conceptual and practical problems. Thus, if a rate of return test is to be used, then the question of asset revaluation arises. The use of historical costs for working assets, typically original cost less depreciation, would tend to understate their value when capacity costs are rising rapidly. If assets are to be revalued, the costs of either (a) exactly reproducing the power system at today prices; or (b) replacing it with an equivalent system, also at today prices; might be used after netting out depreciation to allow for the loss of value corresponding to the economic and functional obsolescence of existing equipment. Significant difficulties of interpretation clearly will occur in the practical application of either approach.

Whichever criterion or combination of criteria is used, it is important that the initial tariffs based on strict LRMC be included in the utility financial forecast. Then these first round tariffs may be adjusted through an iterative process until the chosen parameters of financial viability fall within the acceptable range. Although this process is usually quite ad hoc, some practical guidelines may be effectively used for reconciling strict LRMC and the revenue requirement. The relative adjustments to strict LRMC between major consumer categories like residential and industrial, as well as among the different rating periods (like peak and off-peak) within a given consumer category, will determine the share of the revenue burden to be borne by each user group in a given rating period (NARUC Report 69 1979).

The simplest practical method of adjustment, which also appears to be the most equitable, is to retain the relative structure of LRMC and vary the average rate level by equiproportional changes. In general this procedure will not be economically efficient.

The application of the inverse elasticity rule whereby the greatest (least) divergence from strict LRMC occurs for the consumer group and rating period where the price elasticity is lowest (highest), is the most satisfactory adjustment procedure from the viewpoint of economic efficiency (Baumol and Bradford 1970). In the case of two goods, the following expression applies:

$$(1 - LRMC_1/p_1)/(1 - LRMC_2/p_2) = (1/e_1 + 1/e_{12})/(1/e_2 + 1/e_{21})$$

$LRMC_i$ and p_i are the strict LRMC and price, respectively, of good i; while

$$e_i = (\partial Q_i/\partial p_i)/(Q_i/p_i)$$

and

$$e_{ij} = (\partial Q_i/\partial p_j)/(Q_i/p_j)$$

are the own and cross price elasticities, respectively, of demand (Q) with respect to price (p). The two goods 1 and 2 may be interpreted as either the electricity consumption of two different consumer groups in the same rating period or the consumption of the same consumer group in two distinct rating periods. In practice, a larger number of consumer types and rating periods must be considered and application of the rule will be limited by lack of data on price elasticities and the need to use subjective estimates (Malko and Faruqui 1980). This technique may appear to penalize some customers more than others, thus violating the fairness objective.

Adjustments involving lump-sum payments/rebates or changes in customer and connection charges are also consistent with economic efficiency provided consumers electricity usage is relatively unaffected by these procedures, i.e., consumption depends mainly on the variable charges. The magnitude of the adjustments that can be made may be insufficient however. Another related approach for reducing revenues is to charge strict LRMC only for marginal consumption and reduce the price for an initial block of electricity use. These subsidies on customer charges or on the initial consumption block can also be tailored to satisfy the lifeline rate requirement for poor consumers, but such measure tend to complicate the price structure.

In practice, an eclectic approach involving a combination of all these methods is most likely to be successful.

Other Considerations

There are several additional economic, political, and social considerations that may be adequate justification for departing from a strictly marginal-cost-based tariff policy. The decision to electrify a remote rural area, which may also entail subsidized tariffs because the beneficiaries are not able to pay the full price based on high unit costs, could be made on completely noneconomic grounds, e.g., for general sociopolitical reasons such as maintaining a viable regional industrial or agricultural base, stemming rural to urban migration, or alleviating local political discontent. However, the full economic benefits of such a course of action may be much greater that the apparent efficiency costs which arise from any divergence between actual price and strict LRMC. This possibility is likely to be much more significant in a developing country than in a developed one, not only because of the high cost of power relative to incomes in the former, but also because the available administrative or fiscal machinery to redistribute incomes or achieve regional or industrial development objectives by other means is frequently ineffective.

For the same reason, it is particularly difficult to reform pricing policy where low incomes and a tradition of subsidized power combine to create extreme sociopolitical difficulties in raising prices to anywhere near marginal costs. In practice, price changes have to be gradual, in view of the costs which may be imposed on those who have already incurred expenditures on electrical equipment and made other decisions, while expecting little or no changes in traditional power pricing policies. The efficiency costs of "gradualism" can be seen as an implicit shadow value placed upon the social benefits that result from this policy. Another macroeconomic type argument that electricity price increases may be inflationary is rarely valid, because the costs of electricity use are usually a small proportion of household expenses and of industrial production costs. The overstimulation of demand and lack of funds to expand supply, resulting from low electricity prices are potentially much more serious long-run problems that cannot be ignored.

Metering and Billing and Customer Comprehension

Owing to both the practical difficulties and the economics of metering and billing, the tariff structure may have to be simplified. Another crucial factor is that the tariff structure must be comprehensible to the average customer. Otherwise, individuals will not be able to adjust their consumption according to the price signal. Therefore, the number of customer categories, rating periods, consumption blocks, voltage levels, and so on will have to be limited.

The degree of sophistication of metering (e.g., by time of day) depends on the practical problems of installation and maintenance, and the net benefit of metering -- based on a cost benefit analysis that compares the lower supply costs of reduced consumption with the cost of metering plus the decrease in net consumption benefits (see Appendix A3.3 for details). Thus for very poor consumers receiving a subsidized rate, a simple current limiting device may suffice, because the cost of even simple kilowatt-hour metering may exceed its net benefit. In general, various forms of peak load pricing (i.e., maximum demand or time-of-day metering) would be more applicable to large MV and HV industrial and commercial consumers. For practical details of metering see NARUC (1977).

Most LV consumption, especially for households, is metered only on a kilowatt-hour basis, with the price per kilowatt-hour based on a combined energy and "rolled in" capacity charge (e.g., using appropriate coincidence and load factors). More sophisticated meters, such as time-of-day meters which incorporate synchronous clocks, may be affected by power outages. At the other end of the scale, current limiting devices are easier to tamper with.

Recently, the concept of homeostatic utility control has been proposed in the US, using advanced solid-state technology (including use of microprocessors) to implement sophisticated metering, automatic meter reading, load management techniques and pricing structures (MIT Energy Lab 1979). In contrast, some developing countries may lack technically skilled labor for installation and maintenance of sophisticated meters, or even reliable meter readers. Therefore, choice of appropriate metering is usually very country specific, and is likely to involve many practical considerations.

3.5 Practical Implementation and Results of Modern Pricing Structures

In this section we briefly review the types of tariff structures used to implement the LRMC approach, and the most recent empirical evidence regarding the effectiveness of such modern pricing structures. Finally, the US situation and the implications of the PURPA of 1978 are discussed. Practical applications of the pricing methodology described in this chapter are provided in the case studies of Pakistan and Burma presented in Chapters 2 and 4 respectively.

Types of Tariff Structures

Over the last 50 years, price structures have become increasingly complex as both the techniques for analyzing the structure of supply costs and the metering hardware available to apply these tariffs have become progressively more sophisticated. Since the quantity, quality and price of electricity supplied to each consumer can be, if necessary, individually controlled or at least monitored, a high degree of discrimination and structuring is possible with electricity prices. In theory, a separate tariff could be devised for each customer. In practice, however, as discussed in the previous section, the complexity of the tariff would be limited by the metering capabilities, the problems of billing, and the ability of electricity users to comprehend and react to the price signals provided by the power utility.

The structuring of LRMC with respect to voltage level, geographic area, and customer type have been discussed earlier. This section focuses on how tariffs may be devised and implemented, that vary in relation to the following principal aspects: (a) energy or kilowatt-hour consumption; (b) power demand based on kilowatt or kilovolt-ampere consumption; and (c) fixed charges, including both nonrecurrent and recurrent charges. Structuring of aspects (a) and (b) by time of use and usage level will also be reviewed, as well as interruptible rates, the use of tariff adjustment clauses to correct for power factor, fuel surcharges, and so on. These basis building blocks may be combined in various ways to yield literally hundreds of tariff structures differing in their finer details.

The most common form of tariff is the energy charge based on the customer kilowatt-hour consumption over a given period of time, typically one month. Kilowatt-hour meters that record consumption continuously over shorter periods -- for example, 15-min. intervals or during two different periods of the day -- may be used to implement electricity prices that vary by time of use (TOU). During the peak period, typically, the capacity charge is converted into an equivalent kilowatt-hour charge and added to the energy charge.

Unit charges may also be varied according to the number of kilowatt-hours consumed, yielding two basic types of block tariff structures. Block structures may also be used with kilowatt or capacity charges but this is not a common practice. In the increasing or inverted block tariff, the kilowatt price increases as consumption rises. Incorporation of the increasing block structure in applying the LRMC-based methodology has already been discussed, particularly in the section on social or subsidized prices.

The decreasing block tariff, in which the initial slab of consumption has the highest price followed by successively cheaper blocks has been widely used especially for households and small consumers with only kilowatt-hour metering, where more complex metering would not be economically justified (see the section on metering and billing). The rationale for this policy included arguments that: (a) the utility could recover some of the fixed customer costs through the high priced initial block even though kilowatt-hour consumption was low; (b) the first block corresponded to the high cost of supplying the customers peak period load, whereas additional consumption was mainly caused by off-peak appliance use that could be supplied at relatively low cost; (c) the utility should encourage increased consumption to realize economies of scale in production; (d) price discrimination could be used to extract the maximum revenue from smaller users who had low price elasticities of demand while also encouraging consumption of larger users who were more sensitive to high prices; and (e) if temporary excess capacity existed -- for example, when a new hydrosite was developed -- the new energy could be supplied "costlessly" and, therefore, higher consumption should be encouraged to collect the maximum potential revenues.

All of these arguments ignore the fact that if any slab of the decreasing block tariff is significantly below LRMC, it signals the consumer that electricity is much cheaper than it really is, thus encouraging wasteful consumption. First, if customer costs must be recovered then single or recurring fixed charges should be used. Second, unless there is clear evidence that customers with greater consumption have higher user load factor and consume relatively more off-peak energy at the margin, any additional kilowatt-hours consumed by all consumers will be equally costly to supply. Therefore, there would be little basis for price discrimination according to consumption level. Third, even if economies of scale exist at the aggregate level of the utility, they do not apply in the case of the variable costs to individual customers. In fact, few utilities currently exhibit any economies of scale, and real unit costs of supply in the long run are rising. Fourth, it cannot be generally assumed that the consumption of larger users would be more sensitive to price. Fifth, using up any short-run excess capacity is not costless in the long run, because if demand growth is unduly stimulate, future investments must be advanced. Finally, the decreasing block rate is highly regressive and "unfair," because it penalizes poorer consumers who generally use less electricity but must pay higher prices per unit purchased (see also, earlier discussion of lifeline tariffs).

As explained earlier, the purpose of structuring tariffs by TOU, voltage level, geographic area, and so on, is to convey the LRMC of supply to consumers as accurately as possible. Although peak load or TOU tariffs may also be determined on the basis of accounting costs, the allocation of system capacity costs to different pricing periods in this case is usually quite arbitrary. For example, one method attempts to identify peaking, intermediate and base load generation plant and then allocates the costs of these units to the peak, shoulder and off-peak periods. Another procedure uses the probability of contribution to the peak based on the number of hours in each rating period in which demand exceeds some arbitrary threshold level, dived by the total number of such hours in the year, as the allocation criterion (NARUC 1979). None of these methods satisfy the economic efficiency objective, and therefore, references to peak load and TOU rates in the subsequent discussions imply that these are based on LRMC.

TOU metering (when this is justified) is the best way to apply an LRMC-based pricing structure. The Hopkinson or two-part tariff with separate energy and peak demand charges is used widely, but if the consumer's maximum kilowatt or kilovolt-ampere demand is not measured a the time of the system peak, then he or she could be unfairly penalized. If only kilowatt-hour metering is available, the capacity charge may be levied on the customer's connected kilovolt-amperes; for example, with a current limiting braker or fuse to limit the maximum load. But this is even more questionable, since it requires that the relationship between the consumer's peak demand and connected kilovolt-amperes be accurately known. Interruptible tariffs are an extreme form of peak load pricing in which the customer agrees to be disconnected or shed at short notice when there is a power shortage. These prices have to be low because there is no burden on system capacity. Sometimes the interruptible customer is offered the option of remaining on the system at a time of shortage provided he pays a much higher price. In either case, when demand presses on supply the interruptible tariff increases rapidly either to a high value or to infinity -- if the customer is automatically shed.

Fixed charges are most often related to consumer costs as described earlier. A lump-sum payment may be levied to cover the initial cost of providing the service connection, or the repayment period may be spread over several years to provide relief to customers. Recurrent fixed costs are charged to meet the costs of meter reading, billing, and other repetitive expenses. In some cases, the charge based on a consumer's connected kilovolt-amperes is also called a "fixed" charge, but this is really a proxy for the capacity or kilowatt cost which is a variable charge.

In general, the conversion of strict LRMC into applicable kilowatt, kilovolt-ampere, or equivalent kilowatt-hour charges during different pricing periods requires the knowledge of the customer characteristics such as the load factor, diversity or coincidence factor, ratio of connected kilovolt-amperes to maximum demand, and so on (Munasinghe and Warford 1982).

Tariffs contain power factor (PF) penalty surcharges in excess of the regular price to encourage consumers whose PF drops below some acceptable limit to install capacitative correction. Fuel surcharge or fuel adjustment clauses are also becoming increasingly common. This permits the utility to quickly pass on to the consumer any unforeseen increases in fuel costs, especially of liquid fuels. Ideally, any changes in relative input prices would require reestimation of strict LRMC followed by changes in the tariff structure, but the legislative procedure to achieve the latter may take a long time. A convenient short-run fuel adjustment clause can, meanwhile, provide much needed financial relief (Herman 1974).

Recent Empirical Evidence

Peak load pricing or TOU tariffs have been applied in varying degrees for many years in Europe, and more recently in some developing countries (Mitchell et al. 1978, Munasinghe and Warford 1982). In general, high- and medium-voltage industrial and commercial customers have been faced with separate capacity and energy changes, varying by time-of-day or season. Greater deviations are allowed for low-voltage consumers, because of lifeline rate considerations (as discussed earlier), or simplicity of metering (e.g., demand charge based on maximum load limited by breaker or fuse, demand charge rolled-in to energy charge using a typical user's load factor, etc.). Recurrent flat rate charges are also used recover fixed costs. We note that the conditions imposed on interruptible loads are an extreme form of peak load pricing, where the price charged to these customers during periods of capacity shortage is either very high (if the consumer has the option to continue receiving power at an increased price) or infinite (if the load is automatically shed). Some typical power tariffs recently applied in several countries are summarized in Table 3.1.

The HV and MV tariff structures in European countries where there is substantial thermal generation (England and France), reflect the importance of both seasonal and time-of-day demand peaks, and the variations in energy costs of the least efficient or marginal plants operated during the various rating periods. In the scandinavian countries where hydroelectric generation is dominant, seasonal variations in the availability of water (e.g., the additional costs of storage in the winter months) tend to dictate the structure of prices. The HV and MV tariffs adopted in the developing countries are quite recent and somewhat simpler because of the need to introduce these new pricing structures gradually, the shortage of appropriate metering, and so on. The LV domestic tariff in all countries are also relatively uncomplicated particularly due to use of less complex metering; for example, the demand charges (if any) are levied on the basis of subscribed kilovolt-ampere or maximum size of fuse or breaker. Off-peak nighttime discounts are offered in some thermal generation dominated systems. The generally low level of domestic tariffs, especially of lifeline rates, reflect the importance of socio-political considerations.

Where these modern pricing structures have been in existence long enough to show results, their impact has been favorable. Industrial and commercial customers in europe have responded to peak load prices in many ways, including: (a) changing their pattern of production by increasing activity during off-peak periods and vice versa; (b) using waste heat or combustible residual materials to generate their own energy during peak periods; (c) storing heat and energy for use during peak periods; and so on. The dominant response has been to shift load from the peak to the off-peak period, rather than cutback their total electricity consumption.

Table 3.1

Some Typical Examples of Peak Load Tariffs[a]

Country	HV and MV Consumers Demand Charge $/kW/m	 Energy Charge c/kWh	 Fixed Chge. $/mth	LV (Domestic) Consumers Energy Charge c/kWh	 Fixed Charge $/mth	Percentage Hydro
A. **Europe**						
England (Northeastern Area Electricity Board - 1980)	≥650V April-Oct.0.0 Nov.-Mar:6.33	0700-2400 daily 5.91 2400-0700 daily 2.78	1.0 per kVa	I. Two part 8.48 II. White Meter 0900-2200:9.37 2200-0900:4.0	 2.82 3.75	1 to 2
France (Electricité de France - 1980)[f]	5-30KV 3.21	Nov.-Feb.[c] 0700-0900 and 1700-1900:10,36 Oct.-March 0600-2200:6.08[d] 2200-0600 and Sunday:3.06 April-Sept. 0600-2200:3.87[c] 2200-0600 and Sunday:2.42	-	Breaker Size 3kW:6.45 6kW:6.45 9kW:6.45 12kW:6.45 15kW:6.45 18kW:6.45	 3.90 6.14 10.27 14.37 18.49 22.61	38 to 40
Norway (Water Resources and Electricity Board-1975)	Oct-Apr Contract 1.65 All year Contract 1.89	 0.86 Oct-Apr:0.86 May-Sep:0.44	-	0.85 (reg.) 3.34 (in excess of subs.level)	2.78 per Subs.kW	99
Sweden (State Power Board-1980	70 & 130 kV 0.27 (Contract.) 2.71 (Pk.Load)	May-Aug:1.35 Sept-April:1.56	38-75	0600-2200: 4.37 2200-0660: 2.21	3.88 plus 2.24 to 0.59 per amp. of fuse rating (16-200A)	58 to 60

Table 3.1 (contd.)

Some Typical Examples of Peak Load Tariffs[a]

	Country	HV and MV Consumers: Demand Charge $/kW/m	HV and MV Consumers: Energy Charge c/kWh	HV and MV Consumers: Fixed Chge. $/mth	LV (Domestic) Consumers: Energy Charge c/kWh	LV (Domestic) Consumers: Fixed Charge $/mth	Percentage Hydro
B.	**Developing Countries**						
					0-200VA: -	18.5[e]	
	Indonesia (Perusahaan Unum Listrik Nagara -1980)	12&20kV,200-5000 kVA 2.56[e]	1800-2200:4.64 2200-1800:3.20	-	201-500VA: 4.64 501-2200VA: 6.24 2201-6600VA: 7.20 >6600VA: 9.20	4.48[e] 4.48[e] 4.48[e] 4.48[e]	24
	Kenya (Kenya Power Company-1980	11&33kV,>100 MWh per month 6.74[e]	0800-200 Mon-Frid:3.75 All other times:2.25	53.93	≤30kWh: 3.30 >30kWh: 7.49	2.25	75
	Pakistan (Water and Power Development Authority-1980)	66 & 132 kV 7.11	2.64	-	<40kWh: 3.15 41-200kWh: 4.57 >200kWh: 7.61	-	80 to 82

a. To facilitate comparison, all prices have been converted into constant mid 1980 US$, using: (a) exchange rates prevailing in the year that the tariffs were applicable; and where necessary, (b) the U.S. implicit deflator of GNP utp to 1980.
b. Energy generated from hydroelectric sources 1978-89, as a percentage of total generation.
c. Excluding Synday.
d. Excluding Sunday and Nov.-Feb.:0700-0900 & 1700-1900.
e. Per connected kVA.
f. Includes 17.6% valued added tax.

For example, peak loads of a wide range of european industries have been reduced by amounts ranging form 30 to 90 percent, while the group load factor of 70 percent of industrial load in U.K. increased from 45 to 70 percent from 1961 to 1975 (Burchnall 1977, Pelletier 1977, Pioger 1977). In the case of LV residential consumers, the price signals are not so clearly conveyed due to lack of complex metering. Therefore, these tariffs have been effectively supplemented in several european countries, by a coordinated package of other domestic load management techniques, including storage (space and water) heating during off-peak hours and central control of specific domestic loads (e.g., ripple control) (Birchnall 1977, Mitchell et al. 1978). Some problems have been encountered in Europe, especially France, due to shifting of the peak.

PURPA and Peak Load Pricing in the U.S.

US regulatory bodies and utilities have generally hesitated for a long time to change from conventional accounting or embedded cost approaches to rate making, and adopt more modern techniques based on LRMC. This was mainly due to the steady decline in historical electricity prices, resulting from technological improvements and economies of scale; for example, the average price paid by U.S. domestic consumers fell from 6 cents/kWh to almost 2 cents/kWh between 1930 and 1970, while mean annual consumption increased from about 500 to 7050 kWh per household during the same period. However, the recent worldwide increases in rates of inflation and energy costs have reversed the trend, with the average revenues received from households rising almost 100% to a little over 4 cents/kWh between 1970 and 1978, while mean consumption also continued to increase to about 8850 kWh. These developments have stimulated interest in marginal cost pricing. Thus for example, in 1974 the National Association of Regulatory Utility Commissioners (NARUC) requested that a study of load management and rate design be carried out, the first phase if which was completed in 1977 while further work is ongoing (NARUC 1977, 1979). Meanwhile, the National Energy Act (NEA) passed by Congress in October 1978 which includes the Public Utilities Regulatory Policies Act (PURPA) has been a major step forward in helping to rationalize electricity tariffs in the U.S. (PURPA 1978) In Section 111(d), PURPA establishes federal standards for power pricing (applicable to utilities with retail sales exceeding 500 GWh in a baseline year after January 1, 1976), and the individual state regulatory authorities have to determine by November 1981 whether or not they wish to apply the standards (Section 111(a)). Although a few state regulatory bodies (e.g., California, Michigan, New York, and Wisconsin) have been applying TOU rates for several years, at least to some large consumers (including embedded/accounting cost based TOU rates in Michigan), the majority are still uncommitted in this regard.

While compliance with PURPA is not mandatory (see Section 111(a)), the detailed description of these standards in Section 115, relating especially to Cost of Service, Time-of-Day Rates, and Load Management Techniques, may be interpreted in the spirit of the LRMC approach described earlier, in terms of structuring tariffs, and using marginal costs based on long-run considerations. For example, Section 115(a) describes the methods to be used for determining cost of service:

"Such methods shall to the maximum extent practicable

1) permit identification of differences in cost incurrence for each such class of electric consumers, attributable to *daily* and *seasonal time of use* of services; and
2) permit identification of differences in cost incurrence attributable to differences in *customer, demand, and energy components of cost*.

In prescribing such methods, such State regulatory authority or nonregulated electric utility shall take into account the extent to which total costs to an electric utility are likely to change if
1) *additional capacity* is added to meet peak demand relative to base demand; and
2) *additional kilowatt-hours* of electric energy are delivered to electric consumers".

Note: Author's italics for emphasis.

The Federal Energy Regulatory Commission (FERC) has followed up quickly to implement Section 133 of PURPA which sets out the requirements for "Gathering Information on Costs of Service." While this is only the first step required to calculate the costs of service, after which there yet remains the long drawn-out process of devising and implementing a new tariff, it is nevertheless an essential prerequisite for all that is to follow. FERC has clearly ruled against several utilities which challenged the appropriateness of the marginal cost approach, and has already required 190 utilities to comply with the data gathering and reporting requirements of Section 144, by November 1980 (FERC 1979).

The European experience and preliminary results from ongoing rate studies in the U.S. indicate that the benefits of implementing TOU tariffs (based on LRMC) for large HV and MV consumers could be substantial (Johnson 1976, Reynolds and Creighton 1980). Thus, Mitchell et al. (1978), have roughly estimated that potential long-term savings in both investment and operating costs for the whole U.S. would be about $1.3 to 3.5 billion per eyar, corresponding to the postponement or avoidance of constructing about 28 GW of capacity. Some specific results of TOU pricing experiments in the U.S. are summarized below but should be interpreted cautiously. First, changes in load shape due to TOU rates are difficult to analyze because demand in a given rating period will be affected not only by the price in that period but also by the price changes in other periods. Second, short- and long-run responses may be quite different and cannot be easily separated out. Finally, the effects of many other factors such as climate and user tastes also complicate the analysis.

More specific results indicate that 515 large industrial and commercial users in 5 service areas of California, Michigan, New York, and Wisconsin with minimum demands ranging from 300 kW to 5000 kW responded to TOU prices as follows (NARUC Report 79 1980). About 20 percent of the customers surveyed reduced their peak demand. Over one third of these switched their load to the off-peak period, mainly by shifting equipment use, while the rest reduced their peak demand without shifting to off-peak use, by more efficient operation of air conditioning, ventilation and lighting. 55 percent of users reported no change in peak demand; on average 82 percent of these did not respond to TOU tariffs because their production process was inflexible or the change was not cost effective, while only 18 percent made no effort to change. Finally, 25 percent of the large users increased peak demand because of stepped up production. About 90 percent of all users surveyed indicated that they understood the TOU rates well, and nearly two thirds had made impact studies of the new price structures.

Studies involving TOU metering for residential consumers are now being carried out in several states. Some broad but tentative conclusions have already been reported for Arizona, Connecticut, Ohio, North Carolina, Rhode Island, and Wisconsin, including: reduced kilowatt demand during the peak hour, reduced kilowatt-hour consumption for the entire peak period, more favorable diversity factor for TOU customers, reduction in intermediate period (or shoulder) demand, and no evidence of needle peaking outside the peak period (Miedema et al. 1978, 1980). Another set of results form Wisconsin indicates that an eight to one ratio in the peak to off-peak price would reduce the average peak to off-peak energy consumption during the summer by 24 percent, with the degree of response being positively related to the number of large household appliances owned (Caves and Christensen 1980). While all of these studies report on short-run results, additional long-run consumer responses may be anticipated due to the increasing ability of households to change their patterns of appliance ownership and usage over time.

However, application of TOU prices to LV residential and commercial customers requires more study, to determine whether the significant costs of additional metering is outweighed by the net savings due to improved patterns of consumption (Munasinghe and Warford 1982). For example, a recent British study indicates that on the basis of such a cost-benefit analysis of metering, TOU prices for domestic consumers would not be justified (Electricity Council 1978). More generally, load management techniques such as control of domestic water and space storage heating and cooling have been very effective in Europe and show promise in the U.S. (de Grasse 1977, Hastings 1980, NARUC Report 91 1980). Available load control technologies include radio and ripple control, power line carrier, telephone line, time switches, interlocks, and load limiters, etc. However, the usefulness of these methods in the U.S. has to be verified further in the light of wide difference in energy use patterns, tastes and climate. Meanwhile, even some simple initial steps such as the elimination of promotional decreasing block tariffs (see PURPA, Section 111(d)(2)), would often constitute a significant improvement in rate structuring.

While the more extensive application of peak load pricing based on LRMC is clearly justified in the U.S., the transition period may take many years. First, the truths and myths regarding modern tariff structures must be well understood by customers, utility managers and regulators. The rate hearing process and other means of disseminating information will play a key role in this respect. Second, the application of new tariffs could begin with the larger customers at HV and MV levels, because they are less numerous and may be sensitized more easily. Over the years, more and more customers including LV users would become subject to the new tariffs. Third, the changes in tariff structure should be initially small, to avoid customer resistance due to unfamiliarity or hardship caused by large increases in the total electricity bill. Later, tariffs could be altered more rapidly to approximate LRMC better. Finally, alternative rate structures could be offered simultaneously (i.e., both the conventional and new tariffs), and customers could gradually be won over to the new tariff. We conclude by noting that the implementation and effectiveness of future pricing structures will depend critically on the development of new hardware and techniques. For example, advances in solid state technology may permit cheaper metering and justify more complex tariff structures even at the LV level. Also, as load management techniques improve, centralized control of customer appliances by the utility could be used more effectively, in conjunction with more accurate and practically instantaneous pricing signals. In fact the complexity of the tariff structure would be limited only by the ability of customers to comprehend it. These ideas are embodied in the concept of homeostatic utility control where sophisticated devices including microprocessors would be used to perform a wide range of switching, metering and signaling functions. Such developments envisage an era in which the "new" philosophy of adjusting the demand to meet the available supply would increasingly supplement the existing practice of merely adjusting supply to follow variations in the demand (Burchnall 1977).

Appendix A3.1 Dynamics of Peak Load Pricing

Mathematical Derivation of the Dynamic Version of Marginal Cost Pricing Rule

In this section, we present the dynamic version of the basic (static) rule of pricing presented earlier. The results show that setting the price equal to the marginal cost at every instant of time over the period under consideration will maximize the net benefits of energy consumption to society (for details, see: Crew and Kleindorfer 1979).

The model is formulated in the general context of a peak load pricing case, where capacity constraints could limit the available supply at certain times. This yields a two-part pricing structure in which the marginal costs of supply are higher during the peak period when energy demand begins to infringe on the capacity constraint.

Consider the case of an energy supply system delivering output to society. We seek to find the set of price (and output) levels over some given period of time that will maximize the present discounted value of net benefits or welfare represented by the area under the demand curve minus the area under the supply curve. The problem is formulated with time varying continuously to simplify the presentation, but the result also holds for the discrete time case.

The net benefits of consumption at any given time t are represented by the expression:

$$NB = \int_0^Q p(q,t)\,dq - c(q,K,t) - i(t)$$

where

p = consumer's willingness-to-pay or price curve
q = output or production per unit time period
c = production cost function which represents the available technology for energy supply
K = capital stock or available capacity
i = investment rate

The welfare maximization problem may be formulated as the following optimal control problem:

$$\text{Max} \int_0^T NB.\exp(-rt)\,dt$$

subject to

$$K(t) \geq q(t) \geq O$$
$$I(t) \geq i(t) \geq O$$

$$\text{and } i(t) = \dot{K}(t) + \delta K(t)$$

where

r = discount rate;
I = upper limit on investment rate
δ = rate of capital depreciation, and
T = time horizon under consideration.

Now we may define the Hamiltonian

$$H = NB + \lambda(i - \delta K);$$

and the corresponding Lagrangian function:

$$L = H + m_1(K\text{-}q) + m_2.q + m_3(I\text{-}i) + m_4.i$$

By Pontryagin's maximum principle, the solution to the optimal control problem must also maximize the Lagrangian function at each instant of time over the time horizon under consideration, thus giving rise to the following set of conditions (for details, see: Takayama 1974).

$$(\partial L/\partial q) = 0$$
$$(\partial L/\partial i) = 0$$
and $$(\partial \lambda/\partial t) = r - (\partial L/\partial K)$$

Expanding the first of these conditions yields the desired optimal price:

$$p^*(t) = (\partial c/\partial q) - m_1 + m_2$$

We may interpret the above using the complementarity conditions:

$$m_1(K\text{-}q) = 0, \; m_2 q = 0; \;\text{ and } m_1 \geq 0, \; m_2 \geq 0$$

Case 1. Peak period-capacity constrained: $K(t) = q(t) > 0$.

$$p_1{}^* = (\partial c/\partial q) + m_1$$

Thus, during the peak period, the welfare maximizing optimal price is equal to the marginal operating cost $\partial c/\partial q$ (holding capacity fixed), plus the marginal cost of adding new capacity, m_1.

Case 2. Off-peak period-no capacity constraint: $K(t) > q(t) > 0$

$$p_2{}^* = (\partial c/\partial q)$$

In the off-peak case, only the marginal operating costs enter into the optimal price.

These are the well known marginal cost pricing conditions in a relatively simple form, and it is clear that they hold for a dynamic price path over a given time period just like in the static case developed in the main text.

Allocation of Capacity and Energy Costs Among Peak and Off-peak Users

The simplified geometric representation of a typical electric power generation system is used below to show from a conceptual viewpoint, how a long run marginal cost (LRMC) analysis based on the optimal system expansion plan yields the following idealized conclusions:

1) peak users should pay the peak LRMC of capacity as well as energy;
2) off-peak users should pay only the off-peak LRMC of energy;
3) LRMC of peak capacity = LRMC of base load capacity - net fuel savings due to this base load plant.

Consider an all thermal generation system having the annual LDC shown in Figure A3.1. There are only two types of plant whose linearized cost characteristics are given in the table below, and also in the figure. A more realistic generalised system model would have to consider a number of complicating factors such as a larger number of rating periods and plant types, noncoincidence of the rating periods with the economic crossover points between different plant types, economies of scale and variable heat rates for a given plant type, hydroelectric plant including pumped storage facilities, losses, reserve margins and stochasticity of supply and demand, and so on. The most important difference between this general case and our simpler model is that some capacity costs would have to be borne by consumers outside the peak period. However, the bulk of the capacity costs would still be allocated to peak period users. An elegant exposition of this result is provided in Wenders (1976).

Plant Type	Capacity Cost per kW Installed (annuitized)	Operating Costs per Hour
1) Peaking (e.g., gas turbines, GT)	a	e
2) Base Load (e.g. steam)	b	f

Total cost of 1 kW which is used h hours per year:

GT: $a + e.h$
Base: $b + f.h$

Let H be the hours of operation which corresponds to the crossover point for which GT and base load plant total costs are equal.

Therefore,

$$a + e.H = b + f.H$$

$$H = (b-a)/(e-f)$$

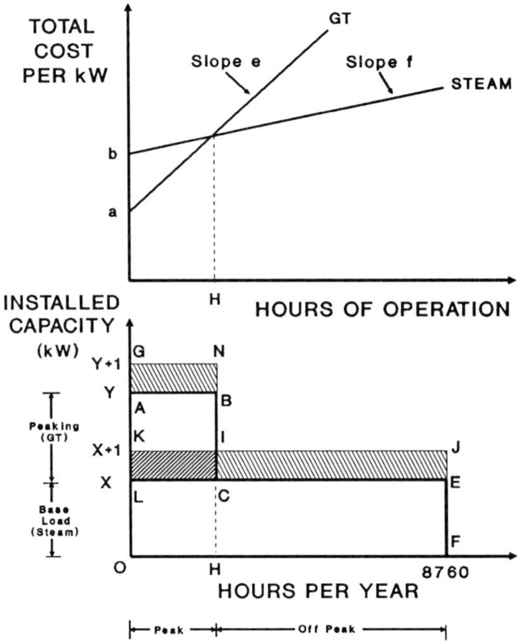

Figure A3.1: Plant Costs and Annual LDC

The most economic use of plant can be determined by examining the LDC, OABCEF:

1) for planned based load operation (i.e., more than H hours per year), use base load plant: X kW;
2) for planned peak operation (i.e., less than H hours per year), use GT: (Y - X)kW.

Total annual costs of meeting the demand depicted by the LDC is:

$$C_0 = X(b + f.T) + (Y-X).(a + e.H).$$

Case 1: Only peak period demand increases by 1 kW (as shown by shaded area AGNB in Figure A3.1).

The optimal system planning response is to increase GT by 1 kW. Total annual cost is $C_1 = X(b + f.T) + (Y + 1 - X)(a + e.H)$. Therefore, increase in cost $C_1 - C_0 = a + e.H$. This is the increase in system costs incurred because of the 1 kW increase in marginal (or incremental) demand during the peak period, and thus the peak period user must pay this cost. The peak costs consist of

capacity charge = a per kW per year
energy charge = e per kWh.

It may be seen that peak users payment = a + e.H = increase in system costs.

Case 2: Only off-peak demand increases by 1 kW (as shown by shaded area CIJE in Figure A3.1).

The optimal system planning response is to add 1 kW more of base load plant. But now there is 1 kW less of GT required than before. Total annual cost C_2 = (X+1).(b + f.T) + {Y - (X+1)} (a + e.H). Therefore, increase in cost:

$$C_2 - C_0 = (b + f.T) - (a + e.H)$$

$$= (b-a) + (f-e).H + f(T-H)$$

Substituting for H

$$C_2 - C_0 = (b-a) + (f-e).(b-a)/(e-f) + f(T-H)$$

$$C_2 - C_0 = f(T-H).$$

Therefore, the increase in system cost incurred due to the 1 kW increase in marginal off-peak demand is equal to the energy cost of operating the base load plant during this period, and thus off-peak users must pay only this energy charge f per kWh. There are no capacity costs incurred by off peak users, since off peak users payment = f(T - H) = increase in system cost.

In particular, we note that the base load capacity cost (b) is exactly offset by the total cost saving due to GT which is not required any more (i.e., capacity cost, a, plus net fuel cost saving (e - f).H, inside the shaded area LKIC). In other words:

Peak capacity cost = Base load capacity costs - net fuel savings
or a = b - (e - f).H.

Case 3: Both peak and off-peak demand increases by 1 kW.

This case is a linear combination of Cases 1 and 2, and therefore consumer charges are:

peak capacity charge = a per kW per year;
peak energy charge = e per kWh;
off-peak energy charge = f per kWh.

Clearly, total peak and off-peak users payment = b + f.T = increase in system cost. These results maybe generalized to include more types of generating units, and rating time periods; e.g., n plan types and rating periods, where these rating periods are chosen to coincide with the economic crossover points between competing types of plant. In this case LRMC prices would be:

0 to H_1	= peak period:	capacity charge a_1 per kW per year, and energy charge e_1 per kWh
H_1 to H_2	= 2nd period:	only energy charge e_2 per kWh
H_{n-1} to T	= nth period:	only energy charge e_n per kWh.

Appendix A3.2 Marginal Costs in Hydroelectric and Mixed Hydro–thermal Systems

The analysis of the LRMC of generation has been discussed so far chiefly in terms of an all-thermal power system. There are several special issues that arise in computing LRMC for all-hydro and mixed hydro-thermal systems, where seasonal variations in LRMC are particularly important. For a more mathematical treatment of the effects of seasonal water flows on the structure of LRMC, see Turvey and Anderson (1977). Again, the most accurate method of calculating LRMC would be to use a computerized model for long-range system planning and to simulate the effect of different load increments on system capacity and energy cost, as described at the beginning of this chapter. In the absence of such a sophisticated planning model, simpler methods of analysis described below must be used to derive the LRMC.

All-Hydro Systems

To present the analysis of an all-hydroelectric system clearly, it is necessary to make the following convenient assumptions:

a. The operation of an entire multi-reservoir, multi-plant system is modeled in terms of an "equivalent" single composite reservoir.
b. The system is analyzed essentially for one year, for example, with long-run capacity costs represented by annuitized investment costs.
c. Deterministic or expected values of water inflows and outflows are used, and the effects of uncertainty are ignored.

The "equivalent" composite reservoir concept is a representation often used in the engineering analysis of hydroelectric systems (Arvanitidis and Rosing 1970a and 1970b, Soares, Lyra and Tavares 1980). Also, although a one-period or annual analysis is no substitute for a complete long-range system model, it illustrates the essential features to be analyzed in the LRMC estimations. Finally, the complications raised by uncertainty may be introduced later. Thus, these simplifying assumptions do not alter the validity of the basic conclusions to be drawn from the model.

Figure A3.2 depicts the typical annual flow of water into and out of the lumped equivalent reservoir. From month 0 to 5, the reservoir is being discharged or drawn down, because this is the dry season, and the water required for generation exceeds the inflow. Therefore, from point A to B the reservoir is being constantly depleted, until at point B the reservoir will be at its lowest level. For a well-planned and well-operated system in a normal year, the reservoir at this time of year will be almost empty, containing only a small safety reserve margin of water. Months 5 to 10 represent the wet season during which water inflow from the catchment area exceeds the desired outflow. The reservoir fills up, consequently, between points B and C. In month 8 the reservoir is full, and excess water must be spilled, since the net inflow is still positive even after the daily generation requirements have been fully satisfied. At point D spilling ceases, and the onset of the dry season again requires a steady draw-down of the reservoir between months 10 and 12. After point A', the annual cycle described above will be repeated.

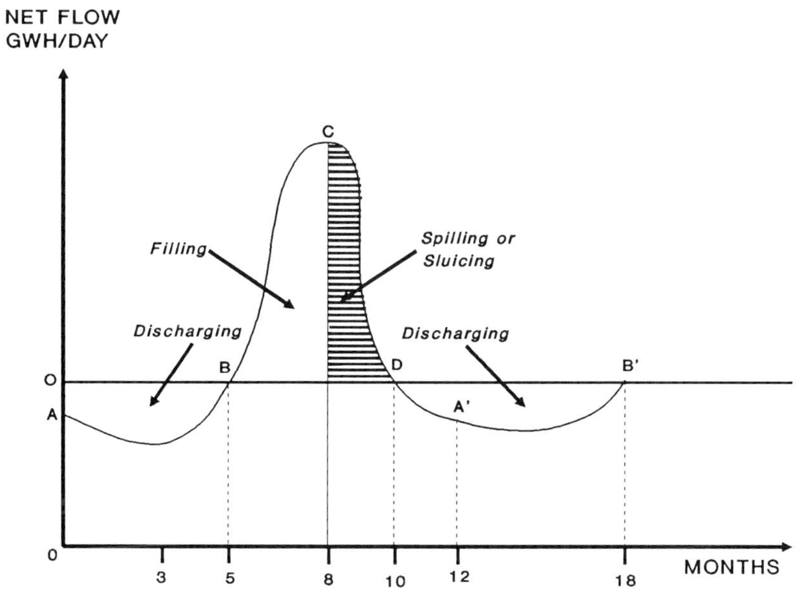

Figure A3.2: Annual Variation of New Water Flows into the Reservoir in an All-Hydroelectric System

If significant spilling occurs over several months and this pattern is likely to continue for many years, then long-run marginal energy costs during this period will be essentially zero. Assuming that there is sufficient plant capacity to meet the peak demand, additional energy may be generated by running through the turbines the water that would otherwise have been spilled. The only marginal costs involved might be a small increase in operating and maintenance costs. In contrast, reservoir capacity is required when the net water inflow is negative, and the corresponding costs represent marginal energy costs during the dry season. Long-run marginal costs of power are incurred, as in the case of the all-thermal system discussed earlier, when additional generating capacity is needed to meet a sustained increment in peak period demand.

In summary, therefore, in an all-hydro system the LRMC of generating capacity incurred during the peak period would be based on the cost of increasing peaking capability, that is, additional turbines, penstocks, expansion of the power-house, and so on. Incremental energy costs chiefly attributable to the dry season would be the costs of expanding reservoir storage. In certain instances it may be justifiable to allocate a fraction of the dam costs to the capacity costs. This depends, however, on the specific case involved, for example, whether or not more storage is needed to firm up the additional capacity. When significant spilling of water occurs (for example, during the wet season), incremental energy costs would be small, usually involving operation and maintenance costs only. Correspondingly, at times when demand does not press on capacity, incremental generating capacity costs may be ignored. If the system is likely to be energy constrained and all incremental capacity is needed primarily to generate more energy because the energy shortage precedes the capacity constraint for many future years, however, then the distinction between peak and off-peak costs and between capacity and energy costs tends to blur. In the extreme case of a system constrained for energy, it may be sufficient only to levy a simple kilowatt-hour charge at all times, because hydro energy consumed during any period, except when spilling, usually leads to an equivalent draw-down of the reservoir. This charge could be determined by applying the AIC method to total incremental system costs. These results are summarized in Table A3.1.

Table A3.1

Marginal Generating Costs in an All-Hydro System

	Peak Period		Off-Peak Period	
Type	**Wet Season**	**Dry Season**	**Wet Season**	**Dry Season**
Capacity Constrained	Capacity	Capacity and energy	None	Energy
Energy Constrained	Total incremental costs required to supply additional kilowatt-hours			

Mixed Hydro-Thermal Systems

The estimation of marginal costs in mixed hydro-thermal generating systems depends critically on the mix of generating plants used at different times. Several basic configurations illustrate the principles governing the calculation of LRMC. First, consider a predominantly thermal system in which all available water is used and thermal generation is required throughout the year. The typical daily pattern of generation for such a system is shown in Figure A3.3. Base-load steam plants represented by the shaded area I would be run continuously during the twenty-four-hour period; gas turbines depicted by the shaded area III would be required to meet the peak period demand. All the water collected during the year would be used to supplement the thermal generation on a daily basis by filling up the unshaded area II of the LDC. The LRMC analysis for this system would closely follow the procedure described earlier in the all-thermal system.

A general principle for determining the value of water used in a mixed hydrothermal system, for example, in area II of Figure A3.3, is analyzed geometrically below. Consider the pattern of plant operation discussed earlier, which is analyzed in Figure A3.4, on the basis of minimizing cost. As described in Appendix A3.1, the gas turbines with relatively low capital and high operating costs, represented by the steepest solid line, are used for peaking duty. The steam turbines having higher capital and lower operating costs, as shown by the flatter solid line, are operated as base-load plants. Since the water is used between these two types of plants, an imputed value for water would be given by the broken line. The position of this line would usually be determined by conditions of operation that would seek to ensure that all available water in storage would be used up during the year to supply the energy represented by the shaded area of the annual LDC. This area also corresponds to the unshaded portion II of Figure A3.3a. The intermediate slope of the broken line indicates that the cost for each kilowatt-hour of water-generated energy lies between the energy costs of peaking plants and baseload thermal plants. In brief, one method of determining the value of water in a mixed hydrothermal system is based on the equivalent costs of the thermal generation it displaces. If pumped storage is involved, the value of the water is the cost of pumping, net of appropriate losses.

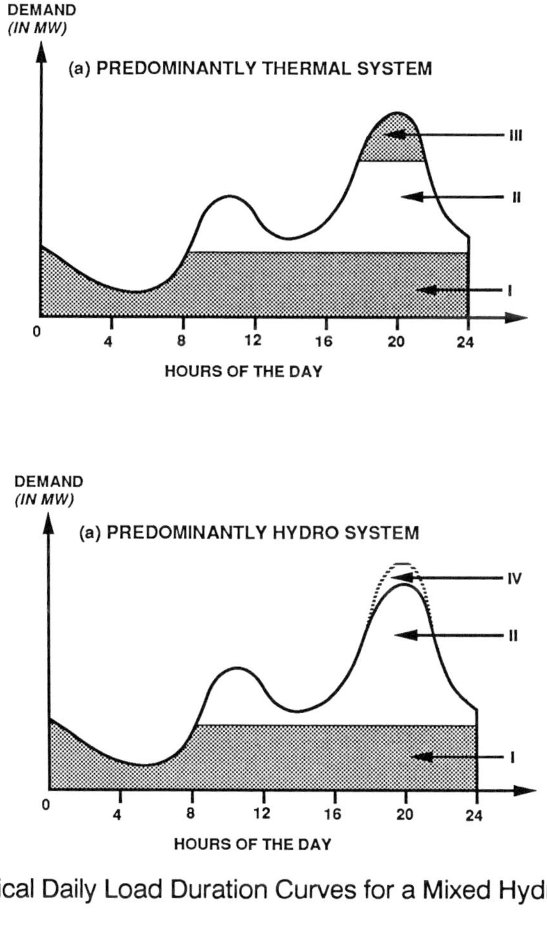

Figure A3.3: Typical Daily Load Duration Curves for a Mixed Hydro-Thermal System

The second configuration is one in which hydro makes a bigger contribution than in the previous case. Here, steam turbines or diesels are operated continuously to supplement limited hydro resources during the dry season. This mode of operation is shown in Figure A3.3b, where the shaded area I represents thermal plants, and the unshaded area II is hydro generation. To minimize the present value of system costs in the long run, the smallest possible amount of thermal capacity must be installed and base-loaded or run twenty-four hours a day in the dry season, with stored water meeting peak demands.

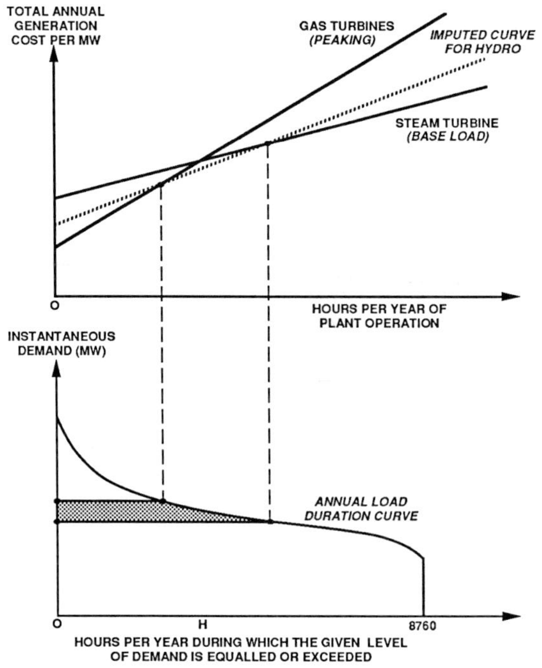

Figure A3.4: Imputed Value of Water in a Mixed Hydro-Thermal System

In this situation, incremental energy must always be supplied by operating the thermal plants, that is, marginal fuel costs must be incurred whenever additional kilowatt-hours are needed, including the wet and dry seasons as well as peak and off-peak periods. Even if more stored water can be run off to supply additional energy during a given period, this implies that there must be greater thermal generation at another time to compensate for the water loss. In addition, since thermal capacity is fully utilized, the marginal costs of the additional capacity divided by the number of hours of operation at full capacity must be added to the marginal fuel costs to obtain marginal energy costs. An increment of demand shown by the broken line enclosing area IV could be met by increasing installed hydro capacity. But the additional energy in area IV must be supplied by augmenting thermal capacity, and this cost has already been charged to marginal energy costs. Therefore, the marginal capacity cost during the peak period would be the incremental cost for a kilowatt of hydro capacity minus the cost of the appropriate kilowatt fraction of thermal capacity used to generate energy.

Other hydrothermal generation configurations are possible, but usually the LRMC for these cases can also be analyzed in the same way, using fairly simple and straightforward considerations. Complications may arise, particularly if the pattern of plant operation is likely to change rapidly in the future. This might occur, for example, if the marginal plant used in the peak period shifts from gas turbines to peaking hydro, or if the reverse situation occurs. Then the value of LRMC would have to be calculated as a weighted average. The weights would depend on the share of future generation by the different types of plants used and on the timing of the transitions.

Appendix A3.3: Peak Load Pricing and the Metering Decision

As discussed earlier in Appendix A3.1, per unit costs of energy supplied during peak periods exceed those of off-peak periods, sometimes by a substantial margin. Table A3.2 provides some typical examples based on electricity utility data.

Table A3.2

Typical Marginal Costs of Supply at Different Periods of the Day (1980 US cents per kWh)[1]

	Fuel Cost	Capacity Cost
Peak Period[2]	6-10	2-3
Off-Peak Period[3]	1-4	

[1] Based on an all thermal system.
[2] Gas turbines. All capacity costs including associated transmission and distribution costs have been allocated to the peak period.
[3] Nuclear, coal-fired steam, or diesel.

Source: Author's estimates.

These higher marginal costs of supply during daily and seasonal peak demand periods should be reflected in tariff structures so that energy users are confronted with them during peak demand periods. If they are not, consumption during peak periods would be larger and during slack periods somewhat lower. This has been illustrated in Figure A3.5, where the demand during the peak period D_{pk} is greater than off-peak demand D_{op}. If a uniform, nondifferentiated tariff P_{av} was used that reflects an average of the baseload and peak period marginal supply costs, the quantity consumed during the peak period would be equal to Q_1. However, if true peak periods costs P_{pk} were charged, consumption would be reduced to Q_3 instead. As a result, overall capacity needs would be lower.

During the off-peak period, consumption would tend to be correspondingly greater (i.e., increase from Q_2 to Q_4) because of the lower price P_{op} relative to the averaged-out price. Another reason for increased consumption during that period would be the partial, intertemporal substitution from the higher-priced peak load to the base load period. Hence the consequence of a peak load tariff would be an overall reduction in needed total supply capacity together with a more efficient, i.e., more uniform, utilization of the remaining capacity. This would lead to overall reductions in marginal costs and, hence, prices to energy users (or to higher profits to suppliers or higher government income, as the case may be). Overall economic efficiency would be increased.

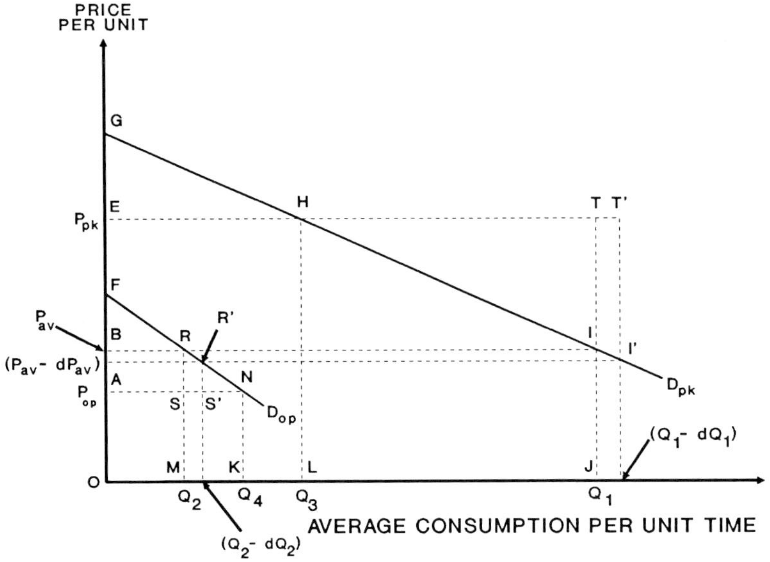

Figure A3.5: Uniform versus Peak Load Pricing and the Metering Decision

Peak load pricing requires time-of-use metering, however, and such metering is usually more costly than simple volume metering only. One of the questions to be addressed, therefore, is whether the additional costs of metering are justified by the resulting savings and increase in the efficient utilization of energy producing equipment.

This is not an issue if the problem is one of adjusting prices to reflect seasonal demand variations. Seasonal tariffs could be instituted and metered by standard volumetric metering equipment without significant additional costs. Seasonal tariffs can be justified on equity as well as efficiency grounds, because consumers would be confronted by the true seasonal supply costs of their consumption. Seasonal price differentials may not lead to major substitutions from peak to off-peak period uses because postponing energy consumption for several months would be rather impractical. The main effects on consumption would come from a genuine reduction in demand in response to the higher seasonal price (i.e., a movement along a given demand curve).

Metering of residential demand by time-of-day is a more difficult problem because the costs of an average sized electric household meter would increase several fold. Thus the use of time-of-day metering devices wold not be economically justified for residential users in low income countries, if the net benefits realized through savings in energy used were less than the increased costs of the more complex metering. The costs of time-of-day metering are much less of a problem for larger users, such as large commercial establishments or industrial plants, because the added costs of metering are small relative to the amount of energy consumed. Most peak-load pricing schemes, therefore, have concentrated on such users, although several developed countries have attempted to apply such tariffs in the residential sector as well.

The degree of sophistication of metering -- for example, by time of day -- should be determined by the net benefit of metering, the practical problems of installation and maintenance, ease of billing, and so on. The economic rationale for making the metering decision depends on the net benefit of metering based on a cost-benefit analysis that compares the lower supply costs of reduced consumption with the cost of metering and billing plus the decrease in net consumption benefits. This procedure is illustrated in the following discussion, using a two-period tariff. An economic analysis of the merits of a multiperiod tariff involving several generation technologies is provided in Renshaw (1980).

Figure A3.5 depicts the analysis underlying the metering decision to implement a two-period, time-of-day tariff for electricity supply. D_{pk} and D_{op} are the demand curves for an average hour during the peak and off-peak periods of the day, each lasting H hours and (24 - H) hours, respectively. If a simple uniform price P_{av} is levied throughout the day, the daily consumption will be Q_1.H and Q_2.(24-H) kilowatt-hours respectively during the peak and off-peak periods. Instead, if a two-period tariff (P_{pk}, P_{op}) is charged during the peak and off-peak hours, the corresponding levels of consumption will be Q_3.H and Q_4.(24 - H). Suppose that the LRMC of supplying a kilowatt-hour during the peak and off-peak periods are MC_{pk} and MC_{op}, respectively, where the former includes both capacity and energy costs and the latter includes only the energy cost.

The net consumption benefit of the uniform tariff is given by:

$$NB_I = [B_I] - [C_I]$$
$$= [(\text{area OGIJ}).H + (\text{area OFRM}).(24\text{-}H)] - [Q_1.H.MC_{pk} + Q_2.(24\text{-}H).MC_{op}]$$

Similarly, the net consumption benefit of the two-period tariff is given by:

$$NB_{II} = [B_{II}] - [C_{II}]$$
$$= [(\text{area OGHL}).H + (\text{area OFNK}).(24\text{-}H)] - [Q_3.H.MC_{pk} + Q_4.(24\text{-}H).MC_{op}]$$

Therefore, the change in net benefit is:

$$\Delta NB = NB_{II} - NB_I$$
$$= [(\text{area MRNK}).(24\text{-}H) - (\text{area LHIJ}).h] + [(Q_2\text{-}Q_4).(24\text{-}H).MC + (Q_1 - Q_3).H.MC_{pk}]$$

A further simplification is possible if it is assumed that the two-period tariff reflects strict LRMC; that is $P_{pk} = MC_{pk}$ and $P_{op} = MC_{op}$. Then, $(Q_4 - Q_2)$.MC = area MSNK, and $(Q_1\text{-}Q_3).MC_{pk}$ = area LHTJ. Therefore, the following is obtained.

$$\Delta NB = (\text{area HTI}).H + (\text{area SRN}).(24\text{-}H)$$

Assuming no change in this average daily cycle during the first year, the annual increase in net consumption benefits is given by:

$\Delta ANB_{l} = 365.\Delta NB$

Let the difference between the two streams of metering and billing costs (installation, maintenance, administration, etc.) associated with the two-period and uniform tariffs be $\Delta C_1, \Delta C_2, ... \Delta C_T$, over the lifetime T of the meter. Suppose that the corresponding stream of net benefits can also be estimated as ΔANB_1, ΔANB_2, ...ΔANB_T. If the condition:

$$\sum_{t=1}^{T} (\Delta ANB_t - \Delta C_t)/(1+r)^t$$

is satisfied, implementation of the two-period price structure would be economically justified.

Even if the decision to install more complex metering is negative, further analysis is helpful. Thus, it is clear that applying the uniform price P_{av} instead of the strict LRMC prices $P_{pk} = MC_{pk}$ and $P_{op} = MC_{op}$ will result in a loss of net efficiency benefits:

$$\Delta NB^1 = (\text{area HTI}).H + (\text{area SRN}).(24-H)$$

The level of P_{av} may be adjusted, however, to minimize this loss of net benefits.

Suppose that a reduction in price from P_{av} to $(P_{av} - dP_{av})$ results in increased consumption given by $(Q_1 + dQ_1)$ and $(Q_2 + dQ_2)$ in periods 1, 2 respectively. The new net benefits may be written:

$$\Delta NB^2 = (\text{area HT'I'}).H + (\text{area S'R'N'}).(24-H)$$

The change in net efficiency benefit is:

$$\begin{aligned} d(\Delta NB) &= (\Delta NB^2 - \Delta NB^1) \\ &= (MC_{pk} - P_{av}).dQ_1.H - (P_{av} - MC_{op}).dQ_2.(24-H) \end{aligned}$$

Using the familiar first-order condition $d(\Delta NB)/dP_{av} = 0$ yields the optimal uniform price that will minimize the loss of net benefits:

$$P^*_{av} = \frac{(H.dQ_1/dP_{av}).MC_{pk} + [(24-H).dQ_2/dP_{av}].MC_{op}}{(H.dQ_1/dP_{av}) + [(24-H).dQ_2/dP_{av}]}$$

This result may be generalized. Thus, suppose the tariff analyst wishes to consolidate n pricing periods of the day, where H_i is the duration, Q_i is the average hourly consumption, and MC_i is the LRMC during period. The optimal uniform price would be:

$$P^*_{av} = \left[\sum_{i=1}^{n} (H_i.dQ_i/dP_{av}).MC_i \right] / \left[\sum_{i=1}^{n} (H_i.dQ_i/dP_{av}) \right]$$

Note that derivatives of the form $H_i.(dQ_i/dP_{av})$ refer to a change in consumption during a given pricing period i due to a uniform change in price across all the pricing periods that need to be consolidated under the single price regime. Generally, in most countries information on demand would be inadequate to determine the magnitude of such derivatives accurately. In this case, a practical approximation might be:

$$P^*_{av} = \left[\sum_{i=1}^{n} (H_i.Q_i).MC_i \right] / \left[\sum_{i=1}^{n} (H_i.Q_i) \right]$$

Here the uniform price is set equal to the weighted average LRMC using the consumption levels in the different pricing periods as the weights.

While such a partial approach to peak-load pricing is preferable to one which disregards the issue completely, it also raises a number of problems. Ideally, the added costs of supplying energy during peak-load periods should be fully charged to the consumers using it during that time. Unfortunately, residential demand fluctuations are usually substantially higher than industrial and commercial ones. Hence the residential sector is a major contributor to the extra peak period costs.

However, as pointed out above, peakload metering often cannot be used in the household sector. This means that peakload pricing schemes can only be partially applied. On equity as well as efficiency grounds the full utility system marginal cost differential between base- and peak-load periods should not be charged against those users that can be metered. To achieve economically efficient consumption patterns among them, they should be confronted with no more than their proportional share of the total added costs of peak-load supplies. For example, if metered users account for, say, 40% of total peak period consumption, their peak period tariffs should be raised to cover 40%, but no more, of the added costs of peak period supply. The other 60% should added to the averaged-out LRMC-based uniform tariffs of the remaining groups. It should be noted that the three prices charged to the various user groups -- base-load and peak-load prices to users with time-of-day meters, average prices to all others -- should be based on the respective long-run marginal costs, i.e., the marginal costs of supply including the annuitized costs of new investments. The reasons for this requirement have been detailed above.

It should also be noted that both per unit marginal capital and energy costs will differ between peak-load and base-load periods. Because of their higher utilization rates, base-load supply installations tend to be more capital intensive but have relatively lower energy operating costs compared to peak-load plants. The opposite is true for peak-load plants that are sensitive to capital costs, but for which per unit running costs are less significant because of their shorter operating hours.

In energy supply systems that have both short-term (i.e., daily and weekly) and seasonal demand variations both seasonal and time-of-day peak-load tariff structures should be combined. If seasonal peaks affect base-load as well as peak-load demands, separate seasonal tariff adjustments should apply to both to reflect the additional costs of base- load and peak-load supply capacity, respectively. If only peak-load demands increase, adjustments should be limited to the peak-load tariffs only.

Two part tariffs, consisting of a fixed capacity charge and volumetric energy charges are widely used in order to reflect the higher costs of peak-load supplies. The fixed portion of such tariffs are usually based on the installed capacity of equipment (size of motors or processing equipment) or the maximum capacity of the supply connection (size of supply pipe, capacity of wiring, etc.). Another alternative is to install metering devices that register the maximum demand in a given meter reading interval. Such meters are less costly than time-of-day metering equipment. While such tariffs are relatively easy to implement and administer and do not require the high time-of-day metering costs they do not really solve the peak-load problem. This is so because the capacity charge is invariant with respect to time. In other words, the charge is the same regardless of the period of time in which the full installed capacity is utilized. No incentive exists to users to reduce consumption during peak-load periods. Hence, while overall installed user capacity might be lower in response to prevailing capacity charges, consumption patterns of installed capacity would not be affected. Time-of-day metering, therefore, is a more preferable demand management and pricing tool.

The discussion in this section has proceeded as if there existed only one basic type of user and marginal cost and two periods of on-peak or off-peak. This is an obvious simplification. Many different user classes exist, and the marginal costs of supply to them will differ as well (for example, by different voltage levels, in the case of electricity supply). Peak demand periods may well be broken down into more discriminating time periods to better reflect variations in the marginal costs of supply.

References

Arvanitidis, N.V. and J. Rosing, "Composite Representation of Multi-Reservoir Hydroelectric Power System", *IEEE Transactions on Power Apparatus and Systems*, vol. PAS-90, February 1970a.

Arvanitidis, N.V. and J. Rosing, "Optimal Operation of Multi-Reservoir System Using Composite Representation," *IEEE Transactions on Power Apparatus and Systems*, vol. PAS-90, February 1970b.

Baumol, W.J. and D.F. Bradford, "Optimal Departures From Marginal Cost Pricing," *Amer. Econ. Rev.*, pp. 265-283, June 1970.

Boiteux, M. "La Tarification des Demandes en Pointe," *Rev. Gen l'Electricite*, vol. 58, 1949.

Boiteux, M. and P. Stasi, "The Determination of Costs of Expansion of an Interconnected System of Production and Distribution of Electricity," in J. Nelson (Ed.), *Marginal Cost Pricing in Practice*, Prentice-Hall, Englewood Cliffs NJ, 1964.

Burchnall, J.A. "A Review of Off-Peak Tariff Incentives and Their Effect in England and Wales," Economic Commission for Europe- Committee on Electric Power, Brussels, Belgium, Report No. EP/SEM. 5/R17, Mar. 1977.

Caves, D.W. and L.R. Christensen, "Residential Substitution of Off-Peak for Peak Electricity Usage Under Time of Use Pricing," *The Energy Journal*, vol. 1, Apr. 1980, pp. 85-142.

Cicchetti, C.J., W.J. Gillen and P. Smolensky, *The Marginal Cost and Pricing of Electricity*, Ballinger Publ., Cambridge MA, 1977.

Crew, M.A. and P.R. Kleindorfer, *Public Utility Economics*, St. Martin's Press, New York, 1979.

de Grasse, R.V., "Electric Storage Heating After Two Years," *Public Utilities Fortnightly*, vol. 23, Jan. 6, 1977.

Donziger, A., "The Underground Economy and the Theft of Utility Services," *Public Utilities Fortnightly*, Nov. 22, 1979, pp. 23-27.

Dupuit, P. "De l'Utilite' et de sa Mesure," *La Reforma Soziale*, Turin, Italy, 1932.

Electricity Council, "Domestic Tariffs Experiment", Load and Market Res. Rep. 121, London, U.K., 1978.

Federal Energy Regulatory Commission, "Procedures Governing the Collection and Reporting of Information Associated with the Cost of Providing Electric Service," Final Regulation, Docket No. RM79-6d, Washington, DC, June 5, 1979.

Francfort, A. and P. Woo, "Lifeline and Incremental Cost Residential Electric Rates," *Public Utilities Fortnightly*, Feb. 17, 1977, pp. 15-20.

Garfield, P.J. and W.F. Lovejoy, *Public Utility Economics*, Prentice-Hall, Englewood Cliffs NJ, 1964.

Hastings, B.F., "Ten Years of Operating Experience with a Remote Controlled Water Heater Load Management System at Detroit Edison," *IEEE Trans. Power App. Syst.*, vol. PAS-99, July/Aug. 1980, pp. 1437-1441.

Herman, G.F.K., "Electricity Tariffs," Economics Dept., Univ. Stellenbosch, South Africa, 1974.

Hotelling, H., "The General Welfare in Relation to Problems of Taxation and of Railway and Utility Rates," *Econometrica*, vol. 6, July 1938, pp. 242-269.

Johnson, C.R. (FEA), prepared testimony before the Connecticut Public Utilities Control Authority, Docket No. 760204, 1976.

Malko, J.R. and A. Faruqui, "Implementing Time-of-Day Pricing of Electricity: Some Current Challenges and Activities," in M.A. Crew (Ed.), *Issues in Public Utility Pricing and Regulation*, Lexington Books, New York, 1980.

Marginal Costing and Pricing of Electrical Energy, Canadian Elec. Assoc., Montreal, Canada, May 1978.

Miedema, A.K. "Overview of DOE Analyses of Residential TOU Data," Research Triangle Institute NC, Apr. 1980.

Miedema, A.K., S.B. White, C.A. Clayton, and D.P. Lifson, "Analysis of Experimental Time-of-Use Electricity Prices," Research Triangle Institute, NC 27709, Dec. 1978.

MIT Energy Lab., "New Utility Managment and Control Systems", Cambridge, MA, 1979.

Mitchell, B.M., W.G. Manning, Jr. and J.P. Acton, *Peak Load Pricing*, Ballinger Publ., Cambridge MA, 1978.

Munasinghe, M., "A New Approach to System Planning," *IEEE Trans. Power App. Syst.*, vol. PAS-99, May-June 1980.

Munasinghe, M., *Economics of Power System Reliability and Planning*, Johns Hopkins University Press, Baltimore MD, 1979.

Munasinghe, M., "The Costs of Electric Power Shortages to Residential Consumers," *J. Consumer Res.*, vol. 6, Mar. 1980, pp. 361-9.

Munasinghe, M., "An Integrated Framework for Energy Pricing in Developing Countries," *The Energy Journal*, vol. 1, July 1980, pp. 1-31.

Munasinghe, M. and M. Gellerson, "Economic Criteria for Optimizing Power System Reliability Levels," *Bell J. Economics*, vol. 10, pp. 353-365, Spring 1979.

Munasinghe, M. and J.J. Warford, "Shadow Pricing and Power Tariff Policy," in *Marginal Costing and Pricing of Electricity*, Canadian Elec. Association, Montreal, 1978, pp. 159-180 .

Munasinghe, M. and J.J. Warford, *Electricity Pricing*, Johns Hopkins University Press, Baltimore MD, 1982.

NARUC Electric Utility Rate Design Study, "Rate Design and Load Control: Issues and Directions," EPRI, Palo Alto, CA, Nov. 1977.

NARUC Electric Utility Rate Design Study, "1978 Plan of Study", EPRI, Palo Alto CA, Dec. 1977.

NARUC Electric Utility Rate Design Study, "Guide to Study Reports," EPRI, Palo Alto CA, Aug. 1979.

NARUC Electric Utility Rate Design Study, Rep. 66, EPRI, Palo Alto CA, June 1980.

NARUC Electric Utility Rate Design Study, Rep. 67, EPRI, Palo Alto CA, July 1980.

NARUC Electric Utility Rate Design Study, Rep. 69, EPRI, Palo Alto CA, Nov. 1979.

NARUC Electric Utility Rate Design Study, Rep. 72, EPRI, Palo Alto CA, Nov. 1979.

NARUC Electric Utility Rate Design Study, Rep. 79, EPRI, Palo Alto CA, Apr. 1980.

NARUC Electric Utility Rate Design Study, Rep. 91, EPRI, Palo Alto CA, June 1980.

Panzar, J.C., "A Neo-Classical Approach to Peak Load Pricing," *Bell J. Economics*, vol. 7, Autumn 1976, pp. 521-30.

Pelletier, Y., "Application des Methodes d'Analyse de Donnees a l'Etude d'Un Echantillon de Courbes de Chargemoyenne Tension," Electricite de France, Paris, Apr. 1977.

Pioger, Y. "Elements de Reflexion Sur la Deformation des Courbes de Chage Clobales et Sectorielles: Evaluation de l'effet induit par la tarification," Electricite de France, Paris, July 1977.

Public Utilities Regulatory Policies Act (PURPA), Rep. 95.1750, House of Representatives, 95th Congress-2nd Session, Oct. 10, 1978.

Reynolds, S.P. and T.E. Creighton Jr., "Time of Use Rates for Very Large Customers on the Pacific Gas and Electric Company System," *IEEE Trans. Power App. Syst.*, vol. PAS-99, pp. 147-151, Jan./Feb. 1980.

Ruggles, N., "The Welfare Basis of the Marginal Cost Pricing Principle," *Rev. Economic Studies*, vol. 17, pp. 29-46, 1959.

Ruggles, N., "Recent Developments in the Theory of Marginal Cost Pricing," *Rev. Economic Studies*, vol. 17, 1959, pp. 107-126.

Soares, S., C. Lyra and H. Tavares, "Optimal Generation Scheduling of Hydro-thermal Power Systems", *IEEE Transactions on Power Apparatus and Systems*, vol. PAS-99, May-June 1980.

Suelflow, J., *Public Utility Accounting: Theory and Applications*, Michigan State Univ. Press, East Lansing MI, 1973.

Steiner, P., "Peak Loads and Efficient Pricing," *Quart. J. Econ.*, Nov. 1957.

Squire, L. and H. van de Tak, *Economic Analysis of Projects*, Johns Hopkins Univ. Press, Baltimore MD, 1975.

Symposium on Peak Load Pricing, in *Bell J. Economics*, vol. 7, Spring 1976, pp. 197-250.

Turvey, R., *Optimal Pricing and Investment in Electricity Supply*, MIT Press, Cambridge MA, 1968.

Turvey, R. and D. Anderson, *Electricity Economics*, Johns Hopkins Univ. Press, Baltimore MD, 1977.

Wenders, J.T., "Peak Load Pricing in the Electric Utility Industry," *Bell J. Economics*, vol. 7, Spring 1976, pp. 232-241.

Williamson, O.E., "Peak Load Pricing and Optimal Capacity under Indivisibility Constraints," *Amer. Econ. Rev.*, vol. 56, Sept. 1966, pp. 810-827.

Chapter 4

ELECTRICITY TARIFFS IN BURMA[1]

4.1 Power Sector Background

Sector Organization and Development

Electric power supply industries have been nationalized in Burma since 1951. The sector is run by the Ministry of Industries No. 2, which controls power, heavy industry, and petroleum. The Electric Power Corporation (EPC) operates a power system with a November 1980 installed capacity of 438 MW, of which 168 MW (38%) is hydro and 56 MW (13%) is diesel. EPC's system accounts for about 88% of the country's total installed capacity. The remaining 62 MW of generating capacity is owned by other industries. Burma's population has low access to electricity (about 7%) when compared to other South-east Asian countries like Thailand (27%) and the Philippines (32%).

Burma has large exploitable hydro-electric resources, of which only about 4 percent has been utilized. There are also limited gas and petroleum resources. The estimates of gas reserves indicate a 20-year supply sufficient to support three fertilizer plants (two existing units plus one more under construction) and six gas-fired turbine power stations (about 100 MW of existing capacity plus a further 100 MW under construction).

Due to the long gestation period for hydro-electric projects, hydro will be used to supply long-term demand (late 1980's onward), while thermal power plants are expected to meet short- to medium-term demand (up to 1985), as shown in the generation expansion program in Table 4.1, the power-energy balance in Tables 4.2 and 4.3, and Figure 4.1. The power-energy balance in Tables 4.2 and contribution of thermal plant up to the late 1980s. Corresponding transmission and distribution facilities are also being built. The main 132 and 220 kV transmission lines planned up to 1990 are given in Table 4.4. The likely system configuration in 1990 is shown in Figure 4.2.

In addition to constructing the necessary transmission and distribution lines to ensure effective use of the existing and committed generation plants, emphasis must be placed on reducing system losses, establishing an effective planning function in EPC and rationalizing tariffs. These actions will also require significant attention being paid to human resource development.

[1] Edited version of a guest lecture presented at the USAID sponsored Energy Management Training Program for senior officials from developing countries, held at the Institute for Technology Policy in Development, State University of New York, Stonybrook, NY, in October 1981.

Table 4.1

EPC Power Expansion Program

Year	Hydel	Gas Turbine	Thermal	Remarks
1980-81				
1981-82		Prome 37 MW		
1982-83		Chauk 13 MW		
1983-84				
1984-85	Sedawgyi 25 MW (130 GWh)		Thaton 12 MW	Sedawgyi dam is under construction and scheduled for completion in 1984. Thaton is not scheduled for interconnection.
1985-86	Baluchaung No. 1 24 MW (177 GWh)			Increase in firm capacity at Lawpita Station of 15 MW.
1986-87				
1987-88	Nyaunggyat 56 MW (165 GWh)			
1988-89	Zaungtu 22 MW (86.7 GWh)			
1989-90	Paungloung 94 MW (718 GWh)			

Source: EPC.

Table 4.2

System Energy Forecast and Generation (GWh)

	Full Capability		Actual						
	(MW)	(GWh)	80	81	82	83	84	85	86
Lawpita	140	1130	720	858	920	985	1055	1130	1130
Sedawgyi	25	130					130	130	130
Nyaunggyut	56	165						113	128
Zaungtu	21.6	87						44	87
Baluchaung (I)	24	177						88	177
Total Hydro Generation			720	858	920	985	1185	1505	1652
Gas Turbine (Gas-fired)			349	350	440	530	547	419	507
Ywama G/T (Oil-fired)			7	12	25	33			
Ahlone & Ywama Steam			62	44	21	12			
Total System Generation			1138	1264	1406	1560	1732	1924	2159

Table 4.3

System Demand Forecast
(MW)

	Actual						
	80	**81**	**82**	**83**	**84**	**85**	**86**
Lawpita	127	130	132	134	137	140	140
Sedawgyi					25	25	25
Nuaunggyut						40	50
Zaungtu						11	21
Baluchaung I						12	24
Total Hydro Demand	127	130	132	134	162	228	260
G/T Gas-fired	55	72	93	118	142	110	119
G/T Ywama	11	12	15	18	-	-	-
Ahlone & Ywama Steam	10	8	6	4	-	-	-
Total System Demand	203	222	246	274	304	338	379

Table 4.4

EPC Transmission Line Program[a]

Year	Section	kV	km
1981/82	Kyunchaung - Monywa	132	193
	Mandalay - Anisakan	132	58
1982/83	Toungoo - Yeni	132	72
1983/84	Mandalay - Sedawgyi	132	64
1984/85	Thazi - Toungoo	230	242
	Chauk - Taungdwingyi	132	176
	Mann - Taungdwingyi	132	97
1985/86	Prome - Rangoon[b]	230	322
	Toungoo - Prome[c]	230	177
	Pyinmana - Taungdwingyi	230	97
	Prome - Taungdwingyi	230	152
1986/87	Thazi - Nyanggyat	132	70
1989/90	Paunglaung - Pyinmana	132	30

[a] Major transmission lines only.
[b] Including construction of a new sub-station to the east of Rangoon.
[c] Program is not finalized yet.

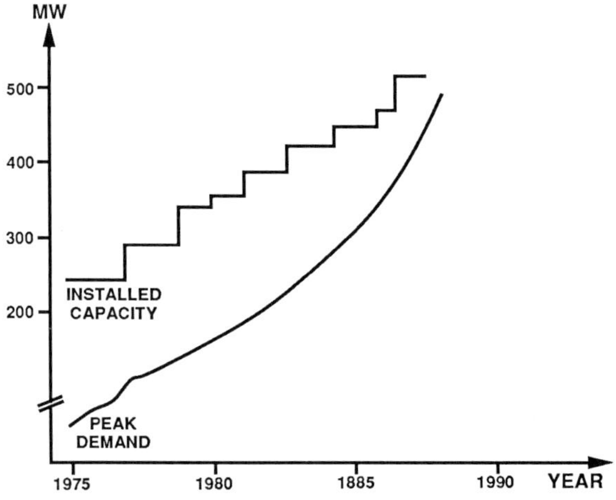

Figure 4.1: Peak Demand Forecast and Installed Capacity

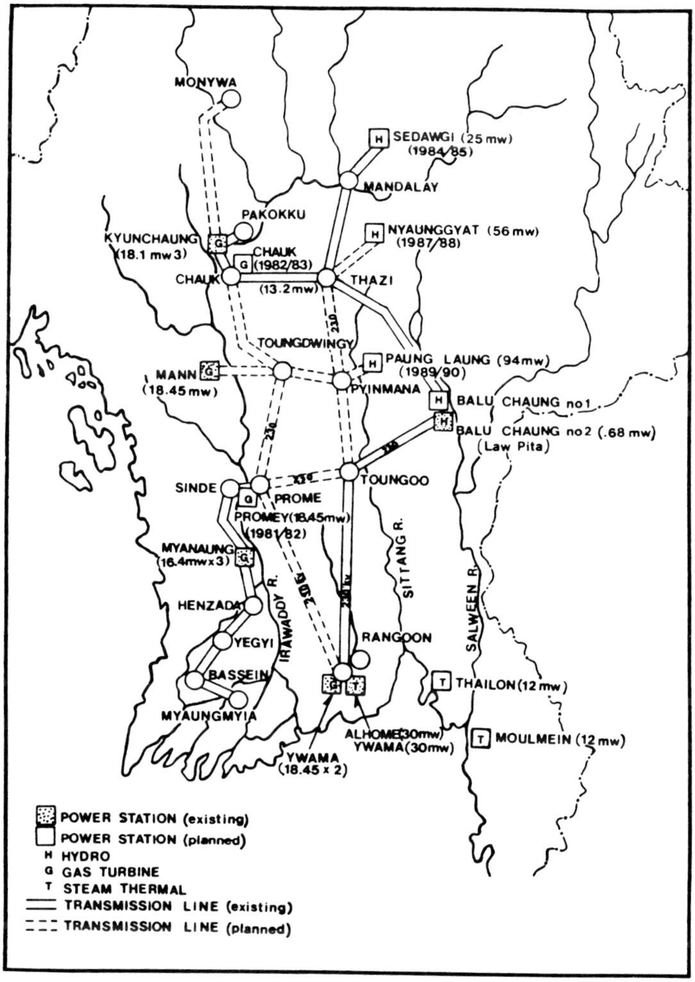

Figure 4.2: Power Supply System (programmed for 1990)

4.2 Strict LRMC for the EPC System

The strict LRMC of supply for the EPC system are derived in this section. The calculations are carried out in constant 1980-81 kyatts using an official exchange rate of 1 U.S. dollar = 6.7 kyatts (KY). Strict LRMC is estimated in shadow prices and these are used as the base results. For purposes of comparison, the calculation is also repeated without shadow prices.

Generation Capacity Cost

The EPC power system utilizes a combination of hydro-electric units and gas turbines to generate electricity. The peak demand, which typically consists of a four-hour period between 6 p.m. and 10 p.m. (Figures 4.3 and 4.4), is met by operating both gas turbine and hydro-electric units over the next 6 years. Approximately 87% of peak-period demand is met by gas turbines, while the balance 13% is met by peaking hydro units. Therefore, generation costs are calculated on the basis of this mix of the two types of peaking plants.

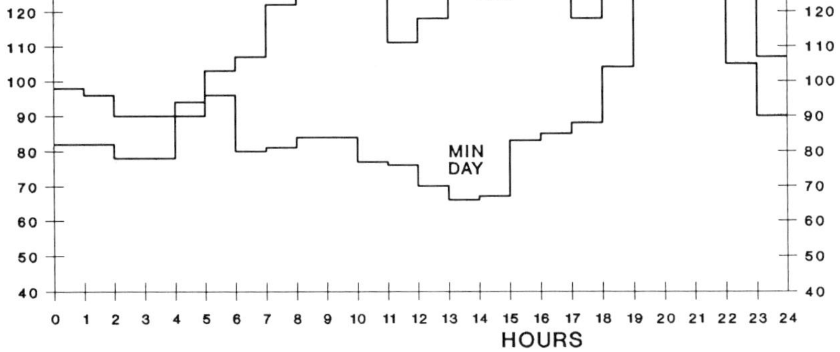

Figure 4.3: Maximum and Minimum Daily Load Curves for the Month of June 1979

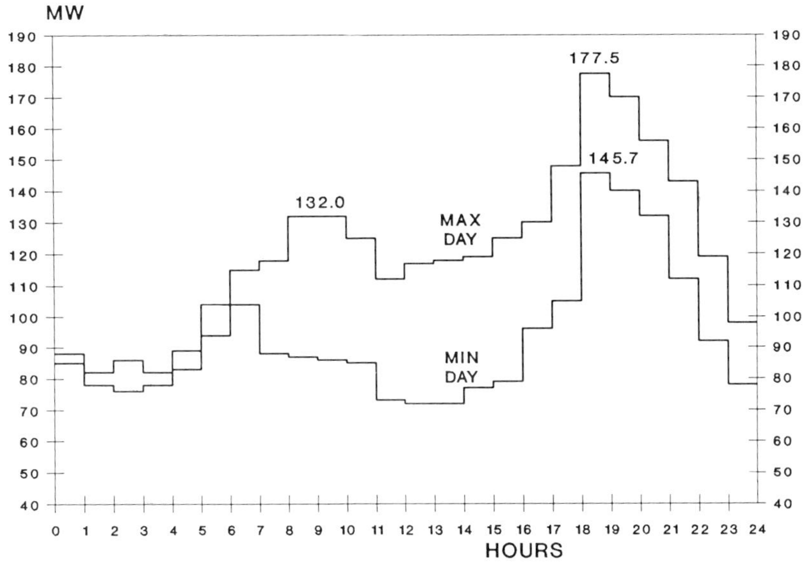

Figure 4.4: Maximum and Minimum Daily Load Curve for the Month of Dec. 1979

The generation capacity costs for the hydro plant and gas turbine are summarized in Tables 4.5, 4.6, and 4.7. The weighted capacity cost per kW installed in 1980-81 border prices = 4,463.96 KY. This measures the long-run marginal cost of meeting a kW increment in peak demand at the bus-bars (including the reserve margin). Operation and maintenance, administrative, and facilities costs associated with generation were estimated as 6% of gas turbine capacity costs, and 2-1/2% of hydro capacity costs. These were also weighted by the factors 0.87 and 0.13 and included later in Tables 4.17 and 4.18.

Table 4.5

**Generation Capacity Costs for Peaking
Hydro - Investments (in Million Kyatts) Sedawgyi and Nyaunggyat Projects
(Excluding Dam Costs but Including Transmission Spur Line Costs)**

Year	Foreign	Local Material	Total Investment Market Prices	Total Investment Border Price	Total Investment Discounted Border Price	Total Investment Discounted Market Price	MW Increment	Discounted MW Increment
1980	14.59	4.64	19.23	18.77	18.77	19.23		
1981	52.03	8.86	60.89	60.00	54.54	55.35		
1982	123.49	38.14	161.63	157.82	130.36	133.51		
1983	90.61	41.83	131.94	127.81	95.99	99.09		
1984	38.48	8.53	47.01	46.16	31.53	32.11	25	17.1
1985							56	34.8
1986								
Total					331.19	339.29		51.9

Note:

Hydro discounted costs (1980-1981) adjusted for 20% reserve margin (i.e., x 1.2);

$$\frac{339.29 \times 10^6 \times 1.2}{51.9 \times 10^3} = 7{,}844.86 \text{ KY/kW in market prices.}$$

$$\frac{331.19 \times 10^6 \times 1.2}{51.9 \times 10^3} = 7{,}657.57 \text{ KY/kW in border prices}$$

Table 4.6

Generation Capacity Costs for Gas Turbines Investments (in Million Kyatts) for 100 MW of Capacity

			Total Investment		Total Investment Discounted	
Year	**Foreign**	**Local Material**	**Market Prices**	**Border Prices**	**Border Prices**	**Market Prices**
1980	248.80	92.70	341.50	332.23	332.23	341.50
Total					332.23	341.50

Note:
Gas turbine discounted cost (1980-81) adjusted for 20% reserve margin:

$$\frac{341.50 \times 10^6 \times 1.2}{100 \times 10^3} = 4098 \text{ KY/kW in market prices}$$

$$\frac{332.23 \times 10^6 \times 1.2}{100 \times 10^3} = 3986.76 \text{ KY/kW in border prices}$$

Table 4.7

Generation Capacity Costs, KY/kW Installed (1980/81) Prices

	Capacity Cost		Weighted Capacity Cost	
Unit	**Market Prices**	**Market Prices**	**Border Prices**[a]	**Border Prices**[a]
Hydro	7,844.86	7,657.57	995.48	1,019.83
Gas Turbine	4,098.00	3,986.76	3,468.48	3,565.26
Weighted Total			4,463.96	4,585.09

[a] The relative weights are 0.13 for hydro and 0.87 for gas turbine.

Transmission and Distribution Capacity Costs

Next, the LRMC of transmission and distribution (T&D) are calculated. Generally, all T&D investment costs are allocated to incremental capacity since the designs of these facilities are determined principally by the peak kW they carry rather than the kWh. The concept of structuring by voltage level may be introduced at this stage. Consider three supply voltage categories: high, medium, and low (HV, MV, LV). Consumers at each voltage level are charged only upstream costs. Therefore, capacity costs at each supply voltage must be identified.

HV Transmission

The planned investment program for HV transmission capacity (66/132 kV) for 1980-86 is summarized in Table 4.8 and columns 1-3 of Table 4.11, while projected HV demand is derived in Table 4.12 and columns 1-3 of Table 4.14. Using the results presented in those two tables and the average incremental cost approach (see Chapter 3), the incremental HV capacity cost can be calculated as: 459.56 x 106/91.55 x 103 = 1133.60 KY/kW.

LV Distribution

The planned investment program for LV distribution capacity (220/400 V) is summarized in Table 4.10 and columns 7-9 of Table 4.11. Peak LV demand is derived in Table 4.12 and columns 7-9 of Table 4.14. Using these estimates, the average incremental LV capacity costs: 147.15 x 106/40 x 103 = 3678.75 KY/kW.

Strict LRMC of Capacity at HV, MV, and LV

The capacity costs at the HV, MV, and LV levels calculated earlier are consolidated in Table 4.15.

Table 4.8

HV Transmission Capacity Costs
(Transmission EHV/HV Investment Plan)

						Discounted Total	
Year	Foreign (1)	Local (2)	Total Market Prices	Total Border Prices (1)+(2)x.9	Discount Factor(5)	Market Prices (6)=(3)x(5)	Border Prices (7)=(4)x(5)
1980	19.30	14.20	33.50	32.08	1	33.50	32.08
1981	27.60	12.60	40.20	38.94	0.909	36.54	35.40
1982	95.70	18.50	147.20	112.35	0.826	94.33	92.85
1983	225.40	13.70	239.10	227.73	0.751	179.56	178.61
1984	163.10	15.00	178.10	176.60	0.683	121.64	120.62
1985	67.50	10.00	77.50	76.50	0.621	48.13	47.50
1986	38.10	3.20	41.80	41.48	0.564	23.39	23.41
TOTAL	637.20	87.20	724.40	715.68		534.42	530.07
Total (1980-1984):						465.57	459.56

Table 4.9

MV Distribution Capacity Costs (MV Investment Plan)

Year	Foreign	Local	Total Market Prices	Total Border Prices	Discount Factor	Discounted Total Border Prices	Discounted Total Market Prices
1980	15.79	3.03	18.83	18.52	1	18.52	18.83
1981	12.50	2.55	15.06	14.80	0.909	13.45	18.69
1982	14.72	3.41	18.13	17.79	0.826	13.70	14.98
1983	14.95	3.50	18.46	18.11	0.751	13.60	13.86
1984	16.76	3.77	20.54	20.16	0.683	13.77	14.03
1985	17.28	3.94	21.22	20.83	0.621	12.83	13.18
1986	17.51	4.04	22.55	21.15	0.564	11.94	12.15
TOTAL	109.52	24.26	133.77	131.36		98.91	100.73
Total (1980-1985):						86.97	88.59

Table 4.10

LV Distribution Capacity Costs (LV Investment Plan)

Year	Foreign	Local	Total Market Prices	Total Border Prices	Discount Factor	Discounted Total Border Prices	Discounted Total Market Prices
1980	1.43	11.75	13.18	12.01	1	12.01	13.18
1981	2.87	20.74	23.61	21.54	0.909	19.58	21.46
1982	3.16	25.31	28.47	25.94	0.826	21.44	23.52
1983	3.66	32.28	35.94	32.71	0.751	24.57	26.99
1984	4.05	34.54	38.60	35.14	0.683	24.00	26.36
1985	4.22	36.69	40.92	37.25	0.621	23.13	25.41
1986	4.61	39.01	43.62	39.71	0.564	22.42	24.60
TOTAL	24.01	200.32	224.33	204.30		147.15	161.56
Total (1980-1986):						147.15	161.56

Table 4.11

Summary of Transmission and Distribution Investment Program (1980-86) in Million Kyatts

Year	HV			MV			LV		
	Cost in Domestic Prices	Cost in Border Prices	Discounted Cost in Border Prices	Cost in Market Prices	Cost in Border Prices	Discounted Cost in Border Prices	Cost in Market Prices	Cost in Border Prices	Discounted Cost in Border Prices
1980	33.50	32.08	32.08	18.83	18.52	18.52	13.18	12.01	12.01
1981	40.20	38.94	35.40	15.06	14.80	13.45	23.61	21.54	19.58
1982	114.20	112.35	92.85	18.13	17.79	14.70	28.47	25.94	21.44
1983	239.10	237.73	178.61	18.46	18.11	13.60	35.94	32.71	24.57
1984	178.10	176.60	120.62	20.54	20.16	13.77	38.60	35.14	24.00
1985	77.50	76.50	47.50	21.22	20.83	12.93	40.92	37.25	23.13
1986	41.80	41.48	23.41	21.55	21.15	11.94	43.62	39.71	22.42
Total 1982-1986			459.56						
Total 1981-1986						86.97			
Total 1980-1986									147.15

Table 4.12

Burma EPC Grid -- Demand and Loss Forecast at the Time of System Peak[a]

Year	Energy Gener- ated	Load Factor (%)	Gener- ation Peak (MW) (1)	Station Use (MW)	HV Loss (MW) (2)	HV Peak (MW) (3) = (1)-(2)	HV Consump- tion (MW) (4)	MV Loss (MW) (5)	MV Peak (MW) (6)=(3)- (4)-(5)	MV Cons. (MW) (7)	LV Loss (MW) (8)	LV Peak (MW) (9)
1979	1027.46	63.4	185.00	3.15	16.65	165.21	16.40	11.90	136.90	27.80	26.18	82.92
1980	1138.09	64.0	203.00	3.45	18.27	181.28	16.40	13.19	151.69	33.50	27.18	91.01
1981	1264.07	65.0	222.00	4.44	19.98	197.58	22.10	14.04	161.44	40.20	26.67	94.57
1982	1406.42	65.0	247.00	4.94	22.23	219.83	30.20	13.27	176.36	48.20	29.48	98.68
1983	1560.16	65.0	274.00	6.85	24.66	242.49	40.80	14.12	187.57	57.30	27.25	102.52
1984	1730.98	65.00	304.00	7.60	27.36	269.04	44.40	13.48	221.16	64.30	29.37	117.49
1985	1924.54	65.0	338.00	10.14	30.42	297.44	48.50	14.94	234.00	81.30	29.01	123.69
1986	2158.03	65.0	379.00	11.37	34.11	333.52	52.90	16.84	263.78	92.90	32.47	138.42

[a] Station use, and HV, MV, and LV losses derived by applying the peak power loss coefficients in Table 4.13 at the points indicated in Figure 4.5.

Table 4.13

Average and Peak Loss Coefficients for Energy/Power (% of Incoming)[a]

Year	Station Use		HV Transmission		MV Distribution		LV Distribution	
	Average	Peak	Average	Peak	Average	Peak	Average	Peak
1979	0.017	0.017	0.060	0.090	0.055	0.080	0.160	0.240
1980	0.017	0.017	0.060	0.090	0.055	0.080	0.155	0.230
1981	0.020	0.020	0.060	0.090	0.055	0.080	0.150	0.220
1982	0.020	0.020	0.060	0.090	0.050	0.070	0.150	0.220
1983	0.025	0.025	0.060	0.090	0.050	0.070	0.140	0.210
1984	0.025	0.025	0.060	0.090	0.040	0.060	0.135	0.200
1985	0.030	0.030	0.060	0.090	0.040	0.060	0.130	0.190
1986	0.030	0.030	0.060	0.090	0.040	0.060	0.130	0.190
Average	0.023	0.023	0.060	0.090	0.048	0.070	0.144	0.214

[a] Peak = Average x 1.5 (except Station Use).

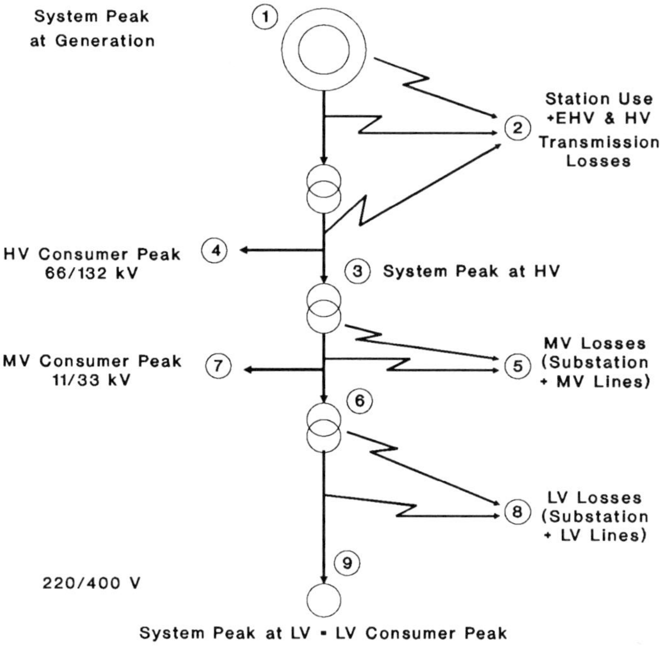

Figure 4.5: Pattern of Load/Demand and Losses in the Power System (see Tables 4.12 and 4.13)

Table 4.14

Summary of Projected Peak Demand at EHV/HV, MV, and LV, 1980-86

Year	HV			MV			LV		
	HV Peak Demand	HV Peak Demand	Dis-counted HV Peak Demand[a]	MV Peak Demand	MV Peak Demand	Dis-counted MV Peak Demand	LV Peak Demand	LV Peak Demand	Dis-counted LV Peak Demand
1979	165.21			136.90			82.92		
1980	181.28	16.07	16.07	151.69	14.79	14.79	91.01	8.09	8.09
1981	197.58	16.30	14.82	161.44	9.75	8.87	94.57	3.56	3.24
1982	219.83	22.25	18.39	176.36	14.91	12.33	98.68	4.11	3.40
1983	242.49	22.66	17.02	187.57	11.22	8.43	102.52	3.84	2.88
1984	269.04	26.55	18.13	211.16	23.59	16.11	117.62	14.97	10.22
1985	297.44	28.40	17.63	234.00	22.84	14.18	123.69	6.20	3.85
1986	333.52	36.08	20.37	263.78	29.78	16.81	138.42	14.73	8.31
Total 1982-1986			91.55						
Total 1981-1986						76.72			
Total 1980-1986									40.00

[a] Assumed discount rate is 10%.

Marginal Energy Costs

The marginal energy costs during the peak and off-peak periods are estimated by looking at the typical pattern of generation plant operation during the next few years. As explained earlier, thermal plant will tend to dominate until some of the long-gestation hydro projects are on-stream from around 1990 onwards. Therefore, incremental energy costs are based essentially on the costs of generating additional kWh during the peak and off-peak periods, using thermal plant.

During peak periods, the marginal machines operated are gas turbines using high speed diesel (HSD) fuel while, at off-peak times, the marginal plants are combined-cycle units running on bunker C/fuel oil. The relevant incremental kWh costs both shadow priced and non-shadow priced at the generators are shown in Table 4.16, together with the assumptions used. These costs have to be increased by corresponding peak and off-peak period loss coefficients in the transmission and distribution networks to obtain the marginal energy costs at the HV, MV, and LV levels, as described in Chapter 3.

Table 4.15

Capacity Costs

	Capacity Cost/kW		Annuitized Capacity Cost[a] KY/kW/Year	
Voltage Level	Border Prices	Market Prices	Border Prices	Market Prices
1. Generation				
Plant I Hydro (13%)	995.48	1,019.83	103.23	105.76
Plant II GT (87%)	3,468.48	3,565.26	456.11	468.83
Weighted Total	4,463.96	4,585.09	559.45	574.59
2. HV	5,019.77	5,085.40	553.18	560.41
3. MV	1,133.60	1,154.72	124.92	127.25
4. LV	3,678.75	4,039.00	405.40	445.10

[a] The assumed lifetimes for HV, MV, and LV capacities are 25, 25, and 25 years, respectively. With an assumed opportunity cost of capital of 10%, the resulting annuitizing factor = 0.1102.

Table 4.16

Marginal Energy Costs at the Generators

Time Period and Plant	Fuel Use (Imp. gal/kWh)	Energy Cost (Kyatt/kWh) Border Prices[a]	Energy Cost (Kyatt/kWh) Market Prices[b]
Peak:	HSD		
Gas Turbines	0.10	0.670	0.270
Off-peak	Bunker C/FO		
Combined Cycle	0.072	0.386	0.144

[a] Based on 1981 traded prices of approximately $35 and $28 per barrel for HSD and Bunker C/FO respectively. Official exchange rate: 1 U.S. dollar = 6.7 kyatts.

[b] Based on 1981 internal prices of approximately $14.1 and $10.5 per barrel for HSD and Bunker C/FO respectively.

Summary of Strict LRMC

Finally, the results obtained in Tables 4.15 and 4.16 are converted to domestic prices and combined with the estimate of losses and expenditures on operation and maintenance to determine strict long-run marginal capacity and energy costs at various voltage levels in Table 4.17 (shadow-priced domestic prices) and 4.18 (non shadow priced-market prices). The underlined values in each table indicate LRMC of peak-period capacity in KY/kW/year at each voltage level.

The shadow priced and non-shadow priced LRMC of capacity at each voltage level differ by only about 5%. However, there is a significant difference (of a factor greater than 2) between the corresponding values for LRMC of energy because of the very low market price for gas paid by EPC relative to its economic opportunity cost as a substitute for petroleum products.

Customer Costs

The customer costs consist of costs of hooking up new consumers, metering, and billing. For convenience, the metering and billing costs have been included in the operation and maintenance, administration, and facilities costs component already included in the previous section.

The typical costs (in 1981 kyatts) of LV customer connection are:

a. *Single-Phase Service*: 1.1 KY per foot of service line, plus 6.5 KY for labor, plus 360 KY for meter (10 to 30 amps).
b. *Three-Phase Service*: 5.2 KY per foot of service line, plus 11 KY for labor, plus 600 KY for meter (10 to 30 amps).

Table 4.17

Summary of Strict Long-Run Marginal Capacity and Energy Costs

Shadow Priced (in Domestic Prices)

	Generation			HV			MV			LV			Energy	
Voltage	Annual Capac.	O&M	Total	Ann. Cap.	O&M	Cum. Total	Ann. Cap.	O&M	Cum. Total	Ann. Cap.	O&M	Cum. Total	Peak[a]	Off-Peak[b]
Generation	621.21	258.6[c]	880.21										0.67	0.386
HV			989[a]	614.64	119.5[d]	1723.14							0.752	0.415
MV						1852.84[a]	138.8	75.3[d]	2066.94				0.809	0.432
LV									2649.92[a]	450.49	247.97[d]	3348.25	1.03	0.491

Table 4.18

Summary of Strict Long-Run Marginal Capacity and Energy Costs

Non Shadow Priced (in Market Prices)

	Generation			HV			MV			LV			Energy	
Voltage	Annual Capac.	O&M	Total	Ann. Cap.	O&M	Cum. Total	Ann. Cap.	O&M	Cum. Total	Ann. Cap.	O&M	Cum. Total	Peak[a]	Off-Peak[b]
Generation	574.6	239[c]	813.6										0.270	0.144
HV			914.15[a]	560.41	109[d]	1583.56							0.303	0.154
MV						1702.76[a]	127.25	69[d]	1899.01				0.326	0.183
LV									2434.63[a]	445[a]	245[d]	3124.73	0.418	0.183

Notes for Tables 4.17 and 4.18

a The losses of incoming peak power and energy equal 11% for HV, 7% for MV, and 22% for LV. Thus, for example, the cost per kW of generation capacity at HV are 880.21/(1 - 0.11) = 989.

b The losses of incoming off-peak energy are 7%, 4%, and 12% at HV, MV, and LV levels respectively.

c See Section 4.2 for details.

d O&M (including administration and facilities costs) per year are 2.2%, 6.1%, and 6.7% of the HV, MV, and LV capacity costs, respectively.

Large industrial and commercial customer costs vary widely depending on location, voltage and kVA.

EPC's official charges for connection are:

a. Flat fee of 10 KY per customer for hookup.
b. General Purpose (Domestic): Meter deposit of 100 KY for first kW and 50 KY for further kW of installed capacity.
c. All Others: Meter deposit of 150 KY for first kW and 100 KY per kW for further kW of installed capacity.

4.3 Proposed New Tariff Structure

Review of Existing Tariff

The existing tariff (see Table 4.19) has several major drawbacks that need to be addressed as soon as possible.

a. All the major tariff categories are of the decreasing block type (with all the blocks well below LRMC), thus encouraging wasteful consumption.
b. Most tariff categories have up to 5 consumption blocks, thus unnecessarily complicating the tariff and confusing consumers.
c. There are effectively no capacity or demand-related charges, especially for large consumers. Only energy or kWh charges are levied, thus distorting the tariff structure and providing little incentive to consumers to improve their load factor.
d. There is no clear-cut distinction between high-, medium-, and low-voltage (HV, MV, and LV) consumers, so the tariff structure does not adequately reflect the different costs of supply at each voltage level.
e. The differences in the Rangoon and out-side-Rangoon tariff schedules appear to be arbitrary and would tend to encourage consumers to move to Rangoon. This is contrary to the government's intentions to decentralize population and economic activity away from Rangoon.
f. Some residential consumers (in Rangoon) have two distinct service inlets, meters and internal house wiring, thus increasing costs. Street lighting is unmetered, so municipal authorities have no incentive to conserve electricity.
g. No provisions exist for power factor or fuel cost surcharges.

The average existing price paid by some typical consumer categories and the corresponding strict LRMC values are shown in Table 4.20. In general, the existing tariffs are well below LRMC, with the greatest distortion being caused by the absence of a capacity charge.

Table 4.19

Existing Electricity Tariffs

Outside Rangoon

1. General Purpose	First	100 kWh	46 pyas/kWh
	Next	300 kWh	42 pyas/kWh
	Excess over	300 kWh	40 pyas/kWh
2. Small Power	First	100 kWh	25 pyas/kWh
	Next	200 kWh	20 pyas/kWh
	Excess over	300 kWh	17 pyas/kWh
3. Industrial			
(a) Other than Pegu, Tharrawaddy, Prome Districts & Loikaw. (minimum of 2000 kWh)			
	First	200 kWh	25 pyas/kWh
	Next	1800 kWh	20 pyas/kWh
	Excess over	2000 kWh	15 pyas/kWh
(b) Pegu, Tharrawaddy, Prome District and Loikaw. (minimum of 2000 kWh)			
	First	200 kWh	22 pyas/kWh
	Next	1800 kWh	17 pyas/kWh
	Excess over	2000 kWh	12 pyas/kWh
4. Bulk	(minimum of 500 kWh)		
	First	500 kWh	54 pyas/kWh
	Next	4500 kWh	44 pyas/kWh
	Excess over	5000 kWh	34 pyas/kWh
5. Street Light	25-watt lamp @		KY 4/-
	40-watt lamp @		KY 5/50
	Every additional 10 watt KY - /50		
	80-watt mercury lamp @		KY 12/-
	Every additional 10 watt KY 1/-		
	40-watt fluorescent lamp @		KY 7/-
	Every additional 10 watt KY -/75		
6. Special	25-watt lamp @		KY 3/-
7. Flat	40-watt lamp @		KY 5/-
	Every additional 10 watt KY -/50		
8. Temporary Lighting	For metered connection same as General Purpose.		
	For point connections		
	40 watt @		KY 1/- per night
	Every additional 10 watt		KY -/25

Table 4.19
(continued)

Existing Electricity Tariffs

Rangoon

1.General Purpose	First	100 kWh	46 pyas/kWh
	Next	300 kWh	42 pyas/kWh
	Excess over	400 kWh	40 pyas/kWh
2. Domestic Power	First	50 kWh	29 pyas/kWh
	Excess Over	50 kWh	19 pyas/kWh
3. Small Power	First	100 kWh	25 pyas/kWh
	Next	200 kWh	20 pyas/kWh
	Excess over	300 kWh	17 pyas/kWh
4. Industrial	First	40 kWh/kW	
	of maximum demand.		
	(minimum of 50 kW)		17 pyas/kWh
	Next	2000 kWh	15 pyas/kWh
	Next	10000 kWh	12 pyas/kWh
	Next	30000 kWh	10 pyas/kWh
	Excess		8 pyas/kWh
5. Commercial	First	40 kWh/kW	
	of maximum demand.		
	(minimum of 50 kW)		40 pyas/kWh
	Next	2000 kWh	24 pyas/kWh
	Next	10000 kWh	18 pyas/kWh
	Next	30000 kWh	16 pyas/kWh
	Excess		14 pyas/kWh
6. Street Light	40- to 50-watt @ KY 6/50 per lamp 60- to 75-watt @ KY 12/- per lamp 100- to 160-watt @ KY 15/- per lamp 160- to 265-watt @ KY 21/- per lamp 265- to 450-watt @ KY 28/- per lamp KY 5/- less for over and above 1,000 lamps.		
7. Temporary Lighting	For metered connections same as General Purpose.		

Note: 100 pyas = 1 Kyatt (KY)

Table 4.20

Comparison of Existing Tariff and Strict LRMC for Typical Consumer Categories

		LRMC					
		Shadow Priced (Domestic Prices)			Not Shadow Priced (Market Prices)		
Voltage Level	Existing Tariff[a] Av. 1980 Price (KY/kWh)	Capacity (KY/kW/mo)	Energy (KY/kWh)	Total[c] (KY/kWh)	Capacity (KY/kWh/mo)	Energy (KY/kWh)	Total[c] (KY/kWh)
HV	0.15 (industrial)	144	0.75(p)[b] 0.42(op)	0.74	132	0.15(op) 0.30(p)	0.40
MV		172	0.81(p) 0.43(op)	0.84	158	0.33(p) 0.16(op)	0.48
LV	0.47 (domestic)	279	1.03(p) 0.49(op)	1.29	269	0.42(p) 0.18(op)	0.85

a Only energy charges levied. For comparative purposes, representative values are given for certain customer classes, since the existing tariff does not explicitly differentiate among voltage levels.

b p = peak; op = off-peak.

c Total cost per kWh = (capacity cost per kW per month/730 x mf) + (0.3) x (peak energy cost) + (0.7) x (off-peak energy cost) where mf = load factor x diversity factor; assumed to be 0.9 for HV, 0.8 for MV, and 0.6 for LV.

Adjustments to Strict LRMC

The proposed new tariff schedule is given in Table 4.21. Because of the wide divergence between the existing tariffs and strict LRMC, the proposed structure may be considered a transitional one that will not be too great a shock to consumers. The principal constraints and considerations used in the adjustment of strict LRMC to achieve this tariff structure were:

a. A target level of about 0.35 KY per kWh for the average tariff level required for EPC's financial viability, i.e., this average price meets the desired increases of 14.5% and 20% in 1981 and 1982, respectively, on the 1980 average tariff level of 0.25 KY per kWh.
b. Limiting increases in prices to levels that will not be too great a shock and cause consumer resistance.
c. Eliminating all decreasing consumption block structures and introducing increasing blocks.
d. Reducing the number of consumption blocks to simplify the tariff.
e. Introducing demand- or capacity-related charges for HV and MV consumers where more complex metering is justified.
f. Distinguishing between HV, MV, and LV consumers in their consumer categories where appropriate.
g. Consolidating the Rangoon and outside Rangoon tariffs into a single schedule.
h. Introducing a power factor (p.f.) surcharge to penalize large consumers and encourage them to improve their p.f.
i. Introducing a fuel cost surcharge to provide temporary relief to EPC in the case of cash flow problems arising from an increase in thermal generation costs due to fuel price escalation.
j. No changes to be made in customer connection charges at this stage.

Table 4.21

Burma -- Proposed EPC Electricity Tariff

Consumer Category: Existing Tariff[a]	Consumer Category: New Tariff	1981-82 Estimated Annual Consumption: Demand (MW)	1981-82 Estimated Annual Consumption: Energy (MWh)	New Tariff: Demand Charge (KY/kW/mo)	New Tariff: Energy Charge (KY/kWh)[b]	Estimated Annual Revenue in 1981-82 (10^3KY)	Average Price (KY/kWh)	Comments
	A. Domestic LV							
General Purpose	A.1 General Purpose							
	<20 kWh	--	96,830	--	0.35	33,891	0.35	Low-income
	20 - 150 kWh	--	85,020	--	0.45	38,259	0.45	lifeline price
	> 150 kWh	--	54,320	--	0.55	29,876	0.55	
Domestic Power	A.2 Domestic Power	--	33,290	--	0.35	11,652	0.35	To be consolidated with A.1 (GP), later
	B. Industrial							
	B.1 HV (66/132 kV)[c]	14.6	114,980	30	0.15	22,503	0.20	Industries operate 24 hr/ day no peak/off peak discrimination required
	B.2 MV (11/33 kV)[c]							
	< 2000 kWh	13.5	76,900	40	0.15 (0.2p) (0.1op)	18,015	0.23	
	2000-4000 kWh	15.5	88,070	40	0.20 (0.25p) (0.15op)	25,054	0.28	No. of blocks reduced from 5 to 3. Further reduction later
	< 4000 kWh	48.4	275,420	40	0.25 (0.3p) (0.2op)	92,087	0.33	
	B.3 LV (230/400 kV)[d]							
	< 100 kWh	2.7	11,820	50	0.25	4,575	0.39	Same comment as for B.2
	100 - 300 kWh	1.1	4,850	50	0.30	2,115	0.44	
	> 300 kWh	3.2	14,050	50	0.35	6,838	0.49	
	C. Commercial (LV: 230/400V)[d]							
Bulk/Commercial	<500 kWh	0.4	2,260	55	0.30	964	0.43	Same comment as for B.2
	500 - 5000 kWh	3.2	18,050	55	0.35	8,430	0.47	
	> 5000 kWh	16.3	92,530	55	0.40	47,770	0.52	Satisfies average price
	D. Street Lighting	--	30,200	--	0.40	12,080	0.40	Increases of 14.5% (1981)
	E. Temporary	--	1,260	--	0.45	567	0.45	& 20% (1982) on average
	F. Departmental (EPC use)	--	660	--	0.20	132	0.20	prices of KY 0.25/kWh
	Total (All Burma)	--	1,000,510	--	--	354,808	0.35	

a Separate tariff schedules exist for Rangoon and Outside-Rangoon. Each consumer category has up to 5 tariff blocks.
b p = peak, op = off-peak (these charges will be implemented when appropriate metering is installed).
c Peak-period demand will be metered.
d Peak-period demand estimated using 0.5 load factor, until metering is available.

Annexure to Table 4.21

1. Power Factor Surcharge

   A. Power factor of 0.85 or greater will be considered normal and not subject to a surcharge. Consumers with power factor less than 0.85 will be liable to the following surcharge on their normal electricity bill:

   p.f. surcharge = 2 x (0.85-actual p.f.) x (normal electricity bill)

   The power factor surcharge is intended as a means of encouraging consumers (especially large industrial consumers) to improve their power factor where appropriate.

2. Fuel Price Surcharge

   A fuel price surcharge will be applied uniformly on all kilowatt-hours sold according to the following formula:

   Fuel price surcharge per kWh = (EPC actual fuel costs - EPC fuel costs based on 1981 prices)/(total kWh consumed)

   The fuel price surcharge is intended only as a temporary measure to relieve EPC's cash flow situation. Any changes in fuel costs should be reflected in changes in the basic tariff schedule as soon as possible.

More specific considerations in certain consumer categories were as follows:

Domestic

Low-Income (approximately 200 KY per month):

Monthly consumption based on following appliances:

3 x 40 W light bulbs x 4 hr/day x 30 day = 14.4 kWh

Therefore, first consumption block set at 20 kWh at 0.35 KY per kWh -- reduced from existing charge of 0.46 KY per kWh. Average electricity bill is 5 k per month, which is affordable.

Medium-Income (approximately 700 KY per month):

Monthly consumption based on following appliances:

6 x 40 W light bulbs x 5 hr/day x 30 days = 36 kWh
1 x 700 W electric iron x 1.5 hr/day x 30 days = 30.5
1 x 120 W small refrigerator x 14 hr/day x 30 days = 50.4 1 x 10 W radio x 3 hr/day x 20 days = 0.9
1 x 120 W fan x 2 hr/day x 30 days = 7.2 (summer only: March to May)

Total = 117.8 to 125.0 kW

Therefore, second consumption block set at 150 kWh at 0.45 KY per kWh, which is about the existing average price. Average electricity bill is about KY 50-54 per month, which is affordable.

Industrial and Commercial

HV: Strict LRMC based capacity cost 144 KY/kW/month at HV (assuming diversity factor of 1.0). But existing capacity charge is negligible. Therefore, transitional tariff capacity charge is KY 30/kW/month.

MV: Strict LRMC based capacity cost 144 KY/kW/month at MV (assuming diversity factor of 1.2). But existing capacity charge is negligible. Therefore, transitional tariff capacity charge is KY 40/kW/month.

LV: Strict LRMC based capacity cost 186 KY/kW/month (assuming diversity factor of 1.5). But existing capacity charge is negligible. Therefore, transitional tariff capacity charge is 50 KY/kW/month for industrial and 55 KY/kW/month for commercial consumers. Government wishes commercial tariffs to be higher than industrial tariffs. All energy costs at HV, MV, and LV are now on an increasing block basis with the price increase limited by the need to avoid consumer resistance.

4.4 Some Remaining Issues

As mentioned, the proposed new tariff is a transitional one that will help to bring the existing distorted and low tariff structure more into line with strict LRMC, as well as with other policy objectives.

Several issues that need to be considered in future tariff revisions are:

1. Widening the peak period to include the daytime peak, especially if this grows rapidly relative to the evening peak.
2. Eliminating the domestic power tariff (and second meter, in some households), and consolidating this with the regular domestic tariff.
3. Further, reducing the number of some consumption blocks from three to two or even one.
4. Increasing the demand charges of MV and HV consumers closer to strict LRMC.
5. The connection costs should be analyzed using better data, and the charges revised to reflect these costs.

In general, the tariff unit in EPC should not be disbanded. It should continue to recalculate the tariff at regular intervals using more up-to-date data, so that revised tariff structures may be introduced when changes in EPC's costs indicate that strict LRMC has diverged significantly from the existing tariff.

Chapter 5

ELECTRICITY DEMAND ANALYSIS AND FORECASTING[1]

5.1 Introduction

This chapter provides an introduction to electricity demand analysis and forecasting, and identifies some of the main problems of practical implementation. There are several interrelated reasons underlying the importance of accurate demand projection. First, the timely and reasonably reliable anticipation of energy needs is vital not only with regard to electricity, but also in the context of overall energy supply. Second, the expansion of power systems usually requires many years to plan and implement. Third, investments in such systems generally are highly capital intensive. If forecasts are too low, shortages may retard development, while the costs of shortages to consumers are usually a large multiple of the volume of energy not supplied. However, if forecasts are too high, large amounts of capital with high opportunity costs might be uselessly tied up for long periods of time. Either of these consequences is costly to an economy; far more costly than the resources that need to be marshaled to undertake detailed and reliable demand studies that could help to avoid such errors.

Before we discuss the details of demand analysis and forecasting, it would be helpful to examine the basic relationship between power and energy. The commonly used unit of energy in heavy current electricity applications is the kilowatt-hour (kWh). The rate of energy per unit time is called power (or sometimes, capacity), which is measured in kilowatts (kW). Therefore, it follows that energy in kWh is the product of power in kW and time in hours. The distinction between power and energy is important because the same amount of energy may be delivered within a small interval of time at a relatively high rate of power flow, or over a longer period of at the lower power level. This point is best exemplified in nuclear fission, by the contrast between the practically instantaneous explosion of an atomic bomb, and the much slower controlled release of power from a nuclear reactor.

The load factor (LF), which is the ratio of average to maximum (or peak) kW over a given interval of time, may be calculated at various levels of aggregation, e.g., for a single customer or the whole system, on a daily or annual basis, and so on. It is important, because the size or capacity (and therefore cost) of power system components are determined to a great extent by their capability to handle peak power flows. Since practically all customers require their maximum kW only during a short peak period during the day, the LF is also a measure of the intensity of capacity use. A load forecast may be made either in terms of the peak power, or the total energy consumed, during a given period (e.g., one year), with the conversion from one unit to the other being made by the use of the LF. In general, the kW peaks of different customers will not occur simultaneously. The diversity factor for a group of consumers is a measure of the divergence of spreading over time of the individual peak loads,

[1] Edited extract from M. Munasinghe, W.G. Scott and R. Dias-Bandaranaike, *Optimizing Rural Electricity Supply*, Energy Department Report, The World Bank, Washington DC, 1982.

and permits the computation of the combined peak load for the group in terms of the disaggregate peak values.

The total amount of both kW and kWh generated at the source will be greater than the corresponding values consumed because of losses in the system (see Chapters 2 and 11). Generation losses are mainly of the station use type, i.e., for driving various auxiliary equipment at the power plant itself. Transmission and distribution network losses are basically resistive losses, including transformer losses. Another type of loss which is important in some countries is theft, and allowances must be made for it in the demand forecast. Since it is also a type of electricity consumption, ideally, the corresponding user benefits and outage costs should also be considered in the optimizing process; in practice this would be extremely difficult to do, because little is likely to be known about the consumers and ethical considerations may make it difficult for the planning authority to give special considerations to such illegally appropriated benefits.

Load forecasts are made for time periods of varying duration. Very short-term demand projections are made on a daily or weekly basis for purposes such as optimizing system operation, and scheduling of hydro units, while short-run forecasts ranging between one and three years are used in hydro reservoir management, distribution system planning and so on. The time horizon for medium-term demand projections is about four to eight years, which corresponds to the lead times required for major transmission and generation projects. Such forecasts, therefore, are especially useful to determine the next steps in building these facilities. Long-range demand projections, usually of a duration of at least 10 years, are extremely important in long-run distribution system expansion planning. This type is the most relevant one for R.E. planning. All the methods discussed below, such as time trend analysis, econometric-multiple correlation techniques, and surveys are used for this purpose.

5.2 Methods of Forecasting

The following are the main methods of making electricity demand forecasts:

1. Time trend extrapolation
2. Econometric multiple regression
3. Field surveys

Time Trend Extrapolation

Trend analysis is a commonly used approach when relevant historical data on electricity consumption (either in the R.E. project area or in a region having similar characteristics), is available. It consists of the extrapolation of past growth trends assuming that there will be little change in the growth pattern of the determinants of demand such as incomes, prices, consumer tastes, etc. These trends are usually estimated by a least square fit of past consumption data against time or by some similar statistical methodology. Depending on the availability of data, they may be estimated either on a global basis for a given area or they may be broken down by consuming sector (e.g., households, commercial enterprises, industry, etc.). Sometimes, ad hoc adjustments are made to account for substantial changes in expected future demands due to specific reasons. For example, this may take the form of projecting on a case by case basis the expected demands of new industrial plants or other economic activities. This combination of overall trend projection together with survey-research type specific adjustments is frequently used.

The main advantages of this approach are its simplicity and modest requirements of both data and analytical skills. Forecasts can be based on whatever data are available. The major disadvantage is that no attempt is made to explain why certain electricity consumption trends were established in the past. The underlying, and usually unstated, assumption is that whatever factors brought about consumption changes in the past will continue unchanged in the future. This is, of course, a rather limiting restrictive assumption, especially in a world in which relative energy prices are changing at rapid rates, and often in a direction opposite to that observed until recently.

Econometric Multiple Regression

Econometric forecasting techniques are usually somewhat more sophisticated, and, in theory, hold out the promise of greater forecast accuracy. Past electricity demand is first correlated with other variables such as prices and incomes, and then future demands are related to the predicted growth of these other variables. However, these methods are frequently nothing more than a special form of trend analysis if the projections of the selected determinants are themselves based on historical trends. The other problem encountered is that of data availability. Usually it is difficult, if not impossible, to obtain the required time series that are needed to produce statistically acceptable (i.e., statistically significant) results. Data series are often short and incomplete especially if the region concerned has only a brief history of electrification. They are also subject to changes in definitions over time and, even more frequently, are in error. Furthermore, the need for "proving" statistically "significant" results is such that long periods of time have to be covered; periods long enough to have experienced significant changes in the underlying structure of the economy. Hence, the confidence in the results must be low, even if they pass the tests of statistical significance. However, the advantages of econometric studies are that they can take a number of important, demand-determining variables, such as price and income, explicitly into account. A specific type of forecasting approach called end-use analysis is used, especially for household consumption, where the numers and types of electricity using devices become important explanatory variables. However, this technique is quite data-intensive and may need to be coupled with field surveys.

The formulation of a typical residential demand model would be based on consumer theory in economics (Nicholson 1978) The direct utility function of a consumer which indicates the intrinsic value he derives from the consumption of various goods may be written:

$$U = U(Q_1, Q_2, ..., Q_n; \underline{Z})$$

where Q_i represents the level of consumption of good i in a given time period (e.g., one year) and $\underline{Z}$ is a set of parameters representing consumer tastes and other factors.

The set of prices P_1, P_2, ..., P_n, for these n consumer goods, and the consumer's income I, define the budget constraint:

$$I \geq \sum_{i=1}^{n} P_i Q_i$$

Maximization of the consumer's utility U subject to the budget constraint yields the set of Marshallian demand functions for each of the goods consumed by the household:

$$Q_i = Q_i(P_1, P_2, ..., P_n; I; \underline{Z}) \text{ for } i = 1 \text{ to } n. \qquad (5.1)$$

Consider the demand function for electricity. Then Eq.(5.1) may be written in the simplified form:

$$Q_e = Q_e(P_e, P_K, P; I; \underline{Z})$$

where the subscript e denotes electricity, while subscript k indicate the substitute form of energy kerosene (e.g., for lighting), and P is an average price index representing all other goods.

Next, assuming that demand is homogenous of degree one in the money variables (i.e., prices and income), we may write:

$$Q_e = Q_e(P_e/P, P_K/P; I/P; \underline{Z}) \qquad (5.2)$$

Thus, starting from consumer preference theory, we may arrive at a demand function for electricity which depends on its own price, the prices of substitutes and income, all in real terms. The effects of other factors, $\underline{Z}$, such as quality of supply, shifts in tastes, and so on can also be explicitly considered. More specific examples of the effects of supply quality on demand are discussed later (for an example from a related utility sector, telecommunications, see Munasinghe and Corbo 1978).

The final specification of an expression such as Eq.(5.2) could vary widely (Taylor 1977, Pindyck 1979). Q_e could be household consumption or per capita consumption; the demand function could be linear, or linear in the logarithms of the variables, or in the transcendental logarithmic form, and could include lagged variables; and $\underline{Z}$ could include supply side constraints such as access to supply, and so on.

Analogously, the industrial demand for energy may be derived from production function theory in economics. For example, consider the output of a particular firm of industry over a given time period:

$$X = F(K, L, M, Q_1, ..., Q_n; \underline{S})$$

where K, L, and M represent the inputs of capital, labor and other nonenergy materials respectively; Q_i is the input of the ith form of energy including electricity, and S is a set of parameters that represents other factors such as shifts in technology, industrial policy, and so on.

To avoid the problem of different types of capital, labor, etc., it is necessary to assume weak separability of the inputs K, L, M and $E_i = (Q_1, Q_2 ... Q_n)$, so that each may be represented as an aggregate, with a distinct, composite price index, i.e., P_K, P_L, and P_M. Furthermore, in order to be able to determine the minimum cost mix of capital, labor, materials, and energy, it is assumed that the individual energy inputs, i.e., oil, gas, electricity, coal, etc., are subject to homotheticity. This makes it possible to use a two-step process. First the mix of fuels that make up the energy input are optimized, and in a second step, the optimal amounts of capital, labor, other material inputs and energy are chosen (see for example, Pindyck 1979).

The problem posed in production theory is the minimization of the costs of producing a given quantity of output X, given exogenous prices of inputs (Shepard

1953). In principle, the solution yields, as in the household case, the electricity demand function:

$$Q_e = Q_e(P_K, P_L, P_M; P_1,..., P_n; X; \underline{S})$$

As before, we may use one of the nonenergy input prices as numeraire, (e.g., P_k) and rewrite Q_e in normalized form:

$$Q_e = Q_e(P_L/P_K, P_M/P_K, P_1/P_K,...,P_n/P_K; X/P_K; \underline{S})$$

The demand for electricity at time t is therefore a function of its own price, the prices of energy substitutes, the prices of nonenergy inputs, and other factors S. Many different choices of variables and specifications of demand function may be used.

In this way demand functions could be developed for other end-use sectors such as agriculture, commerce, and so on. The demand function may be estimated by standard econometric techniques. The estimated equations then could form the basis for future demand forecasts.

The main issues with this approach are:

1. The mechanistic nature of the econometric equations and their extrapolation into the future, which often fails to capture structural shifts in demand growth. Such structural shifts are particularly important in developing countries as a result of the introduction of new technology (e.g., higher quality of electric lighting instead of kerosene lamps).
2. The difficulties of separating out short-run and long-run effects in the analysis of changes in the structure and level of prices.
3. The lack of an adequate data base to make accurate regression estimates. In some areas of developing countries this problem is particularly severe because price and consumption data by consuming sector are usually lacking, income data are unreliable or nonexistent and do not account for changes in income distribution, and statistical data of output and sales of specific industries are often unobtainable because of their confidentiality if the number of producers is small.
4. The inherent limitation of estimation procedures that concentrate on energy prices as demand determinants, but do not account for the costs, availability, life expectancies, and replaceability of the energy-using appliances and equipment that must be utilized with various energy sources. This problem is particularly acute under the assumption of homotheticity for all alternative energy sources, because energy users will usually base their use decisions on total systems costs rather than energy costs only (e.g. total costs of electric versus total costs of diesel-powered irrigation pumps).
5. The erratic behavior of energy prices, unless the analysis relied only on cross-sectional data only (not a practical procedure in developing countries).
6. The fact that specific energy resources are often allocated by governmental fiat, or determined by such factors as availability or reliability of supply, rather than observable market price. Another important factor, may be quality.
7. The problem that demand elasticities, even if they were estimated accurately, are likely to change significantly themselves, rather than remain constant, if expected price changes of energy are large. This is so because for most activities the ability to substitute other resources for energy (including alternative energy resources) is limited.

Field Surveys

Given the limitations of the other forecasting methodologies, surveys potentially provide a direct and reliable tool of demand analysis and forecasting. In essence, surveys consist of a list of more or less sophisticated questions that are put to existing and potential electricity users in order to measure and record their present consumption and future consumption plants. The basic types of questions that might be asked are the following:

1. How much energy (of each type) do you use per month/year?
2. What do you use it for?
3. How much do you pay per unit of energy used (by energy type)?
4. What do you grow, produce or sell and what is the value and value added of each major product?
5. Can you identify specific energy uses with specific outputs?
6. What is your net income (for households)?
7. What are your future expansion plans, and their timing, and what additional energy requirements do they imply?
8. What additional energy-using appliances or equipment are you planning to acquire in the foreseeable future (identify)?

The major problems concerning surveys are the following:

(a) They require substantial amounts of time.
(b) They can be costly.
(c) They require skilled interviewers.
(d) Users may be unable to provide the information asked because they themselves do not know.
(e) Users may be unwilling to provide the information for competitive reasons or because of fear of the consequences of revealing the information, etc.
(f) Users may wittingly or unknowingly give inaccurate answers.
(g) Future energy use plans may be vague, or too optimistic/pessimistic.

Power consumption surveys could also be usefully combined with others, such as general population censuses or similar inquiries (such as income studies, etc.).

5.3 Practical Application

In summary, there is no single universally superior approach to load forecasting. The more sophisticated techniques tend to obscure the basic methodological assumptions as well as weaknesses in the data, and therefore the simpler techniques may be more appropriate in situations where the available information is know to be unreliable. Often, the sum of many disaggregate demand forecasts (by consumer type), estimated using a variety of techniques, can be effectively compared to an independently made global forecast for the entire project area. Table 5.1 lists some of the major factors that are likely to determine the demand for electricity. Estimates of future changes in these determinants are helpful to project demand.

Whenever possible, survey approaches should be used to forecast the demand of large users. This may account for a substantial percentage of the total load. Frequent updating of this information, to account for unexpected changes is an additional important requirement. the determinants of demand for diverse users have to be carefully selected, because they vary from case to case, and the level of disaggregation chosen will depend on the data, as well as on the ultimate purpose of the projection. Demand forecasting is a dynamic process in the sense that the procedure must be repeated often, in light of improved information and techniques of analysis. The building up of a reliable data bank is an invaluable tool in this process of constant revision.

Informed judgement is an essential supplement to the various quantitative methods of electric demand forecasting. An experienced forecaster will be able to make appropriate adjustments based on a number of constraints and complicating factors that cannot be incorporated mechanistically into a numerical algorithm for forecasting electricity consumption.

Load promotion and the attitudes of community leaders can make a significant difference to the growth rate of demand, especially in rural areas. An advance program of information, involving education, information dissemination, meetings, mobilization of groups, and targetting of local leaders, may do much to increase the number of connections and consumption. This will also enable the government to provide firm committments concerning the future timetable for electrification, and eliminate uncertainty. At the same time, caution should be exercised to avoid the temptation of overestimating the importance of such load promotion activities, thereby justifying unviable projects on the basis of unduly optimistic demand projections.

The demand forecast for a given consumer category will depend on the growth rates of the number of connections, the average electricity use per connection, and the user population. Therefore, the evolution of each of these variables needs to be analysed separately. In particular, connection policy is a factor that must be carefully studied. In view of the low levels of income in some areas, high connection charges could act as a significant barrier and negate the favourable effects of low electricity prices. The deterrent effects of excessive connection costs may be inadvertently overlooked, especially in the case of quantitative demand forecasts that rely exclusively on the price variable. On the other hand, heavily subsidised connection charges may help to achieve the high connection rates anticipated in the demand projection, while increasing the risk that the actual consumption per connection would fall far short of the target level assumed in the forecast.

Finally, a distinction should be made between the underlying or potential demand, and the consumption level that is actually achievable. Many practical constraints may prevent the realisation of over-ambitious demand forecasts, including lack of skilled manpower, apathy of the authorities, delays in supplying equipment or unavailability of critical components, diversion of resources for other purposes due to political intereference, and so on. In some cases, institutionally imposed constraints which suppress demand, such as peak lopping or non-connection of consumers, and the presence of captive generation, may lead to a substantial gap between the potential and achievable (or realistic) demand.

Table 5.1

Major Demand Determinants of Electricity Consumption

A. Residential Sector

1. Number of households (or population)
2. Real family income weighted by income distribution
3. Price of electricity
4. Connection charges
5. Availability of service
6. Reliability of service
7. Cost and availability of electricity-using fixtures and appliances
8. Availability and cost of consumer credit
9. Costs of alternative energy sources (kerosene, LPG, natural gas, charcoal, firewood, coal)

B. Commercial Sector

1. Sales or value-added of nonsubsistence commercial sector
2. Price of electricity
3. Connection charges
4. Cost of electricity-using appliances relative to those using alternative fuels for space conditioning
5. Cost of alternative fuels for space conditioning including costs of appliances
6. Reliability of service
7. Working hours of various types of commercial establishments

C. Industrial Sector

1. Price of electricity (demand and energy charges by type of service)
2. Type of industry and its electric intensity
3. Degree of market power of individual industry
4. Relative price of alternative energy sources
5. Reliability of supply
6. Comparative total costs of autogeneration
7. Availability

D. Agricultural Sector

1. Price of electricity
2. Availability
3. Reliability of service
4. Supply costs of alternative energy systems, e.g., diesel-driven irrigation pumps and milling and processing equipment, etc.
5. Comparative total costs of autogeneration

E. Public Services Sector

1. Price of electricity
2. Per capita revenue of local governments
3. Number and size of public schools, hospitals etc.

5.4 Electricity Demand and the Quality of Supply

The importance of supply quality (or reliability) in power system planning was described in Chapter 2. When the supply quality changes, consumption benefits are affected in two ways. In the short run, if the quality drops below the expected level, planned consumption is affected and consumers experience inconveniences and costs (e.g., no lighting for households or loss of output for industries, due to power outage). In the longer run, if consumers anticipate a change in the supply quality, they will adjust their planned demand and behavior patterns. While the first aspect has been investigated in recent papers, the second has not.

Therefore in this section, we develop models to investigate the long-run changes in residential, commercial, and industrial electricity consumption due to variations in the quality of supply. In particular, we show that when the output quality is improved, the planned level of consumption will increase at any given price. This induced demand effect will shift the electricity demand curve outward and increase the consumer surplus, thus providing significant additional consumption benefits. The empirical results described in Chapter 2 (Costa Rica Case Study), indicate that such induced demand shifts are of significant magnitude and will therefore be of critical importance to both demand forecasting and investment planning.

Modelling Background

The demand functions derived and estimated here have the empirically convenient double-log form. In contrast to many earlier studies, they are based on explicit utility maximizing theory. They are also unique in that they explicitly contain the price of substitute energy as an independent variable. In addition, the methodology used here to derive similar double-log functions (which explicitly include the price of substitutes) for goods and services other than electricity.

The dependence of demand on supply quality will vary for different types of electricity consumers. For example, an industrial and a residential electricity consumer could behave very differently from one another when faced with an identical change in supply quality. What is considered to be an adequate quality for one might be far from adequate for the other. If this analysis is taken to an extreme, we could say that every single consumer of electricity behaves differently from every other and that separate demand functions should be derived for each.

In practice, it would be impossible to estimate separate demand functions for each electricity consumer. Even if detailed data were available for each one, there would be insufficient variation in the independent variables to give meaningful results. On the other hand, the fairly extensive literature on aggregation theory demonstrates that, only under unrealistic and restrictive assumptions, is it possible to aggregate individual demand functions into a single function with aggregate variables for purposes of estimation.

A compromise has to be made between the extremes of one aggregate or a large number of individual demand functions. Let us consider three broad categories of consumers: residential, industrial, and commercial. Since there exist similar behavioral characteristics within these categories, we shall assume that aggregation is possible within them. Demand functions will be derived and estimated for "typical" consumers in each of these categories.

Residential Electricity Demand

Let us consider a residential electricity consumer. Let U = U(B,N) be the consumer's utility function, where

B = quantity consumed of all goods and services except energy, and
N = total quantity of energy services consumed from both electric energy services (E) and substitute energy services (S).

In general, N = N(E,S); energy services are a function of electricity services and substitute energy services. The consumer's budget constraint can be written as

$$Y = p_b B + p_e E + p_s S; \qquad \text{where}$$

Y = income
P_b = effective pride of electric energy services
P_e = price of all goods and services except energy, and
p_s = effective price of substitute energy services

The consumer's utility maximizing problem can thus be written as:

$$\text{Max } L = U[B,N(E,S)] + m(Y - p_b B - p_e E - p_s S),$$

where m is an appropriate Lagrange multiplier. The first order conditions reduce to:

$$(\partial U/\partial B)/P_b = [(\partial U/\partial N).(\partial N/\partial S)]/P_s \qquad (5.3a)$$

$$(\partial N/\partial S)/(\partial N/\partial E) = P_s/P_e \qquad (5.3b)$$

Suppose that the consumer utility function can be written in the log-linear form:

$$U = B^{f1}.N^{f2}$$

and also that

$$N = \exp(S^{g1}.E^{g2}),$$

where f1, f2, g1, and g2 are parameters. This function has sufficient generally, but still allows further mathematical analysis. Substituting for the partial derivatives of N and eliminating S gives:

$$E = Kp_s^{t_1} p_e^{t_2} Y^{t_3} \qquad (5.4)$$

where K, t_1, t_2, and t_3 depend on f1, f2, g1, and g2.

Note that $t_1 + t_2 + t_3 = O$, which is a linear restriction on the coefficients. t_1 and t_2 are price elasticities of demand for electricity services, while t_3 is the income elasticity of demand for electricity services. Assuming no money illusion (or that E is homogenous of degree zero in p_s, p_s, and Y), the condition $t_1 + t_2 + t_3 = 0$ is the familiar elasticity condition derived in consumer theory by a straightforward application of Euler's theorem to a general demand function for E.

E and S have been defined as the quantities of electricity and substitute energy derived services, respectively. For substitute energy we will simply assume quantity of service derived to be directly proportional to quantity of such energy consumed. The constant of proportionality can be set to 1 without loss of generality. With this assumption, p_s can be reinterpreted simply as the price of substitute energy. For electricity services, however, the notion of supply quality enters into the argument. Assume that electricity services depend on the actual quantity of electricity consumed, X_{er} (measured in kilowatt-hours), and on the quality of electricity supply, R. The subscripts e and r refer to electricity and residential consumers, respectively. The simplest reasonable functional form for E is $E = f(R). X_{er}$, which indicates that the constant of proportionality f(R) depends on supply quality.

Variations in R effect the price of electricity services, p_e, according to the relationship

$$p_e E = p_x X_{er} = p_x.(E/f(R)),$$

where p_x is the price of electricity (per kilowatt-hour). Solving for p_e and E in Eq.(5.4), we have:

$$X = h(R) p_s^{t_1} p_x^{t_2} Y^{t_3}$$

where

$$h(R) = Kf(R)^{-1-t_2}$$

This equation relates demand and quality of supply. As discussed earlier, when demand shifts with changes in quality, consumption benefits will also change. To estimate these consumption benefits, it is necessary to calculate the area under the (compensated) demand curve for electricity up to the point p_x, the actual price of electricity. This reveals a weakness in the demand function, Eq.(5.4). If $-1<t_2<0$, the area representing total benefits becomes infinite.

One possible way to avoid this weakness, while at the same time making demand function more realistic, is to allow the own-price elasticity of demand (t_2) to vary (Betancourt 1981). It is unreasonable to assume that t_2 remains constant over a wide range of price variations. One simple assumption about this elasticity might be represented by

$$t_2 = k_0 + k_1 \ln p_x.$$

Since we expect $t_2<0$, and since demand for electricity is likely to become more elastic as p_x increases, we expect $k_0<0$ and $k_1<0$.

Substituting for t_2 in the expression for X_{er}, we have

$$X_{er} = h(R)\, p_s^{t_1}\, p_x^{(k_0+k_1 \ln p_x)}\, Y^{t_3}$$

If $G(R) = \ln[h(R)]$, taking logarithms of both sides we have

$$\ln X_{er} = G(R) + t_1 \ln p_s + k_0 \ln p_x + k_1(\ln p_x)^2 + t_3 \ln Y \qquad (5.5)$$

One difficulty associated with allowing price elasticity of demand to vary in the double-log formulation is that the demand equation is no longer homogeneous of degree zero. The homogeneity of the equation depends on the condition $t_1+t_2+t_3 = 0$ or, in this case,

$$t_1 + k_0 + k_1 \ln p_x + t_3 = 0.$$

Since p_x can vary for any estimated values of coefficients t_1, k_0, k_1, and t_3, the left hand side of this equation cannot always be zero unless $k_1=0$. If $k_1=0$, the price elasticity would no longer be variable. Postulating that some variable other than p_x causes the variation in price elasticity will not eliminate the inconsistency. The left hand side of the equation cannot be zero for different values for this variable unless the coefficient associated with this variable is zero and $t_1+k_0+t_3 = 0$.

Although a demand function estimated using the functional form, Eq.(5.5) is not homogenous of degree zero in income and prices, it is still possible to make this condition hold (at least approximately). This can be done during estimation by placing the linear restriction.

$$t_1 + k_0 + k_1 \ln \phi_x + t_3 = 0$$

on the coefficients of the model, where ϕ_x is the mean price of electricity in the sample. This restriction constrains the demand function to be homogeneous of degree zero at the mean price.

Industrial and Commercial Electricity Demand

Consider an industrial or commercial firm consuming electricity and other types of energy. Let $O = O(J,N)$ be the firm's production function, where N is the total quantity of energy services consumed (from both electric energy services E and substitute energy services S), and J is the quantity consumed of all other inputs. As for residential consumers, suppose $N = N(E,S)$.

The producer's optimizing problem can be defined as one of minimizing the cost of producing some quantity of output $\underline{O}$. It can be written in Lagrangian form as:

$$\text{Min } L = p_e E + p_s S + p_j J + m[\underline{O} - O\{J, N(E,S)\}], \text{ where}$$

p_e = effective price of electric energy services,
p_s = effective price of substitute energy services,
p_j = price of other inputs, and
m = Lagrange multiplier.

The first order conditions reduce to:

$$(\partial O/\partial J)/P_j = [(\partial O/\partial N).(\partial N/\partial S)]/P_s \qquad (5.6a)$$

$$(\partial N/\partial S)/(\partial N/\partial E) = P_s/P_e \qquad (5.6b)$$

Suppose that the production function can be written in Cobb-Douglas form:

$$0 = J^{f3}N^{f4}$$

Define $V_i = p_jJ$ as the total returns to all factors except energy, and for the case of residential consumers, suppose

$$N = \exp (S^{g3}.E^{g4})$$

Proceeding exactly as for the case of residential consumers leads to an equation of the form of Eq.(5.5) with V_i in place of Y. Therefore,

$$\ln X_{ei} = H(R) + r_1 \ln p_s + r_2 \ln p_x + r_3(\ln p_x)^2 + r_4 \ln V_i, \qquad (5.7)$$

where X_{ei} is the firm's electricity demand.

Empirical Estimation of Demand Function

One common characteristic of the demand functions, Eqs.(5.5) and (5.7), is the existence of terms of the form G(R) and H(R), representing the effect of the quality of electricity supply on demand. Two problems arise when attempting to estimate a function with such terms. First, the exact functional forms of G(R) and H(R) are unknown. Second, as discussed earlier, the quality of electricity supply R encompasses a variety of different attributes of electricity service. It will be very difficult to quantify such a concept. Both problems might be mitigated, if rather than a continous range of values for R, we consider a set of discrete levels R_0, R_1,....R_p, where R_0 refers to the lowest quality of supply, R_1, R_2,...to successively higher qualities, and we introduce a set of dummy variables D_1, D_2,...D_p, where

$$D_1 = D_2 = D_3 = ... = D_p = 0 \text{ when } R = R_0$$

$$D_1 = 1, D_2 = D_3 = ... = D_p = 0 \text{ when } R = R_1, \text{ and so on.}$$

Introducing an error term u into equation (5.5), we can then write (5.8)

$$\ln X_{er} = (C + a_1D_1 + a_2D_2 ... + a_pD_p)$$
$$+ t_1 \ln p_s + k_0 \ln p_x + k_1(\ln p_x)^2 + t_3 \ln Y + u, \qquad (5.8)$$

where C is a constant, a_1 represents the difference between $G(R_0)$ and $G(R_1)$, a_2 represents the difference between $G(R_0)$ and $G(R_2)$, and so on. Using the regression model, Eq.(5.8), we can estimate $C, a_1, a_2, ... a_p, t_1, k_0, k_1$, and t_3.

From Eq.(5.5), the fractional change in quantity demanded when the quality of supply is changed from R_i to R_j is given by $\{\exp[G(R_j)-G(R_i)] - 1\}$. Using Eq.(5.8), this expression can be represented by $[\exp(a_j - a_i) - 1]$. Although the presentation here has been for the residential consumer category, an identical approach might be applied to industrial and commercial consumers.

The case study of Costa Rica, described in Chapter 2, illustrates how the demand function developed above may be practically estimated and used to optimize power system reliability and planning.

Comparison with other Demand Studies

It is interesting to compare these demand functions with others used in the economic literature to estimate electricity demand. In these studies, residential electricity demand has received by far the widest attention (*The Energy Journal* 1983). The most commonly used residential demand functions are either linear or double-log in form. The dependent variable is usually the per capita demand for electricity (or its logarithm), while the independent variables are usually the ex post average price of electricity, per capita income, and a variety of other variables, including the price of substitute forms of energy and the stock of electricity using equipment.

For some constant level of supply quality, the present demand function, Eq.(5.5), takes a double-log form identical to that used in many of the earlier studies. The independent variables, price of electricity and per capita income, are present as in all previous studies using the double-log form. The price of substitute energy is also present as in many previous studies, while the price elasticity is now no longer constrained to be constant.

Many of the earlier residential demand studies have used demand functions that are not based on utility maximizing theory (Houthakker 1951, Wilson 1971). Sato (1982) has shown that by using additive utility functions, it is possible to derive a system of double-log demand functions with variable elasticities. Betancourt (1981) uses this result to argue that the demand functions employed in his estimations are consistent with maximizing theory. Nevertheless, Sato's derivation has the shortcoming that the additive nature of the utility function does not allow for strong substitutability between goods or services. If the price of substitute energy is to be included in the double-log form, as is done in many estimates of electricity demand, Sato's derivation is no longer relevant and it is not possible to argue that the double-log demand function is consistent with utility maximizing theory.

The demand functions used in this study contain the price of substitute energy and, in contrast with earlier studies, are based on explicity maximizing theory. In fact, the approach used to derive demand functions in this chapter could be used to derive similar functions for goods and services other than electricity. These functions would allow for strong substitutability and would have the empirically convenient double-log form.

References

Betancourt, T., "An Econometric Analysis of Peak Electricity Demand in the Short Run", *Energy Economics*, Jan. 1981, pp.14-29.

Houthakker, H.S., "Some Calculations of Electricity Consumption in Great Britain", *J. Royal Statistical Society*, vol.114, 1951, pp. 351-71.

Munasinghe, M. and M. Gellerson, "Economic Criteria for Optimizing Power System Reliability Levels", *The Bell Journal of Economics*, vol. 10, Spring 1979, pp. 353-65

Munasinghe, M., "The Costs of Electric Power Shortages to Residential Consumers", *Journal of Consumer Research*, vol.. 6, March 1980, pp. 361-9.

Munasinghe, M. and V. Corbo, "The Demand for CATV Services in Canada," *Canadian J. of Economics*, Vol. 11, August 1978, pp. 506-520.

Nicholson, W., *Microeconomic Theory*, Second Edition, Dryden Press, Hinsdale IL, 1978, Chap.4.

Pindyck, Robert S., *The Structure of World Energy Demand*, MIT Press, Cambridge MA, 1979.

Sato, K., "Additive Utility Function with Double-Log Consumer Demand Functions", *Journal of Political Economy*, 1982, pp. 102-24.

Shepard, R.W., *Cost and Production Functions*, Princeton Univ. Press, Princeton NJ, 1953.

Taylor, Lester D., "The Demand for Energy: A Survey of Price and Income Elasticities," in: W.D. Nordhaus (ed), *International Studies of the Demand for Energy*, North Holland, Amsterdam, 1977.

The Energy Journal, Special Electricity Issue, 1983.

Wilson, J.W., "Residential Demand for Electricity", *Quarterly Review of Economics and Business*, vol.11, Spring 1971, pp. 7-22.

Chapter 6

THE ECONOMIC ANALYSIS OF POWER PROJECTS: CASE STUDY OF RURAL ELECTRIFICATION[1]

6.1 Introduction

Since the 1950's, rural electrification has been promoted by enthusiasts, as a major driving force for the development of the rural areas of the third world. The expectations, particularly in the 1960's and 1970's, included rapid economic growth and improvements in the quality of rural life -- especially of the poor, increasing use of electricity for productive activities in agriculture, industry, and commerce, and modernization and other attitudinal changes (Munasinghe and Warren 1979). Such hopes were not surprising, given that R.E. has played a significant role in the developed nations.

On the other hand, skeptics were quick to point out numerous potential problems that might interfere with such optimistic scenarios. In fact, many of their concerns have been realized. Some of the difficulties that continue to plague rural electrification efforts in many developing countries, range from scarcity of capital -- especially foreign exchange, high costs and poor quality of supply, to disappointing load growth, low benefits and productivity gains, and perverse distributional effects -- with benefits favoring the rich rather than the poor.

Because of the substantial investments required to accelerate the pace of rural electrification, government involvement of some type becomes necessary, and most developing countries are pursuing R.E. programs as a part of their economic development efforts. This in turn implies that there are significant advantages to examining R.E. issues in the context of overall national policy. The broad rationale underlying national level planning and policymaking of all kinds is the need to ensure the best use of scarce resources, in order to further socio-economic development efforts and improve the quality of life of citizens. Planning for energy and electric power, including rural electrification in particular, are all components of national economic planning, and therefore share this same general goal.

Meanwhile, because of the many interactions and non-market forces that shape and affect the energy sectors of every economy, decision makers in an increasing number of countries have realized that energy sector investment planning, pricing and management should be carried out on an integrated basis, e.g., within a integrated national energy planning (INEP) framework which helps analyse the whole range of energy policy options over a long period of time (Munasinghe 1980). In summary, R.E. planning must also be closely integrated with overall economic, and energy planning and policy analysis, to meet many inter-related and frequently conflicting national objectives, as effectively as possible.

[1] Edited version of a paper published in *Energy Economics*, vol.10, January 1988, pp.3-17. While the paper focusses on rural electrification, Sections 6.3 and 6.4 on project evaluation and benefit measurement apply more generally to any type of power project.

Using a broad brush approach, R.E. schemes would generally cover regions where agricultural activity (including agro-industries) was dominant, the ratio of labor to capital used in production was high, and incomes were low on average, relative to urban areas. Load densities would also be relatively low, because the number of connections per km of line, the load per connection and the customer load factor tend to be small relative to urban areas. At the same time, costs per connection and per kWh supplied would be significantly higher, operation and maintenance more problematic, and the quality of supply lower. Often small or medium sized towns (population up to 50,000) in a rural area may be included, or suburban areas of larger cities. In fact, from the socio-economic and demographic viewpoint, the rural-urban gradation may often by continuous, and the terminology "regional electrification" is sometimes used to indicate programs covering a mix of rural farms, villages and small towns. From the planning and engineering design perspective, the load and customer density are among the chief parameters which would help to distinguish between urban and rural electric distribution systems.

Electricity can be introduced into rural area in several ways:

1. isolated generators powered by a variety of means including diesel, mini-hydro, etc., serving a single consumer (for example a business enterprise);
2. isolated generators serving several consumers who are connected by a local network, for example a small community; and
3. public supplies from a regional or national grid system.

Typically, these are not distinct ways to introduce electricity into a rural area, -- rather, they might occur sequentially. This procedure involving several phases of electrification can be thought of as an initial building up of demand for rural electricity, by means of autogeneration, to a point where it becomes economically feasible (in most cases) to connect to a central grid (The World Bank 1975).

In this paper, the term "rural electrification" will refer mainly to connections to a central grid, since most R.E. projects being studies are those which are sufficiently advanced to make this economically feasible. This is not to say that the most economical method for electrifying every region in every country once the load has built up, is from a central grid. Even in North America and Europe, where the extent of R.E. is almost 100 percent, many areas continue to be served by local autogenerators as this is the least cost way of providing electricity to such areas.

6.2 Rural Electrification Criteria

Three broad areas of concern that will influence and shape rural electrification policy, may be identified. The first area is centered around the objective of economic efficiency, which implies optimal use of scarce economic resources to maximize output and growth, and therefore, places emphasis on the productive uses of electricity in rural activities. According to this set of criteria, rural areas that showed the greatest promise in terms of industrial, commercial and agricultural growth would be favoured. Electricity prices would tend towards efficient levels based on the long-run marginal costs of supply, with subsidies (if any) being carefully targetted to poverty groups. Cost recovery would be easier under this regime.

A second group of goals are focussed on the satisfaction of the basic needs of citizens, providing electricity services to the poor, and improving the distribution of income and welfare. In this case, the more economically depressed regions of a country would receive greater attention. Pricing policy would rely more on subsidies, and be based mainly on the ability-to-pay of low income groups rather than on supply cost or cost recovery. The third major determinant is based on the principle of cost minimization -- to reach as many rural consumers as possible, in the shortest possible time, using limited resources. With this approach, it would be prudent to pursue R.E. schemes in areas closest to the existing grid, thus keeping costs to a minimum.

In practice, a single objective is unlikely to be dominate, to the exclusion of all others. However, the relative weights attached to these different objectives are likely to vary significantly from country to country, thus giving rise to quite different strategies for R.E. in each case.

Some specific approaches to R.E. include (Smith et al. 1983):

1. Integrated rural development - electricity is treated as only one component of a wider package of infrastructure.
2. Area coverage - comprehensive networking seeks to quickly provide electricity supply to as many customers as possible within the designated project area. More modified approaches may first adopt a backbone system, with increasing numbers of consumers being added on over a phased time period.
3. Grid extension -electrification is based on proximity to existing networks and their relative ease of extension. Each separate branch addition is judged on its own merits.
4. Isolated generation - electrification in areas remote from the national grid, is based on relative cheapness and abundance of local generating sources.
5. Intensification - concentrating on increasing the load factor and connection rates in already electrified areas.

In recent years, an increasing number of developing countries have begun to place greater emphasis on the productive uses approach, encouraging electricity consumption for agricultural pumping, agro-industrial and commercial activity, cottage industries, and so on. There are several important reasons underlying this trend.

First, in a situation of resource scarcity, the greater the mobilisation of finances within the sector, the easier it is to use these funds to expand rural electrification to more remote areas. Emphasis on productive use of electricity will lead to faster load growth, better cost recovery, and more resources for further rural electrification expansion. Second, if equity considerations are very important, other basic needs -- such as access to a safe water supply, better nutrition, and health and educational facilities -- might have a better claim on scarce public funds. In fact, it might be difficult to argue that electricity is a basic need in some countries. A better case can be made for other fuels, such as firewood for domestic cooking in rural areas. At the same time, the benefits of rural electrification might be greatly enhanced when this service is provided together with other rural infrastructural services -- such as transportation, access to credit facilities, and agricultural extension. Areas of productive electricity use are more likely to already have access to these other elements of integrated rural development, thus yielding high economic returns to rural electrification investments. Third, subsidised rural electrification projects are inherently a poor mechanism for achieving long lasting distributional gains, while increasing output and employment through productive electricity use in rural areas is likely to be more effective in this respect -- at least in the long run.

Specific decisions concerning the pace and timing of R.E. schemes, selection of areas to be electrified, investment and cost recovery policy, and pricing and connections policy, require more detailed analysis. For example, projected national rural electrification targets over the planning period will depend on the current extent of R.E., the resources and implementation capability available, and the socio-political pressures for R.E. in rural areas. Similarly, quantitative criteria for ranking areas for electrification may be developed and used in conjunction with other more judgmental factors. Thus where data is available, a formula of the following type could be used for ranking:

$$\text{R.E. Merit Value} = \Sigma\ W_i.C_i$$

where, W_i = weight assigned to characteristic C ; and
C_i = ith characteristic of the rural area under consideration.

In practice, a wide variety of indicators or characteristics and weights may be used, depending on the local conditions and data availability.

While the qualitative links between R.E. and other characteristics of rural areas have been examined in several recent studies, no systematic analyses of the quantitative relationships between such variables and the returns to rural electrification projects appear to have been done. It would be useful to carry out a study of past R.E. programs, and use standard statistical techniques (such as regression analysis, or discriminant analysis) to determine the relative weights or importance of specific rural indicators, in ensuring the success of rural electrification efforts. The results of such an ex-post analysis would be helpful in future area selection and targetting of R.E. investments. However, these relatively simple approaches would only serve as preliminary screening devices, aimed at eliminating the obviously uneconomic R.E. projects and indicating rough priorities for those that remain. The surviving rural electrification schemes would still have to be subjected to detailed project evaluation and cost-benefit analysis, as described below.

At this stage, it is useful to examine how individual projects within an overall national or regional R.E. program are usually defined. A typical rural electrification scheme will invariably include as its core, the primary and secondary distribution networks. However, the scope of the project may be expanded to incorporate upstream investments such as generation sources, transmission, and sub-transmission facilities, as well as downstream components ranging from house-wiring and small electricity using devices to major electric equipment and complementary infrastructure associated with the integrated rural development approach. From the geographic viewpoint also, rural electrification spans the spectrum from individual connections, through specific feeders, single villages and village clusters, up to large areas or regions.

The foregoing suggests that the definition of an R.E. project is therefore likely to be quite case specific, and determined on the basis of convenience in identifying the costs and benefits associated with a given package of investments, in a particular location. However, there is a danger that the aggregate analysis of large areas containing many consumers might mask significant variations at a more detailed level. Therefore, it is advisable (provided the necessary data is available), that a two stage approach be used, in which macrolevel evaluation of a large R.E. project is supplemented by the microlevel analysis of smaller components and subcomponents (especially low load density and thinly populated areas), to help refine the scope of the scheme.

The most important thing would be to establish a clear cut procedure, and then adhere to it as firmly as possible. Ad-hoc, politically motivated interventions in R.E.

policy, especially area selection, should be avoided, since it frequently has disastrous consequences. One of the most important reasons for establishing a R.E. strategy and Master Plan is to minimize undesirable changes of this sort. However, the strategy should be reviewed at frequent intervals to maintain flexibility and the ability to address new problems as they arise.

6.3 Project Evaluation

Projects must be evaluated, because economic resources are not unlimited, especially in the developing country context. The decision to finance an R.E. scheme implies that the scarce resources used for this purpose will not be available for some other project, such as a school or hospital. Moreover, even within a given rural electrification master plan, it will be necessary to rank individual schemes according to some accepted criteria and select the best ones, because the resources available are generally quite inadequate to undertake all possible projects. These criteria are usually closely related to national policy objectives. One of the chief tools for ranking development projects is cost-benefit analysis, described later.

As discussed earlier, some rough choices can be made when regional R.E. priorities are being established at the national level, but the final investment decision on a particular development project is usually made on the basis of a much more detailed, micro-level analysis. In other words, project analysis from a national viewpoint is the preferred method of rationally allocating scarce productive resources, among the different sectors of the economy, as well as individual projects within a given sector.

The successful implementation of any development project usually involves several well defined steps. In order to place R.E. project decisions and economic cost-benefit analysis in its proper perspective, we summarize below the systematic approach used by national governments, aid donors and project financiers in a typical project cycle. The steps include: identification, preparation, appraisal, negotiations and financing, implementation and supervision, and evaluation (Baum 1978).

The Project Cycle

Project identification involves preliminary selection of potential R.E. projects that appear to be feasible and conform to national and sectoral development goals. In the preparation phase which may last one year or more, the engineering-technical, institutional, economic and financial aspects of a proposed R.E. project are systematically studied. At this stage, a potential external financier may provide staff guidance and financial assistance or help the national R.E. authorities obtain other assistance to carry out the studies. In the next stage, project appraisal consists of a detailed review of all aspects of the project. An appraisal report is prepared that comprehensively evaluates the project, in the context of the national and sectoral strategies, as well as the engineering-technical, institutional, environmental, economic and financial issues. In particular, the economic evaluation involved several well-defined stages, including the demand forecast, least-cost alternative, benefit measurement and cost-benefit analysis.

If foreign financial assistance is involved, the appraisal report may be used as the basis for negotiations at which the country and financier discuss the measures required to ensure the success of the project, and the conditions for funding which are usually included in loan agreements. Supervision of the implementation process should be carried out through periodic field inspections and progress reports. Reviews of ongoing R.E. projects help to continually update and improve implementation procedures. Evaluation is the final stage of the project cycle, following disbursement of funds. Project performance audits should be carried out by an independent group involving review of previous project documents and field visits, where appropriate. This analysis yields valuable experience that helps improve the work at all stages of the project cycle for future projects.

Economic Justification

In view of its importance the procedure for the economic justification of an R.E. project is described below, in greater detail. As mentioned earlier, the first stage involves the preparation of a demand (or market) forecast. The least cost alternative, benefit measurement and cost-benefit analysis are inter-related and depend on the economic criteria described below.

Two related criteria are useful in the economic evaluation of investment projects: maximization of net benefits or minimization of costs. Maximization of net benefits is the more general approach, used in the cost-benefit analysis stage, to justify the use of scarce resources in an R.E. project, rather than elsewhere in the economy. To make optimal inter-sectoral allocation decisions in such situations requires the explicit determination of all benefits and costs over the lifetime of the investment. By contrast, a cost minimization approach, used in the least-cost solution stage, eliminates the need to measure the value of the benefits provided. This approach assumes that a given level of demand must be provided in similar quantities and qualities, whatever the supply source. The question then becomes simply one of selecting the lowest-cost method of supplying rural electric consumers. Both these criteria are helpful and used in a complementary way to evaluate R.E. projects.

Following usual cost-benefit practices, the basic objective function to be maximized here is developed in terms of net benefits or income. This leads to maximum economic efficiency in resource allocation. Under certain restricting assumptions mentioned earlier, net benefit maximization is equivalent to minimization of economic costs. Making efficiency, at least initially, the sole criterion for the choice among alternatives has obvious technical advantages. We find that the choice of efficiency as the goal provides us with an unambiguous ranking order among alternative rural electrification schemes. If we ignore the problems that are common to all objective functions, including the evaluation of uncertainty and risk, of unpredictable changes in technology and relative price levels, of the valuation of intangibles and nonmarketed outputs, of positive or negative externalities or spillovers, and the troublesome issue of individual and collective time preferences, any project that yields more net income will be preferable to one that yields less.

However, from the rural consumers viewpoint, it is not really income that we ought to maximize but welfare, or total utility. The result is that an analysis which tells us how to maximize income does not necessarily tell us whether or not we are also maximizing welfare. Nevertheless, the choice of income maximization could be defended if the government would be willing to approach more equitable distribution patterns by redistributing income from the original beneficiaries to those considered to be more deserving.

Efficient Border (Shadow) Prices[2]

In the idealized world of perfect competition, the interaction of atomistic profit-maximizing producers and utility-maximizing consumers given rise to a situation that is called pareto-optimal. In this state, prices reflect the true marginal social costs, scarce resources are efficiently allocated and, for a given income distribution, no one person can be made better off without making someone else worse off. However, conditions are likely to be far from ideal in the real world. Distortions due to monopoly practices, external economies and diseconomies (which are not internalized in the private market), interventions in the market process through taxes, import duties, and subsidies all result in market (or financial) prices for goods and services, which may diverge substantially from their shadow prices or true economic values. Efficiency shadow prices help to establish the actual economic values of project inputs and outputs (Munasinghe 1979).

To derive a consistent set of economic shadow prices for goods and services a common yardstick or numeraire to measure value is necessary. A most appropriate numeraire in many instances is a unit of uncommitted public income at border (shadow) prices. Essentially, this unit is the same as freely disposable foreign exchange available to the government, but expressed in terms of units of local currency converted at the official exchange rate.

The estimation and use of shadow prices is facilitated by dividing economic resources into tradeable and nontradeable items. Tradeables and nontradeables are treated differently. The values of directly imported or exported goods and services are already known in border prices, that is, their foreign exchange costs converted at the official exchange rate. Locally purchased items whose values are known only in terms of domestic market prices, however, must be converted to border prices, by multiplying the former prices by appropriate conversion factors (CF).

$$\text{Border (Shadow) Price} = \text{Conv. Factor x Domestic (Market) Price}$$

$$\text{or BP} = \text{CF x DP}$$

Given the prevalence of non-efficiency considerations like low-incomes and a large non-monetized sector in rural areas, does it still make sense to begin by analyzing the efficiency solution? The answer is: Yes, it does. First of all, it is useful to know by itself which of the various alternatives will result in the largest increase in total net income to the national economy. Second, only if we do know how much income is obtainable can we make an assessment of the "costs" that various distributional or nonquantifiable objectives may have. The efficiency measure provides us with something of a yardstick that can be used to measure the consequences of the latter even if it cannot tell us what heir real value is. In some cases we might find that these costs are unacceptably high. In others it might turn out that the proposed beneficiaries of a specific distributional objective may voluntarily opt for compensatory payments instead, if they find that such payments could make them better off. Knowing the economic losses associated with the realization of non-efficiency goals would greatly facilitate evaluations of potential trade-offs between the multidimensional objectives that usually form part and parcel of any comprehensive national development program, and associated R.E. master plan.

---

[2] see the companion volume: *Energy Analysis and Policy*, Chapter 9, for further details.

What we can conclude is that despite its shortcomings, an income maximization function serves a useful purpose for the evaluation of alternative development policies and rural electric investment decisions. What must be remembered, however, is that finding the most efficient solution by itself does not answer the question: which of the various policies are preferable in terms of overall community welfare? It only provides us with a tool that can be used to calculate the necessary sacrifices in terms of total income that are needed in order to include other and possibly broader social objectives.

The objective function has been formulated in terms of net benefits, which represent the present-value excess of all social benefits over all social costs. The basic rule for accepting a project is that the net present value (NPV) of benefits is positive:

$$NPV = \sum_{t=0}^{T} (B_t - C_t)/(1 + r)^t$$

where B_t and C_t are the benefits and costs in year t, r is the discount rate, and T is the time horizon. The internal rate of return (IRR) is also used as a project criterion. It may be defined by:

$$\sum_{t=0}^{T} (B_t - C_t)/(1 + IRR)^t = 0$$

Thus, the IRR is the discount rate which reduces the NPV to zero, and the project is generally acceptable if IRR > r (some critical discount rate).

Both benefits and costs are defined as the difference between what would occur with and without the project being implemented. As described earlier, for the economic testing, B, C, and r are defined in economic terms and appropriately shadow priced using efficiency border prices. In particular, the shadow price of r is the accounting rate of interest (ARI), which is equivalent to the opportunity cost of capital (OCC) in efficiency prices. However, for the financial analysis of R.E. projects B, C and r may be defined in financial terms.

6.4 Benefit Measurement

If benefits are to be measured accurately, it is important to identify the types of electricity end-use and patterns of demand over the duration of the R.E. project. Therefore, before describing methods of estimating the benefits of rural electrification schemes, it is helpful to begin by analysing the likely evolution of productive activities and lifestyles in rural region, with and without the project (Munasinghe 1987).

Patterns of Energy Use

Figure 6.1 illustrates, as an example, the likely effects of introducing electricity to a rural area that had no such services previously available. To simplify the exposition, the analysis is presented in comparative static terms. As discussed in the chapter on demand analysis, the effects of introducing electricity are likely to be felt over a number of years, as potential users make the necessary investments in electricity-using appliances over a period of time to take full advantage of the new energy source. The three graphs in the figure represent three distinct types of energy use.

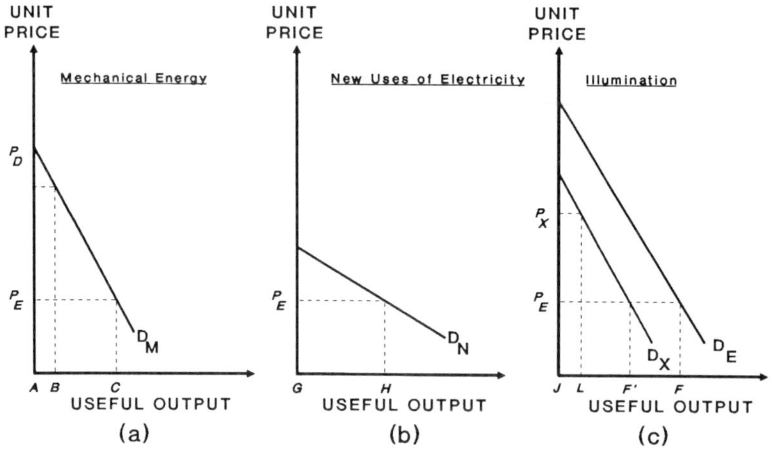

Figure 6.1: Rural Electrification and Energy Use

First, Figure 6.1(a) indicates the demand for an output such as mechanical energy for pumping, where the new electricity supply can substitute for another form of energy that was previously used, with no significant change in the quality of the useful output. The horizontal axis is in units of useful output, e.g. gallons of water pumped, while the vertical axis shows the effective total cost to the customer (including fuel as well as the annuitized and prorated appliance capital costs), per unit of useful output. Since the overall costs of mechanical energy from network-based electricity supply are usually much lower than the costs of operating diesel or gasoline consuming motors, nonelectricity consumers will shift to electricity use, while their demand will also increase from AB to AC, due to the reduction in effective costs. Other examples in this category include kerosene refrigerators, diesel autogenerators of electricity, and so on.

The second category of demand shown in Figure 6.1(b) arises from new uses of electricity that were previously not thought of or considered infeasible because of technical problems or prohibitively high costs. Typical examples include small power tools, television, air conditioning, and so on. The consumption, GH, therefore, represents an entirely new or induced market for electricity.

The third graph, Figure 6.1(c), shows the demand for a service such as illumination or lighting, where electricity will displace another energy source (say kerosene), but with a significant improvement in the final output. The higher quality of output results in the demand curve shifting out from D_X to D_E. If there was no such shift, demand would have increased from JL to JF' because of the lower price of electricity, but with the demand shift, the overall need for lighting would be JF.

Basic Economics of Benefit Measurement

Consider Figure 6.2, which is a static picture of the likely energy use by a typical consumer in the R.E. project area, both with and without the project being implemented. This case is similar to the one shown in Figure 6.1(c), except for the fact that the X-axis here is measured in kilowatt- hours (kWh) or their equivalent. The symbols indicated in the figure are:

Without project condition

D_A = demand curve for an alternative form of energy;
Q_A = quantity of alternative energy consumed, in kWh per month of electricity required to produce the equivalent output;
MC_A = marginal cost of supplying alternative energy; and
p_A = price of alternative energy (subsidised below MC).

With project condition

D_E = demand curve for electricity, shifted outwards due to higher quality of output (e.g., electric lighting);
Q_E = quantity of electricity consumed, in kWh per month;
MC_E = marginal cost of supplying electricity; and
p_E = price of electricity (subsidised below MC).

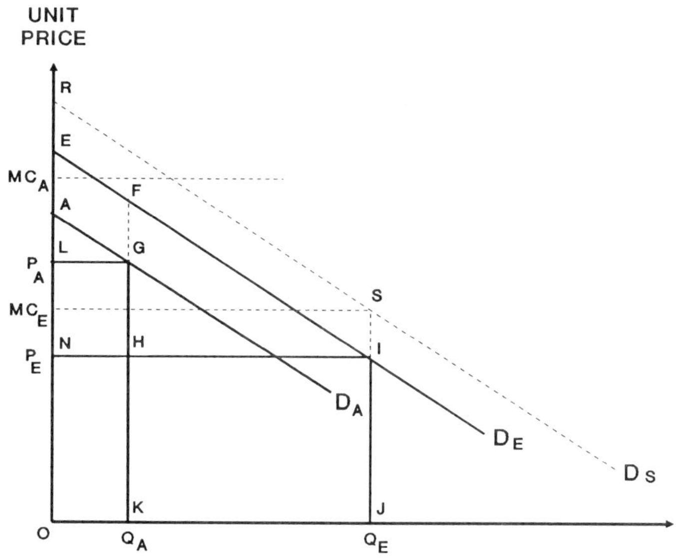

Figure 6.2: Framework for Measuring Net Economic Benefits of Rural Electrification

In conventional microeconomic consumer theory, the benefit derived from consuming a good or service may be measured by the area under the demand curve. Therefore, in the without project situation, [OAGK] is the user benefit of consuming an amount of energy Q_A, where the notation [.....] is used to indicate an area. The corresponding cost of supplying this energy is $MC_A.Q_A$. In the same way, the benefit and cost in the with project condition are [OEIJ] and $MC_E.Q_E$, respectively.

The incremental benefit, defined as the change in benefits brought about by the project, is:

$$IB = [OEIJ] - [OAGK] = [AEFG] + [FHI] + p_E.(Q_E - Q_A)$$

Similarly, the incremental cost is:

$$IC = MC_E.Q_E - MC_A.Q_A$$

Finally, the net benefit due to the R.E. project is:

$$NB = IB - IC$$

$$= \{p_E(Q_E - Q_A) + MC_A.Q_A + [FHI] + [AEFG]\} - \{MC_E.Q_E\}$$

In the above expression, the last term indicates the project costs, and we may write:

$$C = \{MC_E.Q_E\}$$

The remaining part of the expression represents what is usually called the project benefit:

$$B = \{p_E.(Q_E - Q_A) + MC_A.Q_A + [FHI] + [AEFG]\}$$

If we consider the expression for C, the marginal cost element MC_E is the long run marginal cost (LRMC) of supply (see Chapter 3 for details). The expression for B is somewhat more complicated to interpret. The first term in B is the sales revenue corresponding to the additional energy used, while the second is the cost saving due to alternative energy not used. The final two terms are areas representing consumer surplus. In general, these demand curves will shift and consumption levels will change over time. The present value of the stream of (shadow priced) net benefits must be evaluated, year by year, over the lifetime of the project, to yield the NPV described earlier. The other cost-benefit criteria may also be calculated, once the annual stream of net benefits are estimated.

Practical Estimation of Benefits

The project costs C, evaluated in border shadow prices, may be estimated more easily by using the following breakdown:

$$C = (\text{LRMC of supplying R.E. grid} + \text{LRMC of R.E. grid}).Q_E$$

$$= (\text{Total cost of supplying R.E. grid}) + (\text{cost of R.E. grid})$$

The cost of supplying electricity to the R.E. grid, to meet consumption Q_E as well as losses, either from the bulk supply network or from isolated generation, is straightforward to estimate. Furthermore, the cost of the (cheapest) R.E. grid is readily available also, since the least-cost network has already been identified.

Based on the demand forecasts that have been already made, and the corresponding price of rural electricity, the benefit term involving revenues from incremental consumption may be evaluated quite simply. Since the demand projections are usually made with the explicit (or sometimes implicit) assumption that the current price will prevail into the future, in real terms, this is the appropriate price to use when estimating benefits. It is theoretically possible to assume that the future value of p_E will increase or decrease in real terms, but in this case the demand forecast should also be based on the same projected evolution of price. Given the quality of data available, it is rather unlikely that such forecasts of price and consumption can be made with any degree of accuracy. Therefore, the "neutral" assumption of constant real price of electricity is the safest to use. If this R.E. price is well below the level of LRMC (see chapter on pricing policy), then the consumer surplus portion of benefits [FHI] will be relatively large and its estimation becomes more important, as discussed below. On the other hand, in the (rather unlikely) event that price is not very different from LRMC, then the incremental revenues will approximate the full benefits more closely. These revenues are evaluated in domestic market prices, and if the latter are distorted, appropriate shadow prices must be used. One well known approach is to transform market priced values into border priced ones, by multiplying by the appropriate conversion factor; usually the standard conversion factor - SCF, or consumption conversion factor - CCF (Munasinghe 1979).

The second term in the expression for B represents cost savings. Usually, it is reasonably safe to assume that all existing non-electricity using equipment such as kerosene lamps and diesel motors will be replaced, and therefore, equivalent border priced cost savings may be estimated. The rate at which this substitution takes place has to be predicted, and also reflected consistently in the demand forecast. However, there is also the issue of new energy using equipment that might have come into use, in the without project situation. For example, the presence of diesel pump sets suggests that it is profitable for farmers to invest in them, and therefore, more such pumps may be purchased in the future if the R.E. scheme is not implemented. Alternatively, private electricity generation which would be more costly than the R.E. scheme might have emerged. In theory, cost saving benefits may be claimed for all these possibilities, but in practice, such predictions are risky to make. Therefore, the usual approach is to evaluate savings on existing energy using equipment, as a base. Additional cost saving benefits due to substitution of hypothetical new equipment that might have been bought, in the absence of the R.E. project, must be carefully justified on a case-by-case basis.

The third term in the benefit expression, [FHI], is the consumer surplus on incremental consumption of electricity. Additional output due to increased productivity in new activities may be used to approximate some of this willingness-to-pay for rural electricity. For example, the electrification of rural households might result in incremental earnings due to new cottage industries or sewing and tailoring activity. Alternatively, new electric pumps might increase farm yields significantly. The border priced economic value of additional output, net of all input costs including expenditures on electricity, is the appropriate measure of consumer surplus to be used. Similarly, higher net output derived by replacing an existing energy source with electricity, may be used to estimate the consumer surplus area [AEFG]. This benefit is over and above the cost saving benefit arising from the replacement of alternative energy described earlier.

Several typical examples of potentially quantifiable benefits of rural electrification, that may be used to estimate both cost savings and net increases in output, are summarised in Table 6.1. The actual evaluation of such R.E. benefits is described in greater detail in the case study that follows. Some results of quantitative benefit measurement and cost-benefit analysis from recent R.E. schemes are also summarised in Table 6.2. Incremental revenues appear to provide the bulk of the benefits, except in Bangladesh and India, which are both low income countries where the ability-to-pay is a serious problem. The internal economic rates of return (IERR) compare favourably with the opportunity costs of capital (8 to 12 percent) in most countries, except for the case of Egypt where rural electricity prices are heavily subsidised.

Table 6.1

Typical Examples of Rural Electrification Benefits

Quantifiable Benefits: Cost Savings and Increased Productivity

Industrial and commercial uses of electricity

i. Motive power - replacing liquid fuel
ii. Lighting - replacing liquid fuel or gas
iii. Space heating, cooling and refrigeration - replacing liquid fuel, coal, gas, biomass or animal waste
iv. Processing food - replacing liquid fuel, coal, gas, biomass or animal waste
v. Transport - replacing liquid fuel

Household Uses of Electricity

i. Lighting - replacing liquid fuel, gas, biomass, or animal waste
ii. Preparing meals - replacing biomass, animal waste, liquid fuel, coal or gas
iii. Space heating, cooling and refrigeration - replacing biomass, animal waste, liquid fuel, coal or gas
iv. Home appliances (fan, iron, radio, television etc) - replacing batteries, biomass or coal
v. Drinking water - replacing liquid fuel (for pumping)

Agricultural Uses of Electricity

i. Water pumping - replacing liquid fuel, coal, gas or muscle power
ii. Parboiling, heating and drying replacing biomass, coal or liquid fuel
iii. Milling, chaff cutting, threshing, etc. - replacing liquid fuel, hydro and muscle power, coal or biomass

Benefits that Are Difficult or Impossible to Quantify

i. Modernization, dynamism and attitude changes - catalytic effects
ii. Qualify of life, community services, and participation
iii. Income redistribution and social equity
iv. Employment creation
v. Other sociopolitical effects

Table 6.2

Benefit Measurement and Internal Economic Rate of Return in Selected Rural Electrification Projects (1975-84)

Country	Incremental Revenues (% of Total benefits)	Internal Economic Rate of Return(%)
Bangladesh	<5	11
Colombia	55-100	8-13
Egypt	65-85	3-10
India	<10	17-70
Ivory Coast	-	15
Malaysia	100	17
Morocco	100	6
Philippines	50	16-22
Syria	95	8
Thailand	30-60	12-15
Tunisia	60-65	8-12
Yemen (Arab Republic)	100	16-18
Yemen (People's Democratic Republic)	40-85	13-15

Sources: World Bank, Asian Development Bank, Inter-American Development Bank and USAID.

Other Benefits

Promoters of rural electrification often claim that there are significant non-quantifiable gains. These benefits may be conveniently examined in five categories as discussed below, and summarised in Table 6.1. First, it is argued that rural electrification results in widespread modernisation and dynamic growth in rural areas. Most of the productivity gains by households, agriculture and industry, may be quantified as explained earlier. However, if R.E. acts as a catalyst, there may be further unrecognised benefits due to revitalisation and change in attitudes of the entire community.

Second, social benefits could accrue due to overall improvements in the quality of life. For example, electric lights will invariably enhance the ability to study, leading to greater literacy, or the presence of a village center equipped with a television set, radio and other amenities, might help to improve cooperation and the community spirit. Similar gains could occur in other areas affecting social welfare, such as health and sanitation (e.g., safe drinking water), communications, and personal security (e.g., due to better lighting).

Third, rural electrification may be viewed as a effective instrument for improving social equity and income distribution. As described earlier in the section on shadow pricing, social shadow prices may be used to weight benefits and cost to the poor, in order to quantify some of the welfare gains due to income redistribution. However, this is rather difficult in most cases because of several reasons. Data on rural incomes is usually very weak, and therefore it is difficult to identify poverty groups and determine the correct social weights. At the same time, the benefits of R.E. may in fact accrue mainly to the better off sections of the rural community, such as local officials, wealthy landowners, and entrepreneurs. Therefore, targetting the benefits of rural electrification, especially through connections policy and price subsidies (see the chapter on pricing policy), is frequently a much more important practical goal than attempting to quantify distributional benefits, fruitlessly.

Fourth, there may be significant employment impacts following the electrification of rural areas. The direct employment effects due to greater productive uses of electricity will be captured by analysing incremental output, as described earlier. However, intangible benefits include, increased personal satisfaction and family welfare, as well as reduced congestion, crime and social discontent. Another important benefit may arise from reduced rural to urban migration. This may be caused by the perception of greater employment and other opportunities for advancement in hitherto stagnant villages, or because of general improvement in the quality of life.

Fifth and finally, it is sometimes argued that rural electrification may provide a number of other general benefits, viewed from the national viewpoint, such as improving political stability and reducing discontent, improving national cohesion, and reducing urban-rural and inter-regional tensions and inequalities.

The economic basis for claiming additional essentially unrecognised social benefits is analysed in Figure 6.2. Consider the "social" demand curve for electricity D_S, which includes surplus benefits represented by the area [ERSI], due to the factors indicated above. While some of the unquantifiable benefits are included conceptually within the users demand curve D_E, the social surplus benefits are usually not internalised directly within the consumers demand curves, because they are not perceived as benefits by individual users. However, such benefits may be included under the curve D_S which indicates the willingness-to-pay of society as a whole, if they yield identifiable gains to the rural community. Since it is very difficult or almost impossible to quantify these benefits, great care and discipline must be exercised to avoid the temptation of using such claims to justify an R.E. project that otherwise would not have been viable.

6.5 R.E. Project Evaluation in Malaysia

Background

Malaysia is a middle income developing country, with a 1983 income per capita of US$1,860 and a population of 14.9 million. During the decade 1970 to 1980, real gross domestic product (GDP) grew at an average rate of almost 8 percent, while the total energy consumption increased at 8.6 percent. In 1980, total energy use was about 57 million barrels of oil equivalent, of which over 90 percent was derived from petroleum products. The sectoral breakdown of oil use was 43 percent for transport, 29 percent for power generation, 21 percent for industry, and 7 percent for other uses including households.

Malaysia is relatively well endowed with energy resources. Current recoverable oil reserves are estimated to be about 3.5 billion barrels, but if recent domestic consumption growth rates of 8 to 9 percent per annum continue unchecked, the country might become a net oil importer within the next decade. Recoverable non-associated natural gas reserves (mainly off- shore) are greater, and have been estimated at 38 trillion standard cubic feet (scf) or about 6 million barrels of oil equivalent. Planned uses of the gas include: electric power generation, chemical feedstocks, and liquified natural gas (LNG) for export. The usable hydroelectric potential of the country is about 120,000 GWh per year, but less than 15 percent of this lies in peninsular Malaysia, where over 90 percent of the load is located. Over 70 percent of the hydro potential is in Sarawak, and harnessing it would require transmission via undersea cable to the load centres on the peninsula. Recoverable coal resources of 400 to 500 million tons, are quite modest and located mainly in Sarawak. Furthermore, they are relatively low quality lignite with a high moisture content.

Electricity is supplied in peninsular Malaysia by the Lembaga Listrik Negara or National Electricity Board (NEB), an autonomous government power utility. The Sarawak Electricity Supply Corporation (SESCO) and the Sabah Electricity Board (SEB) perform the same functions in the states of Sarawak and Sabah respectively, located on Borneo island. All three organizations fall under the Ministry of Energy, Telecommunications and Posts (METP), which also created an Energy Unit (in 1980) to coordinate overall energy policy and provide infrastructural support to the Cabinet level Energy Committee. Since the rural electrification project described below lies in the peninsula, information concerning Sabah and Sarawak will not be used in the subsequent analysis. References to Malaysia will generally refer to the peninsula, unless otherwise stated.

The production of electricity in peninsular Malaysia, in 1983, is summarized in Table 6.3. About 85 percent of the energy generation and over 75 percent of capacity is oil fired, although the commissioning of over 2500 MW of hydro, gas and coal fired plant will reduce the percentage of oil based capacity to 50 percent by 1990. Further oil substitution is possible depending on a government decision to use gas for power generation in the coming decade.

Table 6.3

Installed Capacity and Electricity Generation in Peninsular Malaysia (1983)

	Installed Capacity (MW)				
Producer	**Total**	**Hydro**	**Steam**	**Gas Turbine**	**Diesel**
NEB	2574	726	1612	100	136
Other	100				
	Energy Generation (GWh)				
NEB	10672	1489	8430	354	399
Other	429	29	26	--	373

There were about 2.4 million households (with an average size of nearly 5 persons) in the peninsula, and the per capita electricity consumption was approximately 770 kWh per year, in 1983. Almost two thirds of the homes are electrified although the degree of electrification varies widely across states, ranging from 44 percent in Pahang to 92 percent in the Federal Territories.

Status of Rural Electrification

R.E. efforts began in the 1950's with the electrification of villages using diesel generators, mainly for security reasons. These early systems were built and operated by NEB's predecessor, the Central Electricity Board. Later, the R.E. systems were sold to private licensees, of whom about 60 still exist. Since the late 1950's each 5 year national development plan has included a progressively larger rural electrification component. In the early 1970's the government felt that the level of rural electrification was too low, and the 1976-80 five year plan had an allocation of M$277 for an accelerated program (US$1 = M$2.3). The Fourth Malaysia Plan (1981-85) provided M$457 for R.E.

Presently, about two thirds of the total peninsular population live in rural areas, defined as those parts of the country lying outside the 73 gazetted urban agglomerations with a population of at least 10,000. About 55 percent of rural households have access to electricity, but the western states and the coastal areas tend to be better served with electricity than the eastern states and inland regions.

There are two principal R.E. programs in Malaysia, with the NEB functioning as the executing agency in both cases. First, the Distribution Department and District Offices of NEB implement an R.E. program (RE-1) funded through the normal expansion of distribution networks. The principal criterion for RE-1 schemes is financial viability. Second, the central government provides earmarked funds to the R.E. Department of NEB, which works in collaboration with the District Managers, to implement a complementary program (RE-2). Criteria for selecting areas to be electrified under RE-2 funding, include: achieving a more equitable distribution of rural electrification by favouring less electrified states, minimizing the distance of the R.E. scheme to the grid, availability of other types of infrastructural facilities, and minimizing costs per consumer connected. About 60 percent of rural electrification funds came from RE-2 and the rest from RE-1, in recent years. The R.E. Department of NEB was set up in 1978, and has helped to overcome a number of constraints to rural electrification, in areas such as: staffing, organization, procurement, way leaves, and so on.

Project Description and Costs

The rural electrification project to be evaluated involves a three year time slice (1983-85) of a five year R.E. program, covering the states of Kedah, Kelantan, Trengganu, and Pahang. The latter program is itself part of the more comprehensive national R.E. program, mentioned at the beginning of this chapter. The project under consideration seeks to electrify about 133,000 new households by end 1986, and raise the average rate of rural electrification in these four states, from about 25 percent to over 50 percent. Details of the number of households to be electrified and the present demographic and socioeconomic profiles of the rural areas are summarized in Tables 6.4 and 6.5.

In 1980, the total population in these states was about 3.3 million, amounting to almost 30 percent of the peninsular total. Almost two thirds of the population live in kampungs or traditional villages, with a mean size of about 577 persons or 133 households. The rate of electrification of homes varied from a low of 16.4 percent in Kedah to a high of 44 percent in Pahang. The main economic activities in rural areas are agriculture (including rice, rubber, and palm oil), forestry, fishing, and cottage industries. Only about 22 percent of all families had a monthly income exceeding M$250 (US$108.7). On average, there were about 5.5 commercial establishments (mainly small shops), and less than one small industrial enterprise per 1000 population.

Table 6.4

Proposed Project (1983-85)

State	Households to be Electrified by End 1986 ($x10^3$)
Kedah	32.2
Kelantan	40.3
Trengganu	23.7
Pahang	36.3
Total	133.0

This background information, and expected future development of the rural areas, may be used to analyze and forecast future demand for electricity. The results are summarized in Table 6.6.

Next, appropriate least cost rural distribution networks are designed to meet the expected load growth. The physical details of the project and the corresponding costs are given in Table 6.7. All costs are in 1982 prices, and the foreign component is net of taxes and duties. A conversion factor for local costs (LCF = 0.85) is used to transform the domestic priced local costs into border prices.

In addition to the electrification component, a number of broader issues were identified, to be addressed during the course of the project. These issues included:

1. Encouraging more productive uses of electricity.
2. Improving the engineering standards and design criteria for greater cost effectiveness, while maintaining an acceptable quality of supply.
3. Strengthening R.E. planning.
4. Implementing an information gathering and feedback system to improve future projects.

Estimation of Benefits

The three principal groups of rural electricity consumers for whom benefits need to be measured, are households, and commercial and industrial users. At this stage, it is convenient to rewrite the equation that defines R.E. project benefits, as follows:

$$B = p_E.Q_E + (MC_A - p_E).Q_A + [FHI] + [AEFG]$$

Table 6.5

Population, Electrification, Income and Occupation Data of Kampungs (Traditional Villages) in Trengganu, Kahang, Kedah and Kelantan

	Trengganu	Pahang	Kedah	Kelantan	Total
Total population of states ($x10^3$)	542.3	770.6	1102.2	877.6	73292.7
Population, Households and Electrification in Kampungs					
Number of kampungs	1973	538	794	386	3691
Population ($x10^3$)	313.2	492.3	7803	545.0	2130.8
Families	64128	100855	260679	128836	554458
Households	58578	88559	235978	109561	492176
Electrified households	23963	39042	38733	22648	124386
(%)	(40.9)	(44.1)	(16.4)	(20.7)	(25.2)
Monthly Income Distribution in Kampungs					
Less than $250 (%)	82.4	58.8	82.5	84.9	77.7
$250-400(%)	13.6	26.7	12.7	11.0	15.6
Above $400 (%)	4.0	14.5	4.8	4.1	6.7
Occupation Profile in Kampungs					
Shops	1512	24.33	4607	3054	11606
Rice mills	124	83	487	489	1183
Oil mills	6	1	2	2	11
Fish, ice and coffee processing	102	17	17	123	259
Saw mills	43	101	63	50	257
Iron works	12	40	41	34	127
Dressmaking, weaving, batik, handicraft	649	116	121	399	1285

Table 6.6

Load Forecast

Year	Households[a] Connected	Sales to Consumers (GWh) Domestic	Commercial and Industrial	Total	Bulk Supply[b] Requirement (GWh)
1983	31195	18.0	84.5	102.5	108.7
1984	75946	43.0	183.5	226.5	239.6
1985	114537	65.0	268.0	336.6	358.6
1986	128674	74.7	281.3	356.0	377.4
1987	135761	81.1	295.5	277.0	399.9
1988	138476	85.9	310.2	396.1	419.9
1989	141245	91.1	325.8	416.9	441.9
1990	144070	96.6	342.0	438.6	464.9
1991	146951	102.4	359.1	461.5	489.2
1992	149890	108.5	377.1	485.6	514.7
1993	152888	115.0	395.9	510.9	541.6
1994	155946	121.9	415.4	537.6	469.9
1995	159065	129.2	436.5	565.7	599.6
1986-2010	159065	129.2	436.5	565.7	599.6

[a] Growth of connections assumed to be 2% from 1987 to 1996.
[b] Bulk supply (at 33 kV) = total sales x 1.06 (based on approximately 6% losses in RE networks).

Table 6.7

Project Components and Costs
(M$ x 10^6)[a]

Item	Foreign Costs Border Prices	Local Costs Domestic Prices	Local Costs Border[b] Prices	Total Costs Border Prices
33 KV OH Lines	27.17	3.94	3.35	30.52
33 KV UG Cables	4.15	2.78	2.36	6.51
33 11 kV Substation	11.13	8.24	7.00	18.13
11 kV Lines	53.00	37.07	31.51	84.51
33 and 11 kV 400 V Substations	25.66	21.27	18.08	43.74
LV Lines	29.02	87.76	74.60	103.62
Service Connection and Meters	22.61	51.56	43.83	86.43
Land and Way Leave	--	15.85	13.47	13.47
Consulting Services	--	3.79	3.22	3.22
Total Costs	172.74	232.26	197.42	390.16

[a] 1982 prices
[b] Border price = domestic price x local costs conversion factor (LCF = 0.85).

Household Consumers

Starting with the domestic consumers and focussing attention on the first term in the above equation, it is relatively straightforward to estimate the incremental revenues, from forecast electricity sales in Table 6.6 and the present average household price of M$0.20 per kWh. Next, the cost saving benefits represented by the second term in the above equation is evaluated, by calculating the foregone expenditures on alternative kerosene lighting. Details of this computation are summarized in Table 6.8.

Table 6.8

Annual Savings Due to Electric Lighting (Border Prices)[a]

Electricity item	Cost(M$)	Kerosene item	Cost(M$)
Wiring costs (M$200 annuitized over 10 years at 10%)	35.4	One Petromax lamp (M$45.0 annuitized over 10 years at 10%)	8.0
M$10 per bulb and a life of 1000 hours	1.4	Fuel cost for 96 kWh equivalent of kerosene at M$ 1.45 per kWh	139.2
Cost of 96 kWh at 17.6 cents per kWh[a]	16.9		
Total cost	53.7	Total cost	147.2
Annual cost saving = M$ 93.5 per household			

[a] Border price = domestic price x SCF = 20 x 0.88 Mcents/kWh.

Further household related benefits may be identified, but are difficult to quantify accurately. For example, considerable surplus benefits appear to be derived from appliances like fans, rice cookers, electric irons, and TV sets, based on the willingness-to-pay for these items. Such benefits may stem from greater convenience and efficiency (e.g., electric versus charcoal iron), as well as better quality and greater hours of use (e.g., electric versus kerosene lighting). Survey data also indicated more general benefits perceived by potential electricity consumers, including improved opportunities for children to study, greater village and household security, and better socioeconomic conditions and opportunities for modernization.

Household activities are also made more productive, due to improved working conditions brought about by better lighting and cooling (with fans) in tropical conditions, as well as better electrical equipment and tools. At the prevailing wage rate, one extra man hour per day would be worth about M$20 per month to a family.

Incomes in the target rural areas are significantly higher than in most Asian and African developing countries. The willingness-to-pay for rural electricity services offered by private operators is quite considerable (Mcents 40 per kWh). Another indicator of the extent of electricity consumption benefits is the use of home autogenerators in the 1-3 kW range, producing electricity at a cost of about M$1 per kWh.

Commercial and Industrial Consumers

As described earlier, most of the business enterprises in the project areas consist of small shops, and a few industries such as rice, oil and saw mills, iron works, and processing plants. A survey of these types of consumers indicated that the benefits derived from increased output made possible by rural electricity supply (e.g., refrigeration), or through the replacement of existing alternative energy sources (e.g., isolated generators), range from M$0.445 to over 0.6 per kWh used.

Economic Cost-Benefit Analysis

The economic cost-benefit calculation, based on the above estimates of costs and benefits, is summarized in Table 6.9. The values in this table are derived as follows. The 1982 border priced investments (from Table 6.6) are divided by an inflation factor of 1.09, to derive the 1981 border priced investments, and then allocated over the period 1983 to 1985. Investments from 1986 to 1995 are 2 percent of the 1983-85 cumulative investment, to meet continued load growth. Annual operation and maintenance costs are estimated as 2 percent of the 1983-85 cumulative investment. The bulk supply costs are equal to the bulk energy supplied (from Table 6.6) multiplied by the LRMC of bulk supply in 1981 border prices. The latter is M$0.192 per kWh, based on the LRMC of bulk supply in 1984 domestic prices (M$0.262 per kWh) multiplied by the standard conversion factor (SCF = 0.88), and divided by an inflation factor of 1.2.

Household incremental revenues in border prices are the product of sales (from Table 6), multiplied by the 1981 average price (M$0.20 per kWh), and the SCF. Total kerosene savings may be computed as the savings per household (from Table 8), multiplied by the number of households connected (from Table 6). The other domestic consumption benefits discussed earlier are not included in the numerical computation, because they cannot be quantified with sufficient accuracy. The chief commercial and industrial electricity consumption benefits are based on the incremental consumption (from Table 6), multiplied by the minimum value from survey data (M$0.445 per kWh), and the SCF.

The internal economic rate of return (IERR) for these streams of costs and benefits is about 20 percent, which compares very favourably with the opportunity cost of capital in Malaysia, of 12 percent. A sensitivity analysis shows that either raising all costs by 10 percent or reducing all benefits by 10 percent, reduces the IERR to about 15 percent. If all costs are increased by 10 percent, while all benefits are also decreased by 10 percent, then the IERR falls to 11 percent. Therefore, the project is acceptable from the viewpoint of economic efficiency, especially since all the benefits were quantified rather conservatively.

Table 6.9

Economic Cost-Benefit Calculation
(1981 Border Prices)

Year	Costs (M$ x 10^6)				Benefits (M$ x 10^6)			
	Investment	Operation and Maintenance[b]	Bulk Supply[c]	Total	Domestic Revenue[d]	Kerosene Savings[e]	Commercial and Industrial	Total
1983	136.22	2.72	20.86	159.80	3.17	2.91	33.12	39.20
1984	115.92	5.04	46.10	167.06	7.57	7.09	71.93	86.59
1985	87.47	6.79	67.89	162.15	11.55	10.69	105.06	127.29
1986	6.80	6.93	72.45	86.18	13.15	12.01	110.27	135.43
1987	6.80	6.93	76.79	90.52	14.27	12.67	115.84	142.78
1988	6.80	6.93	80.61	94.32	15.12	12.93	121.60	149.65
1989	6.80	6.93	84.85	98.58	16.03	13.19	127.71	156.94
1990	6.80	6.93	89.26	102.99	17.00	13.45	134.06	164.52
1991	6.80	6.93	93.92	107.65	18.02	13.72	140.77	172.51
1992	6.80	6.93	98.83	112.56	19.10	13.99	147.82	180.91
1993	6.80	6.93	103.98	117.71	20.24	14.27	155.19	189.70
1994	6.80	6.93	109.41	123.14	21.45	14.56	162.84	198.85
1995	6.80	6.93	115.13	128.86	22.74	14.85	171.11	208.70
1996-2010	0.0	6.93	115.13	128.86	22.74	14.85	171.11	208.70

Internal economic rate of return (IERR) = 20.3%

[a] 1981 border price = 1982 border price (Table 6.6)/1.09 inflation. Investments from 1986-95 are 2% of the 1983-85 total, to meet load growth.
[b] 2% of cumulative investment.
[c] Bulk supply cost = energy supplied (Table 6.6) x 0.192 (LRMC of bulk supply in 1981 border prices).
[d] Domestic revenue (border price) = sales (Table 6.6) x average price x SCF average price = M$ 0.2/kWh; and SCF = 0.88.
[e] Total kerosene savings = annual savings per household (Table 6.8) x households connected (Table 6.6).
[f] Commercial and industrial surplus benefits = sales (Table 6.6) x benefits per kWh (border price). Border priced benefits per kWh = 0.392.

References

Baum, W., "The Project Cycle", *Finance and Development*, vol.15, Dec. 1978.

Munasinghe, M., *The Economics of Power System Reliability and Planning*, Johns Hopkins Univ. Press, Baltimore MD, 1979.

Munasinghe, M., "Integrated National Energy Planning (INEP) in Developing Countries", *Natural Resources Forum*, vol. 4, 1980, pp. 359-73; also available as reprint no. 165, The World Bank, Washington DC.

Munasinghe, M., *Rural Electrification for Development*, Westview Press, Boulder CO, 1987.

Munasinghe, M., and C. J. Warren, "Rural Electrification, Energy Economics and National Policy in Developing Countries", in *Future Energy Concepts*, Institution of Electrical Engineers, London, UK, 1979.

Ray, A., *Cost-Benefit Analysis*, Johns Hopkins Univ. Press, Baltimore, MD, 1984.

Smith, D.V., D.B. Mehta, and P. Hayes, "Regional Rural Electrification Survey", draft report, Asian Development Bank, Manila, Philippines, Oct. 1983.

The World Bank, *Rural Electrification*, Washington DC, 1975.

Chapter 7

RELIABILITY OF ELECTRICITY SUPPLY, OUTAGE COSTS AND VALUE OF SERVICE: AN OVERVIEW[1]

7.1 Introduction

Energy is a *sine-qua-non* in the modern world. Electric power in particular, is considered the key to industrialization, and an engine for growth and development. The energy intensive economic boom during the post second world war period was sustained by abundant and relatively cheap energy, especially oil. However, several developments over the last two decades have shifted the emphasis from the energy supply side, to demand related issues (Munasinghe and Warford 1982, Sanghvi 1983b). Thus, in the electricity sector, significant strides have been made in the analysis and understanding of pricing policy and demand management, conservation, outage costs and the value of service (Munasinghe and Schramm 1983, Sanghvi and Levy 1986, Andersson and Taylor 1986).

The increases in real energy costs of the 1970's provided great impetus to energy demand management and conservation efforts, and the development of effective policy tools, particularly pricing. At the same time, the reliability or quality of electricity supply had also begun to emerge as a key issue, because of the greater dependence of modern economies on power, the tighter resource constraints and reliability margins, and the general climate of greater uncertainty under which power systems had to make investment and operating decisions (sometimes resulting in severe power shortages or conversely, significant overcapacity).

The main objectives of this paper are to provide an overview of key issues concerning the definition, measurement and application of outage costs, and selectively review the related literature. Before discussing reliability, outage costs and the value of service in greater detail, it is useful to outline the broader context in which these aspects must be considered. The lessons of the past few decades suggest that because power and energy are so pervasive throughout a modern economy, it is advantageous to use a broad and integrated framework for analysis, which recognizes crucial energy-macroeconomic links, the interactions between electricity and other forms of energy, and the close relationship between power supply and demand. Such an integrated approach to energy policy analysis is described in Chapter 2.

Within the power sector itself, the supply-demand inter-relationship has been recognized theoretically in the form of the two joint optimality conditions that ensure the economically efficient production and consumption of electricity (Crew and Kleindorfer 1978, Munasinghe 1979). Correspondingly, two key areas for practical decisionmaking in power utilities, concern policies affecting investment-supply and pricing-demand management. From the economic viewpoint, efficient supply is ensured by selecting an investment program and operating the power system in a

[1] Edited version of a paper published in *The Energy Journal*, Special Electricity Reliability Issue, Vol.9, 1988, pp.1-18 (co-author: Arun Sanghvi).

manner that minimizes costs. Efficient electricity consumption patterns are realised by setting the price equal to the marginal cost of supply. The interactions arise because of the cyclical nature of the price - demand forecast - least cost system plan - marginal cost of supply - price relationship. Reliability (as well as outage costs and value of service) are also intimately involved in this process.

Traditionally, power systems have been designed and operated to meet forecast demand, at some target level of reliability that was itself based on historical rules-of-thumb. The more modern approach recognizes that reliability affects both the system costs as well as the cost suffered by consumers due to poor supply quality, and therefore should be subject to a more thorough economic analysis. The reliability optimizing model presented in Chapter 2, shows that it is possible to optimize system investment and operations at the point where marginal benefits arising from improved supply quality are offset by the corresponding costs. Earlier work had demonstrated that for a given price structure, an optimal long-run power system expansion plan and range of reliability levels may be practically determined which maximize the net social benefits of electricity consumption -- thus reflecting an economywide viewpoint (Munasinghe 1979). This planning rule effectively subsumes the traditional criterion of minimizing only the system cost, which is more consistent with the financial viewpoint of a private utility.

In the next section, we provide some basic definitions of outage costs and related concepts such as shortage costs and value of service, and discuss their potential applications in a range of power utility activities. Section 3 reviews the theoretical basis for measuring the costs and benefits of improved reliability and quality of supply, and describes the results of attempts made to actually measure outage costs. Section 4 describes how outage costs have been used in practical utility work and regulation. The final section summarizes our present understanding of the subject, and sets out the principal areas for further analysis.

7.2 Role of Outage Costs, Shortage Costs and Value of Service in Supply Planning and Demand Management

Introduction to Basic Concepts and Definitions

Outage costs represent the economic consequences of service curtailments to the customer, when the demand for electricity temporarily exceeds the available supply capability. These power shortages manifest themselves at the consumer level in several ways, including complete supply interruptions (blackouts), frequency and voltage reductions (brownouts), and sudden sharp fluctuations or surges in frequency and voltage. While all these phenomena will impose costs on consumers, the effects of sudden supply interruptions invariably result in the highest outage costs. Because both power supply and demand tend to vary stochastically, having sufficient redundancy and excess capacity throughout the power system to meet unexpected contingencies is the major safeguard against electricity shortages.

Outage costs are normally estimated by the reduction in customer willingness-to-pay to avoid interruptions in the short- run, assuming that the user does not have sufficient time to change his energy using capital stock. It is important to recognize that outage costs will vary substantially between customers within a class, and between classes of customers. This variation will depend on the degree to which customer applications are dependent upon electricity and the extent to which any production or output can be recovered subsequent to an outage. Two types of outage costs, direct and indirect, are possible, and Table 7.1 summarizes the main characteristics of these two components with respect to the different consumer categories.

Table 7.1

Applications of Outage Costs

	Function	Potential Applications
1.	Generation Planning Bulk Power System	Economic justification of - reliability planning criteria - network protection, communication, and control hardware Optimize reserve margin levels by examining tradeoffs between capital investment and customer outage costs Enhanced resource planning under uncertainty Siting criteria for locating generation plant and transmission lines to address geographical differential in customer outage costs Investment allocation criteria for - the level of investment and allocation by resource type - justification for utility versus customer ownership
2.	Sub-transmission and Distribution	Economic justification for determining - network and line reinforcements - redundancy and performance design requirements for individual transformer and substation locations Budgeting and allocation criteria to: - rank order competing investments based on change in total system versus area outage costs - prioritize substation and feeder maintenance
3.	Operations	Develop shortage management strategies - optimum curtailment levels, priorities, and rate guidelines - service restoration sequence Maintenance scheduling to levelize outage costs instead of reserve margin or LOLP criteria Fuel inventory management and stockpile requirements
4.	Rate Design	Development of customer-based marginal values to replace supply oriented marginal costs for designing: - curtailable and interruptible tariffs - qualified producer/facility rates - real time pricing
5.	Strategic Planning	Cost benefit evaluation criteria for all demand-side management program Market segmentation and opportunity analysis criteria to identify: - need and target for pricing and reliability options - opportunities for additional electrification - customers at risk of bypass

Shortage costs reflect any changes in the customer value of service (willingness-to-pay) following a change in service quality relative to the reliability level they presently receive. Shortage costs are also referred to as long-run outage costs, and include the costs that the customer incurs to change energy using capital stock and other processes, in response to a change in supply quality. For example, if service reliability were to be lowered significantly or if requirements for reliability were to increase, then some customers are likely to find it cost effective to take one or more mitigating actions to cope with future service curtailments. Typically, these adaptive responses (i.e., long-term mitigation measures) include installation of protective switchgear, self generation to backup critical loads, carrying larger in-process inventories for selected items, fuel switching, and relocating to service areas with higher reliability.

Once such adaptation occurs, a service interruption will not result in loss of valued critical loads. Of course, the mitigation measure itself has a certain capital and operating cost (adaptive response cost). Shortage costs reflect both adaptive response costs plus expected outage costs after such adaptation has taken place. Thus, (long run) shortage costs are lower than (short run) outage costs. Otherwise the customer would not find mitigation measures worthwhile to adopt.

To illustrate the linkage between outage cost, shortage cost, and value-of-service, consider the decision facing a customer who must select between two options: so-called "firm" service, and service that offers inferior reliability. If an option offers inferior service quality (e.g., more frequent curtailments), then the willingness-to-pay is reduced as well, and the demand curve shifts downwards. This reduction reflects two main components: (1) a shortage cost, consisting of the cost of any adaptation by the user (e.g., as measured by incremental changes in user capital stock), as well as the change in expected outage costs in the future, conditioned upon such adaptation; and (2) a risk premium component that reflects, among other things, user tolerance for risk as it pertains to the reliability of electric service. In particular, as reliability deteriorates, it can trigger higher anxiety, stress, and other adverse customer response. Additionally, even if a user installs emergency backup capability, it may prove inconvenient to operate and maintain.

The shortage costs plus any risk premium, together define the value of service at a given level of service reliability. For example, if service reliability is increased, then we would expect the value of service to increase by an amount equal to the corresponding change in shortage costs plus the risk-premium (Sanghvi 1983b, EPRI 1986).

Potential Uses of Outage Costs

There are a number of uses of outage costs that have emerged from utility studies. Table 7.2 identifies the range of potential applications.

First, the outage cost concept may be used for setting economic reliability standards in generation, transmission and distribution planning and power system optimization. The engineering-economic marginality condition for optimizing reliability is further explained in Chapter 2.

Second, in the areas of power system operations, disaggregate outage cost information for different customer categories provides an economic basis for developing load shedding priorities during times of capacity shortfalls, and for rationing shortages during energy deficiencies.

Table 7.2

Direct and Indirect Outage Costs

Primary Electricity User	Direct Cost Components (Costs to Household, Firm, Institution, etc.)	Indirect Costs	Remarks
1. Residential	a. Inconvenience, Lost Leisure, Stress, etc. b. Out-of-Pocket Costs • spoilage • property damage c. Health and Safety Effects	Costs on other households and firms	Indirect costs are a minimal if not negligible fraction of total (direct & indirect) costs of a curtailment
2. Industrial, Commercial, & Agricultural Firms	a. Opportunity costs of idle resources for most • labour • land • capital • profits	Cost Component include: a. Cost on other firms that are supplied by impacted firm (multiplier effect) b. Costs of consumers if impacted firm supplies a final good c. Costs (Benefits) to some producers d. Health and safety related externalities	Indirect effects are likely to be minimal capacity related interruptions, but can be a significant component of total costs for longer duration energy short-falls.
3. Infrastructure and Public Services	a. Opportunity cost of idle resources b. Spoilage and damage	a. Costs to public users of impacted services and institutions b. Health and safety effects c. Potential for social costs stemming from looting, vandalism	Indirect costs constitute major portion of total costs of a curtailment.

Third, outage cost data are useful for determining pricing policy. Although the number of tariff options offered to power consumers has increased during the past few years, most customers are still provided with uniform service of the same "high" reliability and quality. However, a single service standard for all ignores the considerable distribution of customer needs for quality, reliability and willingness-to-pay. Under a single service standard, all customers pay for the current level of reliability and quality whether they need it or not. Those requiring less than the system standard will pay less and receive a lower service quality than their uses of electricity might otherwise justify. Application of the single service standard over the long term results in an inefficient allocation of investments (Woo and Toyama 1986, Chao and Wilson 1987).

The imbalance between customer needs, service levels, and price, would present less of a problem if customers of all classes (e.g., residential, commercial, industrial, and agricultural) and sizes were equally distributed over a narrow range of different reliability and quality needs. Potential service and the consequent rate differentials might be insignificant and not worth consideration. However, almost all past and current outage costs studies -- e.g., recent studies in USA (Sanghvi 1983a, Sanghvi 1986), Canada (Ontario Hydro 1980), Brazil (Munasinghe 1979), Costa Rica (Munasinghe 1984), Finland (Finnish Power PCC 1979) and India (Munasinghe 1987) -- tend to consistently show that: (1) the distribution of needs is not randomly distributed but varies with the type of customer, their usage level, and business, and (2) customer needs for quality and reliability are distributed over a wide rather than narrow set of requirements.

Unbundling of service is defined as the ability of utilities to offer customers a choice of options that explicitly trade off service reliability, quality, and price. Although this has not been technically or economically feasible on a large scale until recently, advances in microelectronics, smart metering, and communication and control technologies, now make unbundling increasingly possible. Those selecting less reliable service (characterized by more frequent or longer outages) would pay a lower tariff than those receiving higher quality service. Unbundling service in this fashion may be thought of as another form of product differentiation, or logical extension of the approach on which currently available interruptible load type contracts are based.

Properly designed service options have the potential for enhancing the economic efficiency of the entire system, producing benefits for both the utility and its customers. With unbundled service, reliability options and investments will be targeted and charged to those customers willing-to-pay for their desired levels of service reliability. As a consequence, intra-class subsidies will be reduced and rates will better match the economic cost of supply. Service unbundling also offers the utility a potentially effective strategic marketing tool for responding to the increasing competition in the power markets. Customer preferences for different service options as reflected in outage cost and value-of-service data, provide a basis for meeting these pressures by appropriate market segmentation and product differentiation (Woo and Toyama 1986, Chao and Wilson 1987).

7.3 Outage Cost Estimation

Analytical Approaches to Outage Cost Measurement

While early work on the cost-benefit approach for generation reliability planning employing outage costs was done by the Europeans, some of the more recent theoretical and empirical efforts to measure outage costs and develop value-based planning applications have been made in North America and certain developing nations. In the remainder of this section, we briefly discuss the more important theoretical approaches available for outage cost measurement and how these techniques can be related to the methods used in some of the papers in this special issue. The next section reviews major efforts to estimate and use outage costs and value-of-service.

The theoretical basis for the measurement of outage cost is the loss of consumer welfare as a result of the outage. Stated in this form, however, the theory is too general to be of any practical relevance and hence, several approaches have emerged in the literature which draw upon this fundamental concept. One approach is to estimate outage costs on the basis of observed or estimated willingness-to-pay for planned electricity consumption (Brown and Johnson 1969, Crew and Kleindorfer 1978, Sherman and Visscher 1978). While this method is intuitively appealing, it encounters significant problems in practical application.

First, willingness-to-pay for planned electricity consumption is not an accurate measure of the cost a consumer incurs in an unplanned shortage, since planned and unplanned demand responses correspond to the long and short run demand curves, respectively. If short run demand is very much less elastic than long run demand, then the measurement of willingness-to-pay for planned consumption will considerably underestimate the true outage cost. This would not be a serious problem if the short run demand curve is easily estimated. However, in this case the short run consists of at most a few hours, with the slope of the demand curve varying with the period of the outage, thus making such an estimation practically very difficult. Another criticism of this approach is the assumption that load shedding takes place according to a priority ranking of willingness-to-pay. This may not be tenable in practice.

Another approach to measuring outage costs relies on the principle that electricity is an intermediate good in the production of goods and services so that outage would result in loss of production (see Chapter 8)). Power outages cause losses both in terms of spoilage of raw materials, work-in-progress and finished goods as well as the opportunity cost of idle factors of production. Spoilage can be a very significant problem in chemical processes, steel manufacture, food products etc. Clearly, the duration of the outage would be critical here, as would its timing. Apart from spoilage, production delays and quality deterioration would be other consequences of power outages. This method can be used analogously to measure the outage costs for households. In this case one possible criterion is the opportunity cost of time or leisure foregone (see Chapter 9). The household welfare loss due to power outages has been characterized as a disruption of the household's preferred consumption pattern (Sanghvi 1983). While some work such as doing laundry can be deferred at a very low cost, the interruption of other activities like watching a favourite TV programme, would have a much higher cost attached. Whereas the willingness-to-pay method can only be used to measure the outage costs of aggregate classes of consumers (since it is impossible to estimate individual demand curves), the production cost method can in principle be applied to individual outage cost measurement.

In a situation where the customer suffers a very "high" outage cost, he would find it to his advantage to arrange back-up sources of power to avoid losses from power outages. This intuitive reasoning is confirmed by observing that consumers who incur high outage costs, such as hospitals, hotels and some industries, typically have back-up sources of power. The cost of maintaining this back-up source of power has been proposed as another indicator of outage costs (Munasinghe 1979, Bental and Ravid 1982, Sanghvi 1982). An extension of this approach is seen in recent literature, which has considered the possibility of unbundling the quality attributes of electric power supply, thus allowing the customer to select between different levels of quality (Viswanathan 1985, Chao and Wilson 1989).

Associated with the concept of reliability differentiation, the recent literature also considers other novel service options. One of these is priority insurance where, instead of offering a choice of reliability levels, the utility offers fixed compensation for interruptions but makes no guarantees as to the frequency of the outages (Oren, Smith, Wilson and Chao 1986). This transfers a greater degree of uncertainty to the consumer and provides for the possibility that this risk be assumed by a private insurer. Thus we have a fourth method of determining outage costs since the insurance premium, if actuarially fair, corresponds to the risk preference of a risk averse consumer. This method is appealing from a practical standpoint since it can be easily adapted for measurement using questionnaire or survey type methods where actual data on interruption insurance is not yet available. There is already a fair amount of experience in the use of survey techniques to estimate outage costs for all types of consumers including households (Munasinghe 1979, Doane, Hartman and Woo 1989, Goett, McFadden and Woo 1989, Sanghvi 1986).

Practical Measurement of Outage Costs

Many studies on the measurement of outage costs proceed on the simple but frequently unjustified assumption that output is proportional to the consumption of electricity. In such cases, outage costs can be determined in terms of lost output if the quantity of electricity not supplied due to outage is known (Shipley, Patton and Denison 1972, Kaufman 1975). The recent literature makes it clear that the accurate determination of outage costs requires more sophisticated and disaggregate analysis. The earlier studies give only order-of-magnitude estimates. Two upper bounds to outage costs have been proposed by Telson (1975) along these lines. The larger of these is defined as the ratio of the gross product of a geographical region to the electricity consumed by commercial and industrial consumers in that region. The smaller estimate is the corresponding ratio of wages and salaries to electricity consumption by commercial and industrial consumers. The study by Bernstein and Hegazy (1989), though macro-based, is more sophisticated. They propose an input-output formulation to measure outage costs and apply the technique to Egypt. Previous studies using the input-output approach include an application to Chile (Jaromillo and Skoknic 1973) and a regionalized study by Priestman (1977).

Andersson and Taylor (1986) describe the determination of energy shortage costs in Swedish industry and commerce using a linear programming approach. Available energy is allocated optimally among the various sectors (subject to a set of restrictions governing the operation of the Swedish power industry) and the increase in value-added that would result from availability of an additional kWh is determined. They also describe two surveys done in Sweden in 1969 (Swedish Joint Committee for the Electricity Council 1969) and 1979 (Swedish Association for Electricity Supply Undertakings 1981) and compare the changes in outage costs over the two periods.

The increases in outage costs over the decade, particularly for industry, offices and commerce, is attributed to an increase in dependence on electricity. However, no explanation is offered for the significant decrease in outage costs for households and agriculture which seems paradoxical particularly in view of their observation that the use of home electrical equipment has grown significantly.

Andersson and Taylor also draw attention to an important criticism of survey based studies, i.e., that many of them are based on hypothetical situations. They claim that this leads to an exaggeration of outage costs. Woo and Train (1989) find that, while customers who had experienced outages were more likely to report positive outage costs, such reported costs were in fact low. This lends support to the observation by Andersson and Taylor. Furthermore, some of the empirical evidence cited in Sanghvi (1982 and 1983a) reveals that customers surveyed shortly after an actual outage generally tended to underestimate outage costs. Keane, McDonald and Woo (1989) consider it a major advantage to have been able to use a data set based on customers with significant outage experience.

Sanghvi (1983a) uses the Wisconsin Residential Time-of-Use Electricity Pricing Experiment data base to estimate short-run income-compensated (Hicksian) Linear Expenditure System (LES) demand functions for fixed time periods during various times of the day. These estimates are then used to calculate the losses in consumer surplus during outages of the same duration and time of day. This gives the welfare loss component of short run residential outage costs. The method of estimating domestic outage costs on the basis of survey data and the value of consumers' time, has been applied to households in Brazil (Munasinghe 1979), Costa Rica (Munasinghe 1984) and India (Munasinghe 1987).

Survey and market research based methods predominate in the studies reported in this volume. Doane, Hartman and Woo (1989) attribute the popularity of this approach to the fact that the other common approaches (measurement of market behaviour, use of proxies) are limited by data availability and the inability to obtain cost estimates by time-of-day and season. In Doane et al., households are asked to provide direct cost estimates for a set of outage scenarios distinguished by season, time-of-day and presence/absence of advance notice. They make the important correction for the fact that a large number of customers report a zero outage cost for particular outages (limited dependent variable problem). While they find that outage costs vary significantly by household type, and also by season, time-of-day and duration, the presence or absence of advance notice is found to be insignificant.

A similar result (i.e., insignificance of advance notice) is reported in the study by Goett, McFadden and Woo (1989), who use a probabilistic choice model (multinomial logit) to analyze the selection of hypothetical interruptible service options made by a group of residential electricity customers. However the outage cost estimates reported by Goett et al. far exceed those found in Doane et al. In a subsequent study, Doane, Hartman and Woo (1989) attempt to reconcile this difference in cost estimates by applying, in turn, their previous regression methods and the multinomial logit techniques of Goett et al., to the same group of customers. They conclude that the two estimates are comparable only if an adjustment is made for 'status quo' bias (ie.the premium attached by customers to their existing level of service) present in the choice-modeling approach. In other words, customers attach a much higher cost to a reduction in their existing level of reliability (i.e., willingness to accept (WTA) a decrease in reliability) relative to what they would be willing to pay for an increase in reliability from their existing level (Cummings, Brookshire and Schulze 1986). However, if reliability were actually decreased, the customers would gradually adjust to the new status quo. Thus, failure to correct for status quo bias may seriously overestimate consumer's willingness to pay for alternative reliability levels.

7.4 Application of Outage Costs: Utility and Regulatory Practice

Electric Utility Applications

In the areas of generation reliability planning, several utilities worldwide use the cost-benefit approach for deriving economic reliability standards. Sweden, Norway, France and Italy have pioneered the use of outage costs in optimizing system reliability. More recently, Ontario Hydro and Saskatchewan Power Corporation in Canada, Eletrobras in Brazil, CEB in Sri Lanka and ISA in Columbia (Interconexion Electrica S.A. 1986) have also begun to rely on economic reliability standards for their system planning. New studies are being undertaken by several other utility systems as well. One of the earliest attempts to use outage costs in optimizing the planning of a distribution network was the case of Cascavel, Brazil (Munasinghe 1979). Similar applications to optimize rural electrical systems in Costa Rica and India are reported respectively in Munasinghe (1984), and Munasinghe (1987).

By comparison, consideration of value of service data in reliability planning by U.S. utilities is more limited (Sanghvi 1985b). A notable exception is the case of the Pacific Gas and Electric Company (PG&E) which is evaluating a generation reliability planning criterion based on customer value of service. This reliability criterion, called the Expected Duration of Emergency Actions (EDEA), supplements PG&E's conventional generation planning criteria. Under the EDEA standard, generation capacity for reliability support would be added only up to the point at which the cost of the final increment of capacity is offset by the expected benefits from its use, i.e., the expected reduction in outage cost attributable to that increment of capacity. Resource levels and reserve margin implications of the economically derived EDEA standard are developed by PG&E based upon an analysis that explicitly factors in the impact of uncertainty in key variables (Sanghvi 1985c, Poland 1989). The EDEA concept coupled with this uncertainty analysis represents a significant enhancement of existing reliability planning practice. Of key significance is the fact that EDEA is a probability based measure as opposed to a deterministic criteria (such as reserve margin or outage contingency). As such, it provides a measure of reliability that can be directly related to observable operational measures of the power system.

Transmission reliability planning is typically performed using a contingency analysis approach which defines how the system will cope with a set of failure events. However, in this process little explicit consideration is given to the fact that the probability of the selected single (or multiple) events may be very low, or that the system costs might not be justified by the resulting reliability benefits to users. In this connection, a recent precedent setting case in Florida is worth noting (Florida Public Service Commission 1984). The Florida Power Corporation (FPC) presented to the Florida Public Service Commission (PSC) a conventional single contingency analysis to justify a 500 kV line that would improve service reliability in parts of Western Florida to a forecasted population of 620,000 customers (1.1 million total population) by 1988. The Commission denied this request citing among other things, failure by FPC to satisfactorily justify the benefits of the proposed expenditures.

This project was subsequently approved by the PSC when the FPC demonstrated that project benefits outweighed project costs. To quantify the expected reliability benefits the Company first estimated the decrease in expected frequency, duration, and magnitude of unserved load in the proposed geographic area due to the project. The resulting reduction in annual unserved energy due to the line was based upon a contingency analysis. The unserved energy estimates were further weighted by the probabilities of exposure to the corresponding contingency situations. Estimates of system-wide outage cost for their service territory were obtained as a weighted average of outage costs to the three major customer classes -- residential, commercial, and industrial, as reported in Sanghvi (1982). Another example of practical decisionmaking using reliability concepts in the transmission area was the World Bank's Shaba Rehabilitation Project in Zaire, where the economic benefits of the scheme included an estimate of avoided outage costs due to improved transmission reliability (Munasinghe 1981).

Outage costs have also been used by a number of utilities -- including Gulf Power Company and Kansas City Power and Light Company -- to develop the economically optimal size of coal stockpile to be maintained. A recent EPRI research project has developed a decision analysis model for fuel inventory management which balances the costs of maintaining fuel resources against the risks of running out of fuel. Outage costs are part of the input data in this model (EPRI 1985).

Finally, Chao, Oren, Smith and Wilson (1989) explore the potential benefits of using outage cost differences among customers to devise a pricing scheme based on reliability differentiation. Price differentiation based on the reliability attributes increases welfare as these authors demonstrate by both theoretical argument and numerical analysis. While the use of such schemes of priority service are presently confined to experimental programs, they hold promise for more widespread application in the future. Accurate outage cost measurement would be an important prerequisite for the design and implementation of these pricing schemes.

Regulatory Practice

Value-of-Service concepts have always played an important role in many regulatory proceedings in the US. Proceedings to determine how much of a generation asset should be included in the rate base often includes a "managerial prudence" or "used and useful" criterion. Although specific interpretations may vary, "used" is generally defined to mean that the asset is in service, and "useful" indicates the asset is reasonably necessary to the efficient and reliable provision of service.

In many recent cases, "reserve margin" tests have been applied to evaluate whether a specific asset meets the "useful" criterion. More notably, recent Iowa and New Mexico regulatory proceedings required that optimal reserve margins be determined by explicitly balancing system costs and customer outage costs.

In a recent decision on treatment of uncommitted capacity regarding Public Service Company of New Mexico's (PNM) share of Palo Verde units, the "total cost minimization approach" was a major factor in defining an "inventory policy" (Sanghvi 1985a). Briefly, "total cost minimization" refers to the minimization of capital and operating costs plus expected outage or shortage costs to consumers. Under their policy, any portion of an uncommitted unit that does not contribute to this cost minimum is excluded from the rate base.

In recent plant siting hearings in the US, attention has also been focussed on quantifying the reliability related benefits of new transmission lines. This approach is especially relevant when the line is needed mainly for reliability reasons. In this regard, the Florida Power Corporation docket discussed earlier, represents a precedent setting case.

7.5 Summary and Conclusions

The literature surveyed in this paper portray the shift in emphasis from electricity supply problems to demand related issues, over the past two decades. Reliability and value of service are of prime importance in the analysis of demand behaviour. However, power systems have traditionally been planned and operated on the basis of fixed levels of reliability derived from ad-hoc rules of thumb (eg. the "one day in ten years" LOLP rule). The in-depth studies in this volume confirm the casual empiricism that consumers are differentially affected by power outages. This provides an opportunity to differentiate electric power service on the basis of reliability. The economic rationale for such product differentiation (i.e., improvement of economic efficiency and net welfare) has been clearly established, and different mechanisms for implementing it have been explored. In this respect, recent advances in microelectronics have provided significant new hardware options.

The accurate determination of outage costs, has been shown to be an important prerequisite to the practical implementation of reliability differentiation schemes. Four analytical approaches to the problem of outage costs measurement have been presented and several practical examples of measurement are discussed through the medium of case studies.

An important practical application of outage costs lies in setting economic reliability standards for generation, transmission and distribution planning, and power system optimization. However, while generation expansion planning models such as WASP (IAEA) and EGEAS (EPRI) routinely use outage costs as an input, the implications of using outage costs in power system optimization within the framework of differentiated service has not been adequately explored. Clearly, much more work is needed in this area if the practical usefulness of such methods is to be demonstrated.

The diversity in the estimates of outage costs obtained with different methods must be a cause for some concern. This demonstrates a need to ensure consistency among the different approaches and to explore the reasons why they provide different estimates. While some attempts in this direction are reported in the volume, progress remains limited.

In the studies on priority service and interruption insurance, far less attention has been paid to outages caused by transmission and distribution failures, than is warranted. Distribution related outages usually account for the great majority of annual outages (Munasinghe 1979, Woo and Train 1989). Since distribution outages are generally localized and hardly voluntary as far as the utility is concerned, it is unclear how schemes of load shedding on the basis of prioritized service could be implemented in practice. Whether an outage is localized or not, however, will not affect the implementation of an interruption insurance policy offered by a third party. This is because the insurance premium would reflect the local reliability level, as in the case of an automobile insurance premium that is based on residence location and driver characteristics.

Explicit introduction of outage costs and reliability differentiation into electric power systems will also have important impacts upon regulatory practice. Regulatory bodies have used the criterion that the benefit of capacity addition in terms of reliability improvement should exceed the cost of capacity addition. When different consumers are recognized as having different outage costs, the application of this criterion becomes more complicated. Similarly, regulators would also need to decide how the potential benefits of introducing reliability differentiation (and other novel techniques) would be allocated between the utilities and its different classes of consumers. Finally, interruption insurance schemes may well require the active involvement of a regulatory agency.

References

Andersson, R. and L.Taylor,"The Social Cost of Unsupplied Electricity: A Critical Review", *Energy Economics*, July 1986, pp.139-46.

Bental, B. and S.Ravid,"A Simple Method for Evaluating the Marginal Cost of Unsupplied Electricity", *The Bell Journal of Economics*, vol.13, 1982, pp.249-53.

Bernstein, M. and Y. Hegazy, "Economic Costs of Electricity Shortages: Case Study of Egypt",*The Energy Journal*, Special Electricity Reliability Issue, vol. 9, 1988, pp.173-188.

Brown and M.Johnson,"Public Utility Pricing and Output Under Risk", *American Economic Review*, vol. 59, March 1969, pp.119-28.

Chao, H.P. and R.Wilson, "Priority Service: Pricing, Investment and Market Organization", *American Economic Review*, vol. 77, December 1987, pp.899-916.

Chao, H.P., S. Oren, S. Smith and R. Wilson, "Priority Service: Market Structure and Competition", *The Energy Journal*, Special Electricity Reliability Issue, vol. 9, 1988, pp.77-104.

Crew, M. and P.Kleindorfer,"Reliability and Public Utility Pricing", *American Economic Review*, vol.68, March 1978, pp.31-40.

Cummings, R.G., D.S.Brookshire and W.D.Schulze, *Valuing Environmental Goods: An Assessment of the Contingent Valuation Method*, Rowman and Allanheld, Totowa NJ, 1986.

Doane, M., R. Hartman and C-K. Woo, "Household Preference for Interruptible Rate Options and the Revealed Value of Service Reliability", *The Energy Journal*, Special Electricity Reliability Issue, vol. 9, 1988, pp.121-134.

Doane, M., R. Hartman and C-K. Woo, "Household's Perceived Value of Service Reliability: An Analysis of Contingent Valuation Data", *The Energy Journal*, Special Electricity Reliability Issue, vol. 9, 1988, pp.135-150.

Electric Power Research Institute (EPRI), "Power Shortage Costs: Estimates and Applications," report prepared for EPRI by Faucett Associates, EA-1215, California, December 1981.

Electric Power Research Institute, "Utility Fuel Inventory Model: Version 1-0," ER-4079-CCM, California, August 1985.

Electric Power Research Institute, "Value of Service: Scoping Study", EM-4389, California, January 1986.

Finnish Power Producers Coordinating Council (STYV), "Report on the Value of Non-distributed Energy", August 1979.

Florida Public Service Commission, "Need Determination for Proposed Lake Tarpon-Kathleen 500 kV Line", Florida Power Corporation, Docket No.840044-EI, Order No.13676, Issued September 13, 1984.

Goett, A., D. McFadden and C-K. Woo, "Estimating Household Value of Electrical Service Reliability with Market Research Data", *The Energy Journal*, Special Electricity Reliability Issue, vol. 9, 1988, pp.105-120.

Hall, G., M.Healy and W.Poland, "PG&E's New Methodology for Value Based Generation Reliability Planning", paper presented to the System Planning Committee, Edison Electric Institute, Minneapolis, June 1986.

Interconexion Electrica S.A., "Costos De Racionamiento De Energia Electrica Sectores Industrial, Residencial, Comercio Y Servicios Sistema Interconectado Colombiano", Medellin, Colombia, Oct.1986.

Jaromillo, P. and E.Skoknic, "Costo Social de las Restricciones Energia Electrica", Oficina de Planificacion, Chile, Aug.1973.

Kaufman, A., "Reliability Criterion - A Cost Benefit Analysis", Office of Research, New York State Public Service Commision, Aug. 1975.

Keane, D., L. McDonald and C-K. Woo, "Estimating Residential Partial Outage Cost with Market Research Data", *The Energy Journal*, Special Electricity Reliability Issue, vol. 9, 1988, pp.151-160.

Munasinghe, M., *The Economics of Power System Reliability and Planning*, Johns Hopkins University Press, Baltimore MD, 1979.

Munasinghe, M., "Costs Incurred by Residential Electricity Consumers Due to Power Failures", *Journal of Consumer Research*, vol.6, March 1980, pp.361-69.

Munasinghe, M., "World Bank Research in Electric Power", *Research News*, The World Bank, Washington DC, Spring 1981, pp. 3-10.

Munasinghe, M., "Engineering-Economic Analysis of Electric Power Systems", *Proc. of the IEEE*, vol. 72, April 1984, pp.424-61.

Munasinghe, M., *Rural Electrification for Development: Policy Analysis and Applications*, Westview Press, Boulder CO, 1987.

Munasinghe, M. and M.Gellerson, "Economic Criteria for Optimizing Power System Reliability Levels", *The Bell Journal of Economics*, vol.10, no.1, Spring 1979, pp.353-365.

Munasinghe, M. and G.Schramm, *Energy Economics, Demand Management and Conservation Policy*, Van Nostrand Rheinhold, New York, 1983.

Munasinghe, M. and J.Warford, *Electricity Pricing*, Johns Hopkins University Press, Baltimore MD, 1982.

Ontario Hydro, "Survey on Power System Reliability - Summary of Customer Viewpoints", Report RU80-12, 1980.

Oren, S., S.Smith, R.Wilson and H.Chao, "Priority Service: Unbundling the Quality Attributes of Electric Power", Electric Power Research Institute, Report No. EPRI EA-4851, Palo Alto, CA, 1986.

Poland, W., "The Importance of Including Uncertainties in Economic Generation Reliability Planning", *The Energy Journal*, Special Electricity Reliability Issue, vol. 9, 1988, pp.19-32.

Priestman, D., "The Cost of a Long Term Electrical Energy Shortage Due to Underplanning", paper presented at the Eleventh Annual Pacific Northwest Regional Economic Conference, Eugene, Oregon, May 1977.

Sanghvi, S., "Economic Costs of Electricity Supply Interruptions: US and Foreign Experience", *Energy Economics*, vol 4, no.3, July 1982, pp.180-198.

Sanghvi, S., "Household Welfare Loss Due to Electricity Supply Disruptions", *The Energy Journal*, vol.4, 1983a, pp.33-54.

Sanghvi, S., "Optimal Electricity Supply Reliability Using Customer Shortage Costs", *Energy Economics*, 1983b, pp.129-36.

Sanghvi, S., "Optimum Generation Reserve Study", Final report to Public Service Commission and Office of the Attorney General of New Mexico, 1985a.

Sanghvi, S., "Survey of Reliability Planning Practices at Major U.S.Utilities", Report prepared for Pacific Gas and Electric Co. of California, October 1985b.

Sanghvi, S., "Incorporating Risk Preference in Generation Reliability Planning Using the EUE Criterion", report prepared for Pacific Gas and Electric Co. of California, July 1985c.

Sanghvi, S., "Residential Outage Cost Estimation", final report to Rate Department, Pacific Gas and Electric Company (PG&E) of California, December 1986.

Sherman R. and M.Visscher, "Second Best Pricing with Stochastic Demand", *American Economic Review*, vol.68, no.1, March 1978, pp.41-53.

Shipley, B., A.Patton and J.Denison, "Power Reliability Cost vs. Worth", *IEEE Transactions on Power Apparatus and Systems*, vol.PAS-91, Sept/Oct. 1972, pp.2204-12.

Swedish Association of Electricity Supply Undertakings, "Evaluating Electricity Interruptions", Stockholm, Sweden, 1981.

Swedish Joint Committee for the Electricity Council, "Cost of Interruptions in Electricity Supply", Stockholm, Sweden, Sept. 1969.

Telson, M., "The Economics of Alternative Levels of Reliability for Electric Generation Systems", *Bell Journal of Economics*, vol. 6, no.2, 1975, pp.679-94.

Viswanathan, "Service-Reliability Differentiated Pricing with Special Reference to Electric Utilities", Unpublished Doctoral Dissertation, Stanford University, CA, 1985.

Woo, C.K. and N.Toyama, "Service Reliability and the Optimal Interruptible Rate Option in Residential Electricity Pricing", *The Energy Journal*, vol. 7, July 1986, pp.123-36.

Woo, C.K. and K. Train, "Cost of Electric Power Interruptions to Commercial Firms", *The Energy Journal*, Special Electricity Reliability Issue, vol. 9, 1988, pp.161-172.

Chapter 8

ECONOMIC CRITERIA FOR OPTIMIZING POWER SYSTEM RELIABILITY LEVELS[1]

8.1 Introduction

Reliability standards for electricity supply in most countries are based on past engineering practice and rules-of-thumb. Thus there is a need to develop a more rigorous economic framework which can be used to optimize electric power system reliability. The issue of optimal levels of capacity and electric power reliability has received attention recently as part of the more general economic literature concerning public utility pricing under conditions of uncertainty (stochastic demand and supply) (Brown and Johnson 1969, Crew and Kleindorfer 1976, 1978, Sherman and Visscher 1978). In this case the utility is typically modeled as a net welfare maximizer, where net social welfare equals the difference between expected willingness-to-pay for electricity and either expected supply costs (Brown and Johnson 1969), or the sum of expected supply costs plus the costs of rationing electricity according to consumers' willingness-to-pay (Crew and Kleindorfer 1976, 1978). Such studies indicate that, ideally, both price and reliability should be jointly optimized to achieve maximum net social welfare. We may summarize this complex analysis in a very simplified form as follows: economic efficiency would be maximized if price was set equal to marginal cost, and the capacity level was fixed (thus also determining the reliability level) at the point where the marginal cost of adding to capacity (net of the energy saved) was equal to the marginal consumer benefits realized by the corresponding electric power shortages which were averted.

The short-run marginal cost (SRMC) may be defined as the cost of meeting additional electricity consumption, with capacity fixed. Therefore, in a given period, SRMC would be equal to the marginal energy costs plus shortage costs from unmet demand; during the off-peak hours, since shortages are highly unlikely by definition, shortage costs could be neglected. The long-run marginal cost (LRMC) is the cost of meeting an increase in consumption (sustained indefinitely into the future), in a situation where capacity adjustments are possible. Thus LRMC would include both marginal energy and capacity costs. When the capacity (and therefore the reliability level) is optimal, SRMC and LRMC coincide.

The method of determining optimal reliability described in this paper differs from the earlier approaches. First, in the previous studies, electricity is treated as a good which directly provides consumers with satisfaction. Therefore, shortage costs are measured primarily in terms of the resulting reduction in consumer surplus (plus rationing costs in the case of Crew and Kleindorfer 1976, 1978). However, in this paper, electricity is viewed as being an intermediate product, which is used to produce final goods that are demanded by consumers. Thus we measure outage costs in terms of the effects on the production of final goods and services in various sectors of the

[1] Edited version of a paper published in *The Bell Journal of Economics*, vol.10, Spring 1979, pp. 353-65 (co-author: Mark Gellerson).

economy. The term outage is used to represent all aspects of electric power shortages including supply interruptions, voltage and frequency changes. However, in the case study only the effects of interruptions are considered. If the market prices of productive inputs and outputs are distorted, appropriate shadow prices have to be used (Munasinghe and Warford 1978).

By taking this approach, the problem of having to use consumers' willingness-to-pay for planned electricity consumption as a measure of the costs resulting from unexpected outages may disrupt other activities complementary to electricity consumption. Thus actual short-term outage costs may greatly exceed those estimated on the basis of willingness-to-pay for planned consumption (Turvey and Anderson 1977, Webb 1977, Priestman 1977). Moreover, the method of estimating outage costs described below does not necessarily assume that when outages occur, load shedding takes place according to willingness-to-pay.

Second, giving a rationing or loan-shedding scheme, previous studies maximized net welfare by satisfying joint first-order conditions with respect to prices and capacity, indicating that partial approaches which attempt to "optimize" reliability in the presence of given prices cannot ensure that resulting net benefits are maximized. However, in practice, electricity tariffs are often not readily subject to change. Therefore, this study develops a method of optimizing reliability by weighing the costs and benefits associated with alternative levels of reliability, subject to initially fixed prices. To the extent that this is accomplished, then the net benefits of electricity consumption should at least be increased. Once the optimal reliability level is determined on the first round, tariffs can be revised to reflect any changes in the cost of supplying electricity implied by the new reliability level. With this new level of tariffs and the resulting revised demand forecast, reliability can be re-optimized; and the process can be continued until the self- consistent combination of tariffs and reliability that maximizes the net social benefits of supplying electricity is determined.

Appropriate methods of estimating outage costs for various categories of consumer are presented in Section 8.2. Then in Section 8.3 we develop a model which optimizes the reliability of electricity supply by comparing the supply costs of meeting consumer demand for electricity at different levels of reliability with the corresponding outage costs. Finally, we apply the theory to the case study of Cascavel, Brazil, in Section 8.4, to determine a range of optimum reliability levels for an urban electric power distribution system. Section 8.5 presents the conclusions.

8.2 Estimation of Outage Costs

Residential

The production of household outputs (Becker 1965, Lancaster 1966) which utilize electricity as one of their principal inputs is particularly susceptible to outages. The most important of these outputs are housekeeping, nutrition, and leisure. At any given moment during the day, there is likely to be significant substitutability between electricity-using and nonelectricity-using housekeeping activities. Furthermore, over a given time period (e.g., 24 hours), housekeeping activities temporarily interrupted by an outage may be rescheduled without much difficult. Therefore, outage costs associated with housekeeping may be considered small. In general, nutrition is an activity which could be severely hampered by outages occurring during periods of cooking. However, in the case of Cascavel no significant outage cost was observed because cooking is done almost entirely by gas.

Leisure differs from the previously discussed household activities in ways which make the associated outage costs more significant. First, the enjoyment of leisure is constrained, at least for wage earners, to occur over a relatively fixed period of time. Second, the use of electricity could be considered essential to the enjoyment of certain leisure activities (e.g. TV watching, reading, etc.) during these night time hours; thus, there are very limited substitution possibilities for nonelectricity-using activities during an outage which occurs in this period, at least in the short run. A model of household activity based on the consumer's labor-leisure choice is developed in Chapter 9. This model yields an expression for outage cost which embodies the opportunity costs of other complementary inputs used to produced electricity-dependent leisure, which are made idle by the lack of electricity. Therefore, since the main opportunity cost involved is owing to the lost input of time, the household's net income earning rate provides an estimate of marginal outage cost during the leisure hours. It is assumed that housewives are not constrained to enjoy leisure only during the evening hours, and hence their outage costs are small. Therefore, it is possible to avoid complications arising from having to estimate the value of housewives' time (Gronau 1973).

The results of a survey of residential electricity consumers in Cascavel does confirm that: (a) the chief outage cost imposed on this category of user is the loss of leisure during a critical 1.5 hour period in the evening, and (b) the mean income earning rate for a household is an acceptable monetary measure of the marginal leisure loss due to an outage. Note that a downward sloping demand curve for leisure would imply higher outage costs per unit time for outages of long duration (e.g., 2 hours), involving inframarginal units of leisure.

Generally, the above measure of outage cost is considerably higher than the value of kWh lost, if the latter were priced at the consumer tariff level. In terms of the (short-run) household demand curve for electricity, this underlines the fact that the kWh's lost owing to a random outage are not always marginal. Therefore, the outage cost reflects not just the small wedge shaped area of consumers' surplus lying between the downward sloping demand curve and the horizontal price line, owing to kWh's lost at the margin, but rather a much larger of consumers' surplus foregone corresponding to infra-marginal kWh lost. Moreover, even if the marginal kWh is lost, the resulting outage cost may be significantly greater than the above-described area of consumer surplus, because such an outage is unplanned, and thus leads to the forced idleness of complementary productive factors. The willingness of residential consumers to pay the equivalent of their net income earning rate to avoid evening-time outages, is not surprising in terms of the absolute amounts of money involved. Note that this is a purely monetary measure. If there is diminishing marginal utility of income/consumption, then it is possible to weight the outage costs, depending on the average income level of the household. Such social weights, which should reflect societies income distributional goals, would tend to reduce the disparity between the disutility of outages to low and high income households (Squire and Van der Tak 1975). For example, on this basis, in 1976 a typical electricity-using home in Cascavel with an annual income of Cr$60,000 (i.e., about Cr$28.5 per working hour), would have to pay only Cr$57 per annum more (i.e., about 6 percent of their annual electricity bill of Cr$920), if the incidence of evening outages was halved from 4 to 2 hours per annum. In 1976, US$ 1 = Cr$ 12.5.

The principal advantage of this method for estimating the leisure costs of outages to residential consumers is its reliance on relatively easy-to-obtain income data. For example, as in the case of Cascavel, it is often possible to obtain a good relationship between family income and kWh electricity consumption for a typical sample. Therefore, on the basis of the electric utility company's billing records, the income levels of the electricity-using households may be estimated. However, the empirical test of the value of leisure lost as the result of an outage depends on the short-run willingness-to-pay criterion.

There are still several theoretical problems with this approach. First, traditional work practices such as the 40-hour week, insufficient employment opportunities, union rules concerning hours of work, and so on, may make it difficult to justify equating the wage to the marginal value of leisure. In this case, union pressures were minimal; thus these problems would not be severe. Second, the cost to non-wage earning members of the family is effectively ignored by allowing only the wage earners to represent the household as an income earning unit. Third, there may be some outage expectation, especially in areas where outages are frequent, which tend to reduce outage costs. In certain perverse cases, outages may even provide utility, e.g., enjoyment of a novel situation. Finally, leisure may be enjoyed outside the home, in which case it should be treated separately.

Industrial

Industrial consumers suffer outage costs because materials and products are spoiled, and normal production cannot take place. The cost of spoiled products and materials (SPC) incurred by a given firm over one year, can be measured as:

$$SPC = \sum_{i=1}^{f} [v(d_i) + m(d_i)].(Q/h)$$

where Q = total annual value-added, h = hours of operation in a benchmark year, f = frequency of outages per annum, d = duration of the ith outage, and $v(d_i)$ and $m(d_i)$ = value-added and the value of nonfactor inputs, respectively, embodied in the spoiled products and materials (with both $v(d_i)$ and $m(d_i)$ expressed as a fraction of average value-added per working hour).

Disrupted production results in an opportunity cost in the form of idle capital and labor (IFC) which can be measured as

$$IFC = \sum_{i=1}^{f} [\epsilon.D_i + \epsilon'.\gamma(d_i)].(Q/h)$$

where ϵ and ϵ' are the fractions of normal output not produced during the outage and restart periods, respectively, and $\gamma(d_i)$ is the restart time for an outage of duration of d_i. A restart period is required because often production cannot begin immediately after the supply of electricity is restored; machines must be cleaned, furnaces reheated, and so on. In most cases, $\epsilon = \epsilon'$, and therefore we adopt this simplifying assumption.

If slack production capacity exists, then a fraction λ of lost value-added can be made up by using the productive capacity more intensively after the outage than during normal working hours. Therefore, a part of the outage cost (RC) can be recovered: RC = λ.IFC, where λ varies with the time an outage occurs and its duration. However, an average value of λ can be calculated. In fact, in Cascavel $\lambda = 0$ for most firms.

Firms which do not produce 24 hours per day also have the option of making up some fraction p of lost output by working overtime. In most cases it is unlikely that a profit-maximizing firm would have an economic incentive to do so. However, contractual obligations or other short-run constraints on a firm's behavior might lead to a decision to work overtime to make lost output. The hours of overtime production necessary to accomplish this are given by

$$h_0 = \rho \sum_{i=1}^{f} [v(d_i) + \{d_i+\gamma(d_i)\}.(1-\lambda)\epsilon]$$

and the opportunity cost of overtime production (OTC) can be measured as

$$OTC = h_0.(\underline{w}/w).(L/h)$$

where L is annual labor value-added, $\underline{w}$ and w are overtime and normal wage rates, respectively, and it is assumed that there is no opportunity cost of using capital during overtime production periods. In cases of wage distortion, shadow wage rates could be used to represent the opportunity cost of labor during both normal working hours and overtime.

Thus, allowing for the possibility of both slack productive capacity and overtime production, a general measure of the costs of outages which affect a firm or industry over a given time period (OC) can be determined:

$$OC^{I} = (Q/h).\sum_{i=1}^{f} \{m(d_i)+[1-\rho(1-(\underline{w}/w).(L/Q))][v(d_i)+\{d_i+\gamma(d_i)\}(1-\lambda)\epsilon]\}$$

In most cases, L = 0 and ρ = 0 so that

$$OC^{I} = (Q/h).\sum_{i=1}^{f} \{m(d_i)+v(d_i)+\{d_i+\gamma(d_i)\}\epsilon]\}$$

A survey of the 20 principal industrial users of electricity was conducted in Cascavel. On the basis of this survey, and using the methodology described above we determined the costs of outages of various durations for each of the firms. These costs were weighted according to the firm's share of value-added in the industry and added, so as to give one set of observations relating outage costs and durations for each industry. Finally, the outage cost functions presented in Table 8.1 were estimated for each industry.

Table 8.1

Industrial Outage Cost Functions

No.	Industry	OC Function Type[2]	Estimated Coefficients[1] a (x 10^{-4})	b (x 10^{-4})	Remarks[3]
A.1 & B.3	Mechanical and Metallurgy	Linear	0.0013	3.9	D
A.2	Nonmetallic Minerals	Linear	0.077	3.5	D
A.3	Wood	Linear	0.094	3.9	b
A.4	Vegetable Oils	Piecewise-	0.037	2.1	d < 0.5 hours, A
		Linear	3.6	1.9	d ≥ 0.5 hours, A
A.5	Food and beverages	Piecewise-	1.22	2.30	d < 0.5 hours, D
		Linear	7.07	2.02	d ≥ 0.5 hours, D
		Linear	0.66	0.72	N
A.6	Other	Log	7.2	2.9	d ≥ 0.01 hours, D
B.2	Telephone	Linear	0.002	1.1	A
B.3	Water Treatment	Linear	-0.27	1.1	d ≥ 2 hours, D

1 Adjusted $R^2 > 0.95$ for all estimations

2 Linear and log refer to functions of the type $C/Q = (a + b.d)$ and $C/Q = (a + b.\text{Log}_{10}d)$, respectively, where C = outage cost, Q = annual value added, and d = outage duration (hours).

3 D = Daytime Operation
N = Night-time Operation
A = 24 Hour Operation

Other

The provision of public illumination is constrained to occur during the hours of darkness, and the service is difficult to supply without electricity. Thus an outage which affects public illumination imposes a cost in the form of reduced community well-being since some of the benefits of public illumination, e.g., security and improved motoring safety, are foregone. One can argue that these foregone benefits are worth at least as much as the net supply cost which the community would have incurred for public illumination during the outage periods. It is unlikely that the decisions concerning marginal investments on public illumination would be made on the basis of explicit cost-benefit analyses. However, if street lighting is either much too inadequate, resulting in reduced security for citizens at night, or conversely far too extensive, leading to increased costs and municipal taxes, etc., then there will be public pressure on the city government to rectify the situation. During a one-hour outage, such a minimum estimate of net cost would equal the annuitized value of capital equipment and routine maintenance expenditures per hour of use of the particular street lights affected by the outage; electricity costs are not included, since they are not incurred during outages.

Because of the lack of knowledge of the production process that goes on in hospitals, and the difficulty of measuring outputs, the measurement of outage costs for hospitals is more difficult and less exact that in the case of industrial firms discussed earlier. Two hospitals (80 beds and 200 beds) were surveyed to estimate the opportunity costs of both productive factors which are made idle (e.g., electricity-using equipment and labor) and intermediate products, such as blood and medicines, which might spoil because of outages. Estimating the costs resulting from possible losses of life is a task exceeding the scope of this study; therefore, such costs are not considered here.The principal outage costs were found to occur during the night period (i.e., 1900-0600 hours) as a result of idle labor and capital. The average cost was estimated to be Cr$0.67 per hospital bed per hour of outage.

Outage costs for government offices and commercial customers were found to be minimal because work could often continue by daylight, and because there was usually sufficient slack during the normal hours of work for jobs delayed by any outage to be made up. Supermarkets and hotels reported minor amounts of spoilage for long outages, i.e., over five hours; however, such outages are extremely rare. The kWh consumption of rural consumers in the vicinity of Cascavel was less than one percent of the total in 1974, and their share of demand is unlikely to exceed 5 percent of the total even by 1996. Therefore, this category of consumer will not be considered in detail here.

8.3 Optimization Model

Let the generalized measure R characterize the reliability level of the electric power system over the time horizon being considered, and let D be the corresponding demand for the service assuming a given price level. A simple expression for the net benefit of electricity consumption at given levels of reliability R, demand D, and assuming fixed tariffs may be written:

$$NB(R,D) = TB(D) - OC(R,D) - SC(R,D)$$

where TB(D) is the total benefit from consumption at the level of demand D if there were no outages, OC(R,D) are the outage costs, and SC(R,D) are the supply costs. Note that OC also depends on consumers' expectations of reliability which are related to the actual reliability level. Then we have

$$\frac{d(NB)}{dR} = -\frac{\partial(OC+SC)}{\partial R} + \frac{\partial D}{\partial R}\frac{\partial(TB-OC-SC)}{\partial D}$$

For purposes of the simulations carried out in this study, we assume that $\partial D/\partial R = 0$ (i.e., for the range of reliability levels considered here, variations in R will have negligible effects on predicted demand) and that variations in R do not significantly affect consumers' expectations of reliability. Then the effects of an increase in reliability $|\Delta R|$ on net benefits may be written:

$$\Delta NB = -\Delta OC - \Delta SC = |\Delta OC| - |\Delta SC|$$

where $\Delta OC = (\partial OC/\partial R)\Delta R$, and so on, and use has been made of the fact that $\partial OC/\partial R < O$ and $\partial SC/\partial R > 0$. Therefore, to maximize the net economic benefits of supplying electricity to society, the reliability level should be increased as long as the corresponding decrease in outage costs exceeds the increase in supply costs, and vice versa.

We define the measure of reliability of electricity supply to any group of consumers year t as

$$R_t = 1 - OE_t/TE_t,$$

and over a period of T years, as

$$R = 1 - \left[\sum_{t=0}^{T} OE_t/(1+r)^t\right] / \left[\sum_{t=0}^{T} TE_t/(1+r)^t\right]$$

where OE_t represents the electric energy not supplied because of outages in year t and TE_t represents the total electric energy that would have been supplied in year t if there were no outages, and r is the discount rate. The measure R is analogous to the loss of energy probability (LOEP) criterion used in generation system planning. For a review of the different reliability indices in current use, see Chapter 2.

Consider Figure 8.1, which shows three possible evolutionary paths of system reliability for the given demand forecast starting from a relatively high level and extending over some planning time horizon; i.e., O to T. Path 2 represents the case in which expenditures on the power system are sufficient to maintain the medium level of reliability R^2. In paths 1 and 3 the reliability is adjusted to two different target levels, i.e., R^1 and R^3. The paths vary about the trend line because of the lumpy nature of the investments corresponding to each system expansion plan.

The present discounted values of the annual streams of future system supply costs (i.e., investment and maintenance) and outage costs (net of marginal generation costs of kWh not supplied because of outages) associated with expansion path i are denoted by SC_{it} and OC_{it} and total costs $TC_i = OC_i + SC_i$. The expansion path i which minimizes TC_i also yields the optimum reliability level R_m. Since the system design process involves discrete jumps in R rather than continuous variations, it is better to consider of a band of optimum reliability levels around R_m, rather than a unique value. Furthermore, the impact of the new reliability on price and demand should be examined iteratively, as discussed earlier. In this formulation the measure R_i is an aggregate scalar quantity used for convenience to simply characterize a given system expansion path i. The process of choosing the optimum system design is carried out on the basis of an economic cost benefit analysis, and is quite independent of the actual measure of reliability.

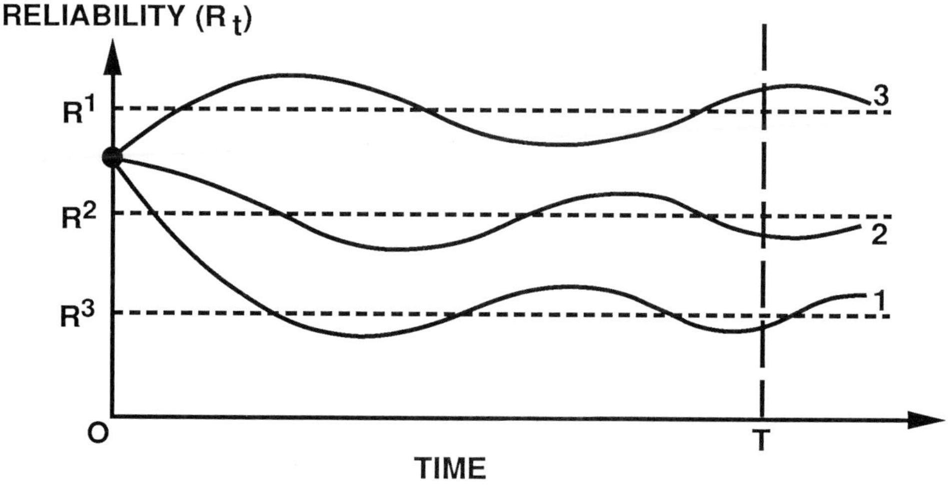

Figure 8.1: Characterization of Alternative Expansion Paths According to the Reliability Level

The above cost-benefit simulation model is sufficiently general to optimize any part of a power system, i.e., at the generation, transmission, or distribution levels. However, in the illustrative case study presented next, only the distribution network is varied, whereas the reliability of power delivered to the substation is exogenously fixed (i.e., generation and transmission reliability are held constant). A given improvement in reliability may, of course, be realized more economically by strengthening generation and transmission rather than distribution, but this is outside the scope of the case study.

8.4 Case Study

The case study involves the optimization of the electric power distribution system servicing the Cascavel urban area, by comparing the costs and benefits of changes in the reliability level. The analysis is characterized by the high level of disaggregation required for distribution system planning. Twenty categories of consumers are identified: 4 residential income classes, 10 industrial sectors, 3 commercial categories (including government offices), public illumination, hospitals, and schools. The city is overlaid with a rectangular grid which divides up the urban area into 247 distinct cells (0.5 X 0.5 km^2), and the planning time horizons spans a 21-year period (1976-1996). The technical and engineering considerations by which the supply side costs were determined are briefly summarized below (for details, see Munasinghe and Scott 1978).

First, a demand forecast is made assuming a constant future real price level disaggregated by consumer category, by geographic cell, and by year, over the plan period. For each consumer type, electricity consumption is correlated with one or more other explanatory variables which are relatively easy to predict. For residential consumers the relationship between kWh use and household income is combined with the projected growth of population and income, to forecast electricity consumption. In each industrial sector the future load is based on the value added per kWh and projected growth of value added. Similarly, the key relationships for commercial users, hospitals, and schools are kWh consumption per unit area of floor space, per hospital bed, and per student, respectively. These loads are allocated among the different geographic cells, for the plan period, starting with the existing pattern of consumers in 1976 available from the billing records of the local electric utility company, and using the Cascavel long-range urban development plan and zoning regulations to determine the future distribution of demand (Plano Diretor do Desenvolvimento 1975).

Next, several alternative distribution systems are designed to meet this load, each using a different target reliability criterion. Thus we distinguish four basic system designs to be called the low, medium A, medium B, and high reliability plans, respectively. Finally, the engineering analysis leads to a determination of the total supply costs, and the expected annual frequency and duration of outages (by cell, over the plan period), corresponding to each basic system plan.

The chief global (or city-wide) characteristics of the basic alternative system plan are summarized in Table 8.2 and shown in Figure 8.2. With respect to the four basic system designs, as the overall reliability level (R) increases, outage costs (OC) decline steadily, whereas supply costs (SC) remain fairly stable until there is a sharp rise in SC between the medium B and high reliability plans. Plan 2B is slightly cheaper although more reliable than Plan 2A, because of better design optimization. However, for this range of reliability, SC may be considered essentially constant. Outage costs per kWh lost (OCK) tend to rise with increasing system reliability.

Table 8.3 indicates that both total energy use (TE) and outage costs are most significant for residential and industrial consumers. For the range of reliability levels considered in this study, the main industrial group appears to be the one most affected by outages, in terms of OCK, followed by residential, service industrial, hospitals, and public illumination. OCK for the main industrial category tends to increase as reliability increases (and the average outage duration decreases), whereas OCK does not vary much with system reliability for the other categories of users.

More detailed analysis indicates, that residential consumers suffer outage costs of about CR$16-22 per kWh lost, over the range of reliability levels considered. This outage cost rises only slightly with income since those with higher incomes consume greater amounts of electricity. Industrial outage costs per kWh lost vary between Cr$37-76 for the wood industry down to Cr$12-35 for other industries. The outage costs results for industrial consumers are consistent with results from other studies. A range of values US$0-7 per kWh lost have been reported for different industries by others. However, because of the wide variations in methodology, different countries of application, and timing of these studies, the results are not strictly comparable (Jamaica Public Service Co. 1975, Jaramillo and Skoknic 1973, Julius 1976, Kaufman 1975, Swedish Committee on Supply Interruption Costs 1968, Taiwan Power Company 1975, and Telson 1975). Our residential outage costs are much larger than in previous studies because in the latter, the value of foregone leisure was ignored. An exception is the Swedish Committee on Supply Interruption Costs (1968), which gave values of residential outage costs similar to ours; once again the results are not directly comparable.

Table 8.2

Global Characteristics of the Alternative Distribution System Plans[1]

System Plan	Reliab. R	Outage Rate[2] OR(%)	Outage Cost OC (10^6 Cr$)	Supply Cost SC (10^6 Cr$)	Total Cost OC+SC (10^6 Cr$)	Outage Cost per kWh Lost OCK (Cr$/kWh)
Basic Designs						
1 (Low Rel.)	0.9935	0.650	150.1	57.5	207.6	14.0
2A (Med. Rel.)	0.9969	0.314	77.2	58.6	135.8	14.9
2B (Med. Rel.)	0.9981	0.191	49.5	57.8	107.3	15.7
3 (High Rel.)	0.9988	0.116	28.4	76.9	105.3	14.8
Hybrid Designs						
4 (Med./High Rel.)	0.9982	0.184	44.5	59.9	104.4	14.6
5 (Med./High Rel.)	0.9983	0.166	44.3	65.2	109.5	16.1
6 (Med./High Rel.)	0.9984	0.159	39.3	67.4	106.7	14.9

[1] Present discounted values of quantities, over the period 1976-2006; discount rate = 12%.

[2] $OR = 100 * (1 - R) 100 * \frac{OE}{TE}$

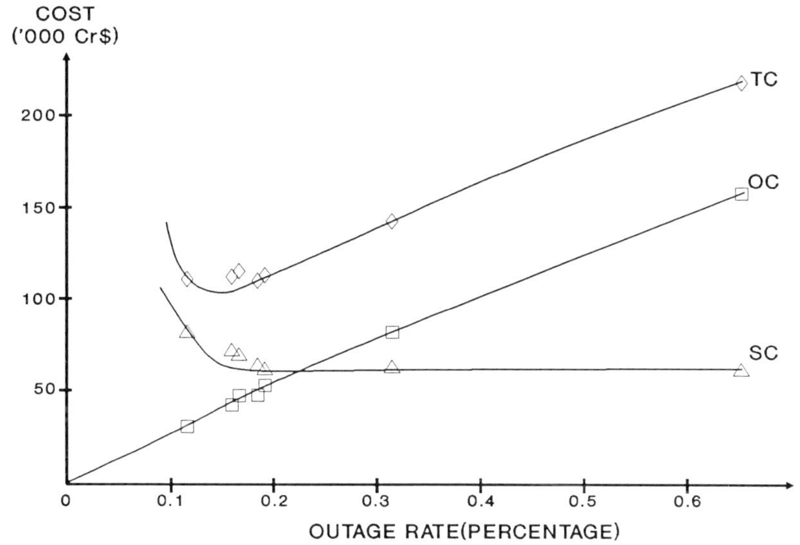

Figure 8.2: Optimization of the Reliability Level for a Distribution System

Table 8.3

Energy Use and Outage Costs by Major Consumer Category[1]

Consumer Category	Total Energy TE (GWh)	Outage Cost OC (10^6 Cr$)				Outage Cost/kWh Lost OCK (Cr$/kWh)			
		Plan 1	Plan 2A	Plan 2B	Plan 3	Plan 1	Plan 2A	Plan 2B	Plan 3
Residential	484.1	68.2	34.8	20.5	9.9	17.9	18.2	18.6	17.5
Main Industrial	371.8	77.4	40.6	27.8	17.8	31.4	33.5	30.0	26.5
Service Industrial	224.7	4.1	1.6	1.0	0.6	3.4	3.2	3.8	3.6
Public Lighting	60.4	0.2	0.1	-	-	0.4	0.4	0.4	0.4
Hospitals	11.7	0.3	0.1	-	-	5.9	5.9	5.9	5.9

[1] See Table 8.2, footnote 1.

As the results in Table 8.2 and Figure 8.2 indicate, by progressing from Plan 1 to Plan 3, there is a steady improvement in overall reliability, a corresponding decrease in global outage costs (OC), and a sharp increase in the distribution system supply costs (SC) between Plans 2B and 3. Thus the optimal distribution system design, where the total economic cost of distributing electricity in Cascavel (OC + SC) is minimized, will likely occur somewhere in the region of Plans 2B and 3. Sensitivity analysis indicates that these results are relatively unaffected by a 10 percent change in the demand forecast, variations in the discount rates between 8 to 14 percent, or variations in shadow wage rate from 0.8 to 1.0 times the nominal wage rate.

Next, using the geographically disaggregate outage costs results, several hybrid distribution system designs are considered, based on the principle of providing high reliability service for high outage cost neighborhoods, and so on; the results are summarized in Table 8.2. The results for hybrid plans 4, 5, and 6 indicate that Plan 4, in which the main industrial zone is supplied at high reliability, while all other areas obtain medium B reliability service, has the lowest total costs. Clearly, further fine tuning of the optimization procedure is possible by examining different parts of the city in greater detail and varying the design of the distribution system in these areas. However, for the purposes of this study, which is to demonstrate the application of the new methodology, we go no further than to identify an approximate range of reliability levels $0.13 < OR < 0.16$ within which the optimal distribution system plan should lie.

8.5 Conclusions

The study demonstrates that it is possible to optimize the design of a long-range distribution system so as to minimize the total costs (i.e., sum of outage and system costs) to society. Since the economic methodology is quite general, it would be possible to treat generation and transmission system reliability in the same way; i.e., optimize the design of the whole electric power system. In addition, the method of measuring industrial outages can be used to identify those industries that are very sensitive to outages, and this information could be used to minimize the economic costs of load shedding.

The case study of Cascavel utilized data that were readily available, mainly from the local utility company, as inputs in the computer models used to determine the demand forecast, the alternative distribution system designs, and the resulting system and outage costs. Results indicate that the principal outage costs are incurred by industrial and residential consumers; and, this fact is reflected in the optimum design of the distribution system, with the high population density core city area and the industrial zone receiving the highest reliability service. Further fine tuning of the system is possible by varying the design in specific areas to obtain a better optimum.

If the optimum plan is adopted, which corresponds to the minimum TC achievable after fine-tuning the procedure, it would result in a net saving of about Cr$8 million (US$0.65 million) in present value terms for Cascavel over the next 30 years, assuming that the alternative would have been Plan 3 in which the existing high level of reliability was maintained into the future over the whole city. Of this net saving, Cr$15 million would be the reduction in system costs whereas outage costs would increase by Cr$7 million. If electricity revenues were reduced by Cr$15 million, we find that the consumers living in the medium reliability areas would pay on the average at least Cr$0.01 less per kWh consumed than with Plan 3. This assumes that the system cost saving of Cr$15 million is allocated to the 1,340 GWh (in present discounted value terms) consumed by users outside the industrial zone. If the tariff of main industrial consumers within the industrial zone were raised to reflect their higher reliability level, the average tariff for other areas of the city could be reduced further. Of course, if tariffs were based on marginal costs, the restructuring of prices would be different. In summary, however, the overall change of Cr$ 0.01 per kWh in the mean price caused by reliability level optimization alone is small compared to the average 1976 residential tariff of about Cr$0.82 per kWh consumed.

Because every city has its specific characteristics, the results of the Cascavel case study cannot be used to draw wide-ranging conclusions regarding generally desirable, distribution system reliability levels for other cities in the developing countries. However, the methodology remains valid and could be applied elsewhere.

References

Becker, G., "A Theory of the Allocation of Time", *Economic Journal*, September 1965, pp. 493-517.

Crew, M.A. and Kleindorfer, P.R., "Peak Load Pricing with a Diverse Technology", *The Bell Journal of Economics*, vol. 7, No. 1 ,Spring 1976, pp. 207-231.

Crew, M.A. and Kleindorfer, P.R., "Reliability and Public Utility Pricing", *The American Economic Review*, vol. 63, No.4, September 1973, pp. 634-650.

Jamaica Public Service Company, "Reliability Dollar Value Analysis" and "The Selection of a Generation Reliability Criterion for JSPC", November 1975.

Jaramillo, P. and Skoknic, E., "Costo Social de las Restricciones de Energia Electrica", Oficina de Planificacion, Chile, August 1973.

Julius, D., "The Economic Cost of Power Outages: Methodology and Application to Jamaica", EW&T Department, World Bank, August 1976.

Kaufman, A., "Reliability Criterion - A Cost-Benefit Analysis", Office of Research, N.Y. State Public Service Commission, August 1975.

Lancaster, K., "A New Theory of Consumer Behavior" *Journal of Political Economy*, 1966.

Munasinghe, M., *The Leisure Cost of Electric Power Failures*, World Bank Staff Working Paper No. 285, World Bank, Washington DC, June 1978.

Munasinghe, M., *The Economics of Power System Reliability and Planning: Theory and Case Study*, Johns Hopkins University Press, Baltimore MD, 1979.

Munasinghe, M. and Scott, W., "Long-Range Distribution System Planning Based on Optimum Economic Reliability Levels", Paper No. A78576-1, *Proceedings of the IEEE Power Engineering Society*, Summer Meeting, Los Angeles, August 1978.

Munasinghe, M. and Warford, J., "Shadow Pricing and Power Tariff Policy", in *Marginal Costing and Pricing of Electricity*, State of the Art Conference on Marginal Cost Pricing, Montreal, April-May 1978.

Plano Diretor Do Desenvolvimento, Prefeitura Municipal de Cascavel, Cascavel, Parana, Brazil 1975.

Priestman, D., "The Cost of Long-Term Electricity Energy Shortage Due to Underplanning", Eleventh Annual Pacific Northwest Regional Economic Conference, Eugene, Oregon, May 1977.

Sherman, R. and Visscher, M., "Second-Best Pricing with Stochastic Demand", *The American Economic Review*, vol. 63, No. 1, March 1973, pp. 41-53.

Squire, L. and Van der Tak, H., *Economic Analysis of Projects*, Johns Hopkins University Press, Baltimore MD, 1975.

Swedish Committee on Supply Interruption Costs, "Costs of Interruptions in Electricity Supply", A.O. Trans. 450, The Electricity Council, UK, September 1968.

Taiwan Power Company, "Evaluation of Generating Capacity Reserve", Taiwan, 1975.

Telson, M.L. "The Economics of Alternative Levels of Reliability for Electricity Generation Systems", *The Bell Journal of Economics*, vol. 6, No. 7, Autumn 1975, pp. 679-694.

Turvey, R. and Anderson, D., *Electricity Economics*, Johns Hopkins University Press, Baltimore MD, 1977, Chapter 2.

Webb, M., "The Determination of Reserve Generating Capacity Criteria in Electricity Supply", *Applied Economics*, March 1977, pp. 19-31.

Chapter 9

COSTS INCURRED BY RESIDENTIAL ELECTRICITY CONSUMERS DUE TO POWER FAILURES[1]

9.1 Introduction

The impetus for this study arose from the need to establish appropriate standards of reliability for electricity supply. As the incidence of electric power failures or outages increases, the economic costs to society of such outages also rise. On the other hand, reliability of electricity supply may be improved with a corresponding decline in the incidence of failures, but in this case the costs associated with the electric power system increase. Ideally, from society's point of view, power system reliability should be optimized at the level where the sum of the economic outage costs and the system supply costs are at a minimum.

The foregoing discussion suggests that reliability should be increased to the point where the marginal social costs of improvements to the power system are equal to the marginal social costs of outages (or electricity interruptions) that are averted (Munasinghe and Gellerson 1979). Comparison of social costs and benefits, a hitherto neglected area, is particularly relevant in the developing country context where investment resources are scarce and large sums are devoted to electric power. The supply-side costs may be determined from fairly straightforward technical engineering considerations based on power system planning methods (Anderson 1972). However, on the demand side, the costs to society of reduced levels of reliability of electricity supply are far more difficult to determine, and have not been much investigated (Telson 1975).

The principal categories of electricity users include industrial, commercial, and residential consumers. As the first two types (and most other categories of user) produce an output having a market value, one measure of outage costs would be the disruption of the stream of output caused by the electric supply interruption, such as due to spoilage of materials, factors of production made idle, etc. Expenditures on standby electricity generating equipment could also be considered outage-related costs, although in this case it would be difficult to determine at what level of outages (i.e., frequency and duration) users would feel that it was worthwhile to install a captive generating plant. In the case of households that generally do not produce a marketable output, a different approach must be used.

In this paper, an explicit theoretical framework for measuring residential outage costs is presented and empirically tested using the results of a survey of household electricity consumers in the city of Cascavel, Brazil (population approximately 90,000 in 1976). Although the model or variations thereof are relevant to most countries, simplicity of application has been emphasized because of data constraints in developing countries. In brief, it is argued that:

---

[1] Edited version of a paper that appeared as Staff Working Paper No. 285, The World Bank, Washington DC, June 1978; and was subsequently published in the *Journal of Consumer Research*, vol.6, March 1980, pp.361-9.

1) The principal outage cost imposed on a household is the loss of leisure during the evening hours when electricity is essential, because during the daytime there is sufficient slack in the execution of household activities that are interrupted by the outage, such as cleaning and washing, to permit rescheduling of these functions without much inconvenience.
2) The marginal monetary value of this lost leisure is equal to the effective net income earning rate on the basis of the consumer's labor-leisure choice.

9.2 Residential Electricity Users

It is convenient to adopt the approach of treating the household as a productive unit that uses inputs, such as the household members time and market goods (electric appliances, electricity itself, etc.) to produce outputs, such as leisure and nutrition, that provide utility (Becker 1965, Lancaster 1966, and Gronau 1973). The cost of an unexpected outage may be estimated in terms of inputs spoiled or made idle, i.e. opportunity costs, as in the case of the industrial and commercial user categories. The effects of known or planned power failures are neglected because costs may be minimized by rescheduling activities, etc. Similarly the effects of brown-outs due to voltage drops and frequency fluctuations, may be ignored.

Even though the household can be logically viewed as a productive unit, the measurement of residential outage costs is not straightforward, because most of the outputs are consumed within the household, e.g., leisure, and are thus not valued in the market, whereas the value of inputs into this production process, especially household member's time, is also often not determined in the market. Household outputs that utilize electricity as one of their principal market inputs are of special interest, particularly with regard to household keeping, nutrition and leisure.

Of particular relevance to housekeeping activities is the possible flexibility of rescheduling household chores, thereby minimizing outage costs. Typical interruptions in electricity supply will affect the functioning of household electrical appliances used for ironing, washing, vacuuming, etc. However, sometimes there is the possibility of substitutability between such electricity-using and other nonelectricity-using housekeeping activities. In addition, over a given period of time, e.g., one day, there will be some degree of substitutability between housekeeping and other nonhousekeeping activities in the home. Thus, the outage cost resulting from idle resources used for housekeeping will be small. Nevertheless, there will be some cost, including the psychological costs of disruptions in the normal routine, as long as it is assumed that household producers rationally allocate their time to various kinds of production before the outage.

Food preparation is an activity that could be hampered by outages at particular times of day, especially if electricity is essential for cooking and/or lighting. For interruptions of relatively short duration, the outage costs may involve only the inconvenience of a delayed meal, but in other cases food may have to be supplied from outside the home, and paid for as an unexpected expense. In areas where outage expectation is high, households may incur additional indirect outage-associated expenses by purchasing nonelectric appliances, such as small kerosene stoves for standby purposes, i.e., in addition to their electric stoves. On the other hand, where cooking is carried out exclusively through the use of other fuels (kerosene, gas, or firewood), outage costs are likely to be small. In Cascavel, no significant outage cost was observed because cooking is done almost entirely by gas.If the expenses associated with nonelectric cooking facilities exceed the costs of purchasing and using an electric stove, and if the former were preferred to the latter specifically because of the low reliability of electricity supply, then the difference in costs between the two alternatives would also constitute an indirect outage cost.

Leisure differs from the housekeeping and food preparation activities in ways that make the associated outage costs more important. First, the production and enjoyment of leisure in most households is constrained, particularly for wage earners, to a relatively fixed period of time, usually in the evening. It is assumed that housewives are not constrained to enjoy leisure only during the evening hours, and because of the substitution possibilities between evening and daytime leisure their outage costs are small. Even if electricity-dependent daytime leisure activities such TV watching were interrupted by outages, there would be sufficient substitutability with nonelectricity-dependent leisure or household activities to minimize the outage costs. Thus, it is possible to avoid complications arising from having to estimate the value of housewives' time (Gronau 1973).

Second, in the case of leisure activities, such as television viewing, reading, etc., which would require the use of electricity during the nighttime hours, there would be limited substitution possibilities for nonelectricity-using activities during an outage, at least in the short-run. As the loss of leisure activities is likely to be the most significant component of residential outage costs, a model of household activity based on the consumer's labor-leisure choice is developed here, to estimate the welfare loss associated with interruptions in electricity supply.

In addition, power outages will impair the functioning of equipment such as refrigerators, air conditioners, and heaters, which may be only partially associated with the types of household activity just discussed. In these cases, the effects of the disruption would be significant only for outages of longer duration (e.g., several hours), which generally occur far less frequently. However, besides the loss of services from such equipment, the damage to or reduced life span of motors and equipment due to voltage variations would have to be taken into account.

Households may also incur indirect outage costs by purchasing standby generators, storage batteries, kerosene lamps, and so on. In this respect, the consequences of outages may be inequitably distributed, because wealthier consumer will be better able to afford voltage-boosting devices and alternative energy sources, thereby reducing direct outage costs. On the other hand, poorer users may be less dependent on electricity, and therefore less susceptible to outages.

9.3 Value of Foregone Leisure

Consider a typical household that maximizes utility over some period of time (D days). Utility, U, is expressed as a function of leisure type S, which cannot be enjoyed without electricity, leisure type V, which is electricity independent, and income, I (net of expenses incurred to enjoy the leisure), which represents all other consumption.

$$U = U(S,V,I) \qquad (9.1)$$

Next, leisure type S is specified as a function of the inputs -- time t (hours), electricity consumption e (kWh) flow of services z from the stock of electricity using equipment (TV), and other inputs represented by x:

$$S = S(t,e,z,x) \qquad (9.2)$$

Similarly, leisure type V could be specified as a function of time Θ, and other inputs represented by m;

$$V = V(\Theta,m) \qquad (9.2b)$$

Furthermore, e and z may be written as functions of the stock of electricity using capital k (the value of appliances annuitized over their useful life), and of time t that is a measure of intensity of use. Therefore, $e = e(k,t)$ and $z = z(k,t)$.

The household budget constraint may be written:

$$I = w(H - t - \Theta) - pe - bk - cx - rm \qquad (9.3)$$

where,

w = effective net income earning rate (per hour);
H = maximum feasible number of hours of work in the time period;
p = mean price per kWh of electric energy
b = equivalent income foregone per unit of k in the time period;
c = cost per unit of other inputs x used;
r = cost per unit of other inputs m used.

Initially, we examine the longer-term situation, e.g., a time period of one year. Maximizing utility subject to the linear budget constraint is equivalent to unconstrained maximization of the expression:

$$L = U(S,V,I) - \lambda[H - t - \Theta - (I+pe+bk+cx+rm)/w] \qquad (9.4)$$

As shown in Appendix A9.1, manipulating the first order conditions for maximizing, yields an expression involving the marginal rate of substitution of income for electricity-dependent leisure, keeping utility constant, as derived from the consumer's long-run decision.

$$MRS_{I,S}dS = wdt + pde + bdk + cdx \qquad (9.5)$$

$MRS_{I,S}$ measures the marginal monetary value of electricity dependent leisure, i.e., it is the income increase that will compensate the consumer just enough for this type of leisure foregone at the margin, to remain on the same indifference curve (Willig 1976).

Now consider the effect of an unexpected outage during the evening hours when leisure type, S, is being enjoyed by the household. Ex-post and over the short-run period of the outage, it is argued that e and z depend only on t, i.e., assuming a putty-clay type of relationship between k, e, and z. For example, in the short-run electricity use may be linearly proportional to time. Therefore it is possible to write:

$$e = \phi(t)$$

Using the above expressions to evaluate Equation (9.5), and replacing dS by δS, and so on, we get:

$$MRS_{t,S}\delta S = [w + p.(\partial\phi/\partial t)].\delta t + c.\delta x + b\delta k \qquad (9.6)$$

Thus the left-hand side of Equation (9.6) represents the overall welfare decrease due to an incremental loss of leisure δS resulting from the unexpected outage of duration δt, while the right-hand side is a measure of the value of the inputs that are required to produce the leisure.

However, the households electricity bill will also be reduced by the amount $p(\partial\phi/\partial t)\delta t$ due to kWh not used during the outage. Therefore, the net incremental welfare loss or outage cost due to an unexpected outage of duration δt may be written:

$$OC^R = MRS_{l,S}\,\delta S - p\,(\partial\phi/\partial t).\delta t$$

$$= w\delta t + c\delta x + b\delta k \qquad (9.7)$$

As also derived in the Appendix, the marginal rate of substitution of income for electricity-independent leisure, keeping utility constant ($MRS_{l,V}$) is:

$$MRS_{l,V}\delta V = w\delta\Theta + r\delta m \qquad (9.8)$$

where δV has been used instead of dV, and so on.

Equation (9.8), is analogous to Equation (9.6), and measures the welfare change due to an incremental change in the availability of electricity-independent leisure δV, in terms of the value of the inputs, i.e., time δt and other input δm, which are required to produce this leisure.

As mentioned earlier, the empirical estimation of residential outage costs is particularly difficult because of the nonmarketable nature of the household outputs produced by using electricity. From a practical point of view, data on wages and income are far easier to obtain than information on the use of appliances and other inputs. Fortunately, in most instances the dominant term in Equation (9.7) would be the wage term.

Take, for example, a television set that is one of the more expensive items of electrical equipment commonly used for leisure. Assuming a purchase price of \$250 and a six-year lifetime (independent of usage and/or outages), and a discount rate of 10 percent, the annuitized value is \$57.50 a year. For a set that is used four hours a day, on the average, the corresponding value of $b.\delta k$ is 4 cents per hour of outage, which is likely to be a negligible amount, even compared with minimum income earning rates for relatively poor households using electricity. Thus, the difficulties of obtaining data on ownership of electrical appliances, or other inputs used during leisure time, will most often far outweigh any resulting refinement to the basic estimate of outage costs derived purely on the basis of more readily available income data.

Therefore, in practice the following expression may be used as a good approximation of Equation (9.7):

$$\delta OC^{R}/\delta t \approx w \qquad (9.9)$$

Arguing analogously, that the wage term is likely to dominate the second term on the right hand side of Equation (9.8), it is possible to derive another approximate relationship:

$$MRS_{l,V}.(\delta V/\delta \Theta) \approx w \qquad (9.10)$$

Therefore, the incremental monetary values of both electricity-dependent and independent types of leisure per unit time are roughly equal to the wage (or income earning) rate.

In practical terms, the theoretically derived link between the two types of leisure and the income earning rate is intuitively reassuring, because electricity-dependent leisure and actual working hours may not be direct physical substitutes for each other. For example, the normal workday could extend from 9:00 a.m. to 5:00 p.m., whereas electricity-dependent leisure may begin only after 7:30 p.m., i.e., the hours of darkness. Therefore, the marginal trade off between electricity-dependent leisure and extended working hours may have to occur via an adjustment in the intervening electricity-independent leisure period.

As discussed earlier, a principal practical advantage of this method of estimating the outage costs of residential consumers on the basis of foregone leisure, is its reliance on relatively easy-to-obtain income data. Often it may be possible to obtain a good correlation between family income and kWh electricity consumption for a typical sample of residential consumers by using the utility company's billing records and information from household budget surveys. In this way the income levels of electricity-using households could be estimated.

Still, this method of estimating residential consumers outage costs may sometimes lead to incorrect estimates for five reasons. First, it is assumed that workers can vary their hours of work to equate their wage with the marginal value of their leisure time. Traditional work practices such as the 40-hour week, union restrictions on hours worked, or insufficient employment alternatives might prevent this. If workers are unable to work as much as they wish, their wage will overestimate the value of lost leisure. A related point is that the daytime wage rate may not be a good proxy for the value of leisure; the marginal wage rate corresponding to the leisure hours may be more appropriate, and in some cases this may be the overtime rate of pay. We note that in Cascavel, unemployment is very low due to rapid economic growth in the area, and union pressures are minimal.

Second, the cost to nonwage earning members of the family is effectively ignored, by allowing only the wage earner to represent the household as an income-earning unit. For example, the value of the housewives' time may be estimated in terms of the average earning rate of women workers, although the corresponding outage cost may not be very accurate because of the greater work-leisure substitution possibilities of housewives, as well as differences between the productivity of housewives and women workers.

Third, residential consumers may develop outage expectations, presumably because of the frequency of such occurrences in the past, so that the possibility of interruptions in electricity supply will be considered when labor-leisure decisions are made. The cost of the outage will then be less than in instances where there is no such outage expectation.

Fourth, is some leisure is enjoyed outside the household that is affected by the outage, then ideally, this case should be treated separately. Finally, in certain unusual situations, there is the possibility that outages themselves may provide some consumption benefits to residential consumers, e.g., perhaps the enjoyment of a novel situation.

These difficulties indicate that the estimate of outage costs inferred from the consumers labor-leisure choice ought to be independently verified. The short-run outage cost is likely to exceed the willingness-to-pay for the lost kWh, based on the households longer-run demand curve. Hence, empirical verification of Equation (9.9) must depend on the determination of the consumers willingness-to-pay in the short-run, to avoid an unexpected outage, as described in the next section.

9.4 Survey Results

A random sample of 27 residential consumers selected from the 1976 list of approximately 9,000 electricity-using households in Cascavel were subjected to an in-depth survey to determine their opinion of the quality and value of electrical service, electricity consumption, income, leisure-time habit patterns, and the value they placed on foregone leisure during an outage.

The complete questionnaire is provided in Appendix A9.1. Unfortunately, the sample size was limited by the time and manpower resources available to carry out the survey. In view of the difficulty of obtaining meaningful answers to questions on outage costs involving hypothetical willingness-to-pay, considerable time had to be devoted to explaining the problem to the consumers, and putting them at their ease. Care had to be taken to avoid giving the impression that the interviewers wished to actually increase or decrease their electricity bill, obtain information on incomes for tax purposes, etc. At least two adult members of each home were required to be present, to obtain replies that were representative of the household. Consequently, a team of three persons (including a translator) took one week to complete the 27 interviews.

All households in the sample indicated that electricity was an essential service in their daily lives; in general, the poorer consumers felt that they were getting good value for their money, whereas some of the higher income households thought that the price of electricity services was too great. A direct estimate of outage costs was obtained by asking consumers how much extra they would be willing to pay to avoid outages of different durations at various times during the day (Questions 4 and 5). Consumers indicated no willingness to pay extra to avoid daytime outages. The direct amounts that consumers were willing to pay in order to avoid evening outages was found to be linearly proportional to outage duration, on the average, over the range 3 to 80 minutes, and the estimate of outage cost is based on values in this range. For outages for less than a few minutes many people indicated that they would sit out the inconvenience rather than pay, and for outages over 1.5 hours the outage cost per unit time tended to also fall off quite rapidly, as households were unwilling to pay large sums of money, i.e., the impact on income became important.

Indirect estimates of outage costs were obtained by determining what additional amounts each household considered reasonable: (a) for payment to the utility company, if the existing (known) incidence of outages was halved; and (b) as a compensatory refund on their electricity bill, if the existing outage rate was doubled (Questions 3, 9, and 12).

The evening time outage cost estimates in Table 9.1 are average values of the direct and the two indirect estimates, for each household. Questions 4, 9, and 12 were not all asked consecutively, to minimize the possibility of consumers giving consistent answers to later questions, based on rational extrapolation from the earlier questions. In fact, good agreement was obtained between the various estimates of outage costs for each household, and this was taken as an added sign of reliability of the estimates. As expected, in general the indirect estimate of outage cost from Question 9 tended to be lower than the estimate from Question 12; these two estimates most often straddled the value of the direct estimate of outage cost.

Table 9.1

Residential Consumer Survey Results

Net Income Earning Rate (Cr$ per Hour)[a]	Outage Cost (Cr$ per Hour)	Net Income Earning Rate (Cr$ per Hour)[a]	Outage Cost (Cr$ per Hour)
7.4	6	27.7	20
7.4	7	30.8	35
7.4	13	40.3	30
10.5	12	40.3	30
11.1	15	45.5	50
14.2	10	48.0	60
15.4	20	55.1	45
18.5	25	64.6	55
18.5	35	64.6	60
19.7	20	73.9	85
21.5	15	99.1	60
21.5	25	116.3	133
24.6	15	166.2	150
24.6	40		

[a] US $1 = Cr $12.35 (end 1976).

The daily leisure period during which electricity was considered essential had a fairly uniform duration of about 1.5 hours (standard error = 12 minutes), and occurred somewhere within the broader time interval 19.30-22.30 hours. Approximately 60 percent of the latter interval was devoted to TV watching, 25 percent to dining and the remainder to other activities, such as reading and conversing. The differences between households were minor; for example, the standard error for the TV watching period was only about 10 percent of the mean. Only one household in the sample did not possess a TV set; however, in this case about 50 percent of their leisure time was spent listening to the radio, and special TV programs, such as football games, were watched on the neighbors set. Thus, these results confirmed the view expressed by local municipal authorities, that practically all electricity-using homes owned a TV set. The mean household size was 5.1 persons, with a standard error of 1.4 persons, and the average time at which families went to sleep was about 23.00 hours, with a standard error of 45 minutes.

Assuming that all the usual conditions of the ordinary least squares (OLS) model are fulfilled (Johnston 1972), a basis unrestricted equation of the form,

$$OC_i = \beta_1 + \beta_2 Y_i + U_i \quad (9.11)$$

is fitted to the observations, where OC_i, Y_i and U_i are the outage cost per hour, net income earning rate per hour, and random disturbance term, respectively, for the ith household. Net income earning rate = gross income rate - tax (estimated from known tax rates). Total household income was in most cases earnings of a single wage earner. The estimated coefficients are β_1 = 3.355 and β_2 = 0.896 (t = 15.12), yielding a good fit with R^2 = 0.901, while the sum of the squares of residuals (SSR) = 3,228.2. The original observations and the fitted line are shown in Figure 9.1.

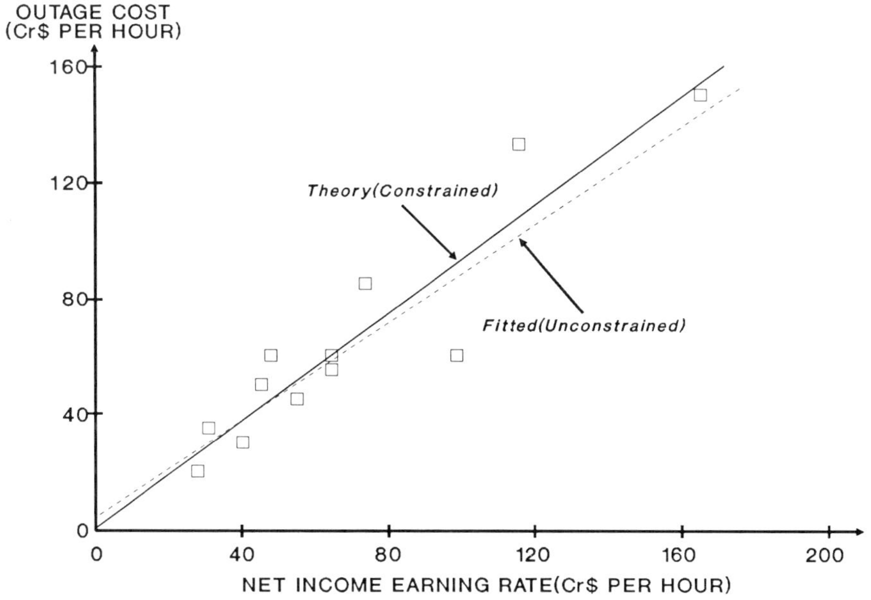

Figure 9.1: Outage Costs Versus Net Income Earnings Rate for Residential Consumers

The null hypothesis to be tested is that β_1 = 0 and β_2 = 1 jointly, giving us the restricted model:

$$OC = Y + U \tag{9.12}$$

This corresponds to the solid line in Figure 9.1, where the new SSR = 3650.2. A Chow test (Chow 1960, Fisher 1970) was performed to ascertain whether the free Equation (9.11) fits the observed values significantly better (in terms of the sum of squares of the residuals), than the restricted Equation (9.12) derived from the theoretical model. From the results of the regression F = 1.634; but $F_{25}^{2}(0.20) = 1.72$, and therefore, the null hypothesis H_0: $\beta_1 = 0$ and $\beta_2 = 1$ is accepted at the 20 percent level, i.e., we accept the restricted equation as the correct one. This statistic is distributed as F with 2 and 25 degrees of freedom: F = [(3650.2 - 3228.2)/2] / [3228.2/25] = 1.634. Acceptance at the 20 percent level implies acceptance at all lower levels, i.e., 10 percent, 5 percent, etc., for a right-tail test.

The slope (β_2) of Equation (9.11) is less than unity. A possible explanation is that the ratio of nonwage to wage income tends to rise as the income level increases. This nonwage income (e.g., property or investment income) may not enter into the consumers labor-leisure decision. Therefore, since net total income (irrespective of source) is plotted along the X-axis of Figure 9.1, rather than net wage income, this would tend to reduce the slope of the fitted line, especially for the higher income levels. Unfortunately, it was very difficult to obtain further information concerning the breakdown of income by source.

Saturation effects could also occur at high income levels, which would tend to impose a ceiling on the willingness-to-pay. Similarly, at the low end of the income scale, there could be a minimum level of outage costs (greater than zero), although it is doubtful whether such poor households would have electricity service. Both of these effects would tend to reduce the slope β_2. The impact of the other (neglected) terms in Equation (9.7) would be to increase the slope of the solid line.

The foregoing analysis has focused on the outage costs per hour of outage, by income level. These results may be examined from a different viewpoint by using the survey data presented in Figure 9.2, a graph of the average annual electricity consumption versus the net annual income of households, based on an independent municipal survey of 701 residential electricity consumers, carried out by city government in 1976. As expected kWh consumption increases monotonically, with evidence of saturation at the upper end of the income scale.

For the rest of this analysis, the population is divided into four income classes: lower (L), lower-medium (LM), upper-medium (UM), and upper (U). The net income and electricity consumptions per annum (Figure 9.2), by income class, are shown in Columns 2 and 3 of Table 9.2. Corresponding outage costs per hour of leisure-time outage are shown in Column 4, assuming that these costs are equal to the mean income earning rate. Next, using Figure 9.2, and the daily pattern of household electricity consumption (i.e., 75 percent of the kWh usage, during the period 1930-2230 hours),the outage costs per kWh lost due to leisure-time outages are calculated (Column 5). It may be seen that while outage costs per hour rise steadily with income level, the outage costs per kWh lost are much more stable; but there is a slight slump in outage costs per kWh lost for the highest income class, due to the saturation effects in Figure 9.2, discussed earlier.

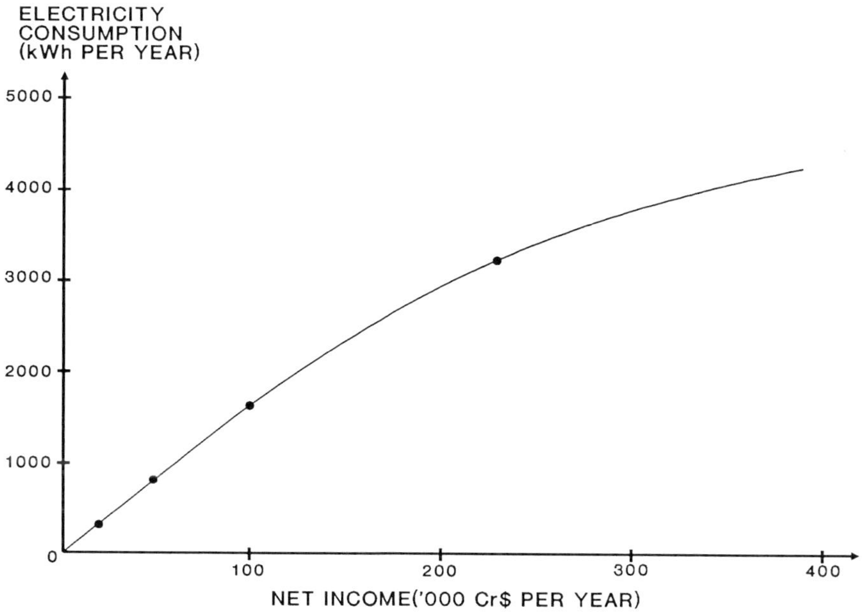

Figure 9.2: Average Electricity Consumption versus Net Income of Households in 1976

Table 9.2

1976 Household Income, Electricity Consumption and Outage Costs by Income Class

Income Class	Net Income[a] (1976 Cr$ x 10^3 per annum)	Electricity Consumption (kWh/Annum)	Outage Costs (OC) Efficiency Prices (Cr$ per Hour of Outage)	Efficiency Prices (Cr$ per kWh lost)	Social Prices (Weighted) (Cr$ per Hour of Outage)	Social Prices (Weighted) (Cr$ per kWh lost)
L	18.4	307	8.71	38.83	32.05	142.87
LM	47.7	794	22.59	38.94	32.05	55.26
UM	97.5	1631	46.16	38.74	32.05	26.88
U	228.6	3283	108.24	45.12	32.05	13.36

[a] US$ 1 = Cr$ 12.35 (end 1976) and 2112 working hours per year.

The analysis thus far has ignored income distribution considerations. Therefore, the results in Columns 4 and 5 have been evaluated in terms of so-called efficiency prices, a terminology derived from cost-benefit analysis (Little and Mirrlees 1974, Squire and van der Tak 1975). However, by treating an outage cost like foregone consumption/income, it may be argued, on the basis of diminishing marginal utility of consumption/income, that the outage costs of poorer households should be weighted more heavily. This type of weighting may be termed social pricing.

Consider a very simple weighting function of the form $W(C) = (C'/C)^N$, where C is the household income, C' is some critical (or benchmark) income level, and N is a parameter. For convenience, we choose N = 1 and C' = Cr\$67,700 per annum = mean income level of all Cascavel households (including homes not consuming electricity). This gives us the social weights 3.679, 1.419, 0.694, and 0.296 to be applied to the outage costs of the L, LM, UM, and U income groups respectively.

The results of social pricing are shown in Columns 6 and 7 of Table 9.2; the situation is inverted with respect to the efficiency pricing regime. The choice of the social weighting function has ensured that the outage costs per hour are constant at Cr\$32.05, while the outage costs per kWh lost during leisure-time vary widely, between Cr\$13 and Cr\$142.

9.5 Conclusions and Discussion

The result of the Cascavel survey confirm that the chief cost imposed on this category of users is the loss of leisure during the evening hours, approximately over a 1.5 hour period. Furthermore, we conclude that during this critical period the mean income earning rate for a household is an acceptable monetary measure of the marginal leisure loss due to an outage. Although the sample size is small, the strong correlation between outage costs and income justifies this conclusion.

Generally, this measure of outage is considerably higher than the value of kWh lost, if the latter were priced at the consumer tariff level (Webb 1977, Priestman 1977). In terms of the (short-run) household demand curve for electricity, Equation 9.9 underlines the fact that the kWh lost due to a random outage are not marginal. Therefore, the outage cost reflects not just small wedge shaped areas of consumers surplus lying between downward sloping demand curve and the horizontal price, owing to kWh lost at the margin, but rather a much larger amount of consumers surplus foregone, corresponding to inframarginal kWh lost. The willingness of residential consumers to pay the equivalent of their net income earning rate to avoid evening time outages is not surprising in terms of the absolute amounts of money involved. For example, on this basis, in 1976, a typical electricity-using household in Cascavel with an annual income of Cr\$60,000 (about Cr\$28.5 per working hour) would have to pay only Cr\$57 more (about 6 percent of their electricity bill of Cr\$920 per annum) if the incidence of evening outages was halved from four to two hours per annum.

If income distribution considerations are ignored (i.e., efficiency pricing), then outage costs per hour of outage rise with income, whereas outage costs per kWh lost are stable. If outage costs are weighted according to the linearly diminishing marginal utility of income, i.e., social pricing, the situation is reversed; outage costs per kWh lost decrease with increasing income.

The principal advantage of this method for estimating the leisure costs of outages to residential consumers is its reliance on relatively easy-to-obtain income data. For example, as in the case of Cascavel, it is often possible to obtain a good relationship between family income and kWh electricity consumption for a typical sample, and thus on the basis of the electric utility company's billing records, the income levels of the electricity-using households may be estimated. Ultimately, the empirical test of the value of leisure lost due to outage depends on the willingness-to-pay criterion.

Appendix A9.1

A. Derivation of Outage Costs

Consider the expression (using the notation of Section 9.3):

$$L = U(S,V,I) - \lambda\,[H - t - \theta - (I + pe + bk + cx + rm)/w]) \qquad (A9.1)$$

The first order conditions necessary for maximization are:

$$\frac{\partial L}{\partial t} = \frac{\partial U}{\partial S}\left[\frac{\partial S}{\partial t} + \frac{\partial S}{\partial e}\cdot\frac{\partial e}{\partial t} + \frac{\partial S}{\partial z}\cdot\frac{\partial z}{\partial t}\right] + \lambda\left[1 + \frac{p}{w}\cdot\frac{\partial e}{\partial t}\right] = 0 \qquad (A9.2)$$

$$\frac{\partial L}{\partial k} = \frac{\partial U}{\partial S}\left[\frac{\partial S}{\partial e}\cdot\frac{\partial e}{\partial k} + \frac{\partial S}{\partial z}\cdot\frac{\partial z}{\partial k}\right] + \frac{\lambda}{w}\left[p\frac{\partial e}{\partial k} + b\right] = 0 \qquad (A9.3)$$

$$\frac{\partial L}{\partial x} = \frac{\partial U}{\partial S}\cdot\frac{\partial S}{\partial x}\cdot\frac{\lambda}{w}c = 0 \qquad (A9.4)$$

$$\frac{\partial L}{\partial I} = \frac{\partial U}{\partial I} + \frac{\lambda}{w} = 0 \qquad (A9.5)$$

$$\frac{\partial L}{\partial \theta} = \frac{\partial U}{\partial V}\cdot\frac{\partial V}{\partial \theta} + \lambda = 0 \qquad (A9.6)$$

$$\frac{\partial L}{\partial m} = \frac{\partial U}{\partial V}\cdot\frac{\partial V}{\partial m}\cdot\frac{\lambda}{w}r = 0 \qquad (A9.7)$$

$$\frac{\partial L}{\partial \lambda} = -\left[H - t - \theta - (I + pe + bk + cx + rm)/w\right] = 0 \qquad (A9.8)$$

Multiplying equations (A9.2), (A9.3), and (A9.4) by dt,dk and dx, respectively, summing the three resulting equations, and rearranging terms yields:

$$\frac{\partial U}{\partial S}\left[\left\{\frac{\partial S}{\partial t} + \frac{\partial S}{\partial e}\cdot\frac{\partial e}{\partial t} + \frac{\partial S}{\partial z}\cdot\frac{\partial z}{\partial t}\right\}dt + \left\{\frac{\partial S}{\partial e}\cdot\frac{\partial e}{\partial k} + \frac{\partial S}{\partial z}\cdot\frac{\partial z}{\partial k}\right\}dk + \frac{\partial S}{\partial x}dx\right] + \frac{\lambda}{w}\left[(w + p\frac{\partial e}{\partial t})\,dt + (p\frac{\partial e}{\partial k} + b)\,dk + c\,dx\right] = 0$$

Simplifying further:

$$\frac{\partial U}{\partial S} ds + \frac{\lambda}{w} (wdt + pde + bdk + cdx) = 0 \qquad (A9.9)$$

Dividing Equation (A9.9) by Equation (A9.5) yields:

$$MRS_{I,S} \cdot dS = wdt + pde + bdk + cdx$$

where

$$MRS_{I,S} = \frac{\partial U}{\partial S} / \frac{\partial U}{\partial I} = - \frac{dI}{dS} \Big|_{U \text{ constant.}}$$

Next, multiplying Equations (A9.6) and (A9.7) by dΘ and dm, respectively, summing the two resulting equations, and rearranging terms as before, yields:

$$\frac{\partial U}{\partial V} \left(\frac{\partial V}{\partial \Theta} d\Theta + \frac{\partial V}{\partial m} dm \right) + \lambda d\Theta + \frac{\lambda}{w} rdm = 0$$

This may be rewritten:

$$\frac{\partial U}{\partial V} dV + \frac{\lambda}{w} (wd\Theta + rdm) = 0 \qquad (A9.10)$$

Once again, dividing Equation (A9.10) by Equation (A9.5), yields:

$$MRS_{I,V} dV = wd\Theta + rdm$$

B. Residential Consumer Survey Questionnaire

The information from this survey will be used by COPEL to analyze and improve the quality of electricity services to customers. All responses will be treated as strictly confidential.

1. What was your average monthly electricity consumption during the last three months?

   ______kWh per month ________ Cr$ per month

2. Do you feel that:

   a. Your electricity supply is good   Yes   No
   b. There are many electric power outages   Yes   No

3. a. On average, how many minutes of unexpected outages per month have you experienced during the last three months?

______________________ minutes per month

b. During what hours is electricity essential for the enjoyment of your leisure?

____

c. Approximately what fraction of outages occurred during these critical hours?

______________________ %

4. If an unexpected outage occurred during these critical hours while you were enjoying your leisure (e.g., watching TV, listening to the radio, reading, having dinner, etc.), how much extra would you be willing to pay to avoid:

a. A one-minute interruption ______________ Cr$
b. A five-minute interruption ______________ Cr$
c. A thirty-minute interruption ______________ Cr$
d. A sixty-minute interruption ______________ Cr$
e. A ninety-minute interruption ______________ Cr$
f. A two-hour interruption ______________ Cr$

5. If an unexpected outage occurred at any other time (e.g. while housekeeping) how much extra would you be willing to pay to avoid:

a. A one-minute interruption ______________ Cr$
b. A five-minute interruption ______________ Cr$
c. A thirty-minute interruption ______________ Cr$
d. A sixty-minute interruption ______________ Cr$
e. A ninety-minute interruption ______________ Cr$
f. A two-hour interruption ______________ Cr$

6. Do you feel that:

a. Electricity is an important service Yes No
b. The service is too expensive Yes No

7. On the average, how do you spend your evening/night-time leisure hours?

   a. Watching TV ______________________ hours
   b. Listening to the radio ______________________ hours
   c. Reading ______________________ hours
   d. Having dinner ______________________ hours
   e. In conversation ______________________ hours
   f. Going out ______________________ hours
   g. Other ______________________ hours

8. When do you normally go to sleep?____________

9. If we were to reduce the incidence of unexpected outages to half its present level, how much extra would you be willing to pay on your monthly electricity bill?

   ______________________Cr$ per month

10. How many persons live in this household?

    ______________________persons

11. a. What are your principal sources of income?

    ____

    b. What was the average monthly income (gross of taxes) of your household during the last three months?

    ______________________Cr$ per month

12. If the level of unexpected outages were to double what reduction in your monthly electricity bill would you consider to be fair?

    ______________________Cr$ per month

References

Anderson, D., "Models for Determining Least-Cost Investments in Electricity Supply", *The Bell Journal of Economics*, vol.6, Spring 1972, pp.267-301.

Becker, G., "A Theory of the Allocation of Time", *Economic Journal*, vol.75, Sept.1965, pp.493-517.

Chow, G.C., "Tests of Equality Between Subsets of Coefficients in Two Linear Regressions", *Econometrica* vol.28, July 1960, pp.591-605.

Fisher, F.M., "Tests of Equality Between Sets of Coefficients in Two Linear Regressions", *Econometrica*, vol.38, March 1970, pp.361-6.

Gronau, R., "The Intrafamily Allocation of Time: The Value of Housewives Time", *American Economic Review*, vol.63, Sept.1973, pp.634-50.

Johnston, J., *Econometric Methods*, McGraw-Hill, New York, 1972.

Lancaster, K., "A New Theory of Consumer Behaviour", *Journal of Political Economy*, vol.74, April 1966, pp.132-57.

Little I.M.D., and J.A. Mirrlees, *Project Appraisal and Planning for Developing Countries*, Basic Books, New York, 1974.

Munasinghe, M., *The Economics of Power System Reliability and Planning*, Johns Hopkins University Press, Baltimore MD, 1979.

Munasinghe, M., and M. Gellerson, "Economic Criteria for Optimizing Power System Reliability Levels", *The Bell Journal of Economics*, vol.10, Spring 1979, pp.353-65.

Munasinghe, M., and J. Warford, "Shadow Pricing and Evaluation of Public Utility Projects", Report No. GAS 14, Energy Department, World Bank, Washington, D.C., 1977.

Priestman, D., "The Cost of Long Term Electrical Energy Shortage Due to Underplanning", Paper presented at the Eleventh Annual Pacific Northwest Regional Economic Conference, Eugene, Oregon, 1977.

Squire, L. and H. van der Tak, *Economic Analysis of Projects*, Johns Hopkins Univ. Press, Baltimore MD, 1975.

Telson, M. L., "The Economics of Alternative Levels of Reliability for Electricity Generation Systems", *The Bell Journal of Economics*, vol.6, Fall 1975, pp.679-94.

Webb, M., "The Determination of Reserve Generating Capacity Criteria in Electricity Supply", *Applied Economics*, vol.22, March 1977, pp.19-31.

Willig, R.D., "Consumers Surplus Without Apology", *American Economic Review*, vol.66, Sept.1976, pp.589-97.

Chapter 10

RURAL ELECTRIFICATION: INTERNATIONAL EXPERIENCE AND POLICY IN INDONESIA[1]

10.1 Introduction

Although the developing countries are turning increasingly to rural electrification (R.E.) to improve the well being of their vast and relatively disadvantaged rural populations, the extent to which R.E. programs should pre-empt scarce resources is still a topic of considerable controversy. Critics have been quick to point out numerous problems that have emerged, ranging from the chronic shortages of capital (especially foreign exchange), high costs and poor quality of supply, to disappointing load growth, low benefits and productivity gains, and perverse distributional effects -- with benefits favoring the rich rather than the poor.

Meanwhile, rural electrification has been promoted by enthusiasts, as a major driving force for the development of the rural areas of the third world. The expectations, particularly in the 1960's and 1970's, included rapid economic growth and improvements in the quality of rural life -- especially of the poor, increasing use of electricity for productive activities, and modernization.

Table 10.1 indicates some selected statistics from a set of representative developing countries. Because of data constraints, these numbers are merely indicative estimates. Given the wide differences in the characteristics of the developing countries, with populations varying from a few hundred thousands to over one billion, GNP per capita ranging from under US$150 to several thousands per annum, and a great variety of resource endowments and sociopolitical structures, it is rather difficult to generalize. However, some interesting points may be deduced from the table.

At first glance, there may be some evidence to support a commonly held hypothesis -- that the extent of rural electrification might be related to the GNP per capita, because higher incomes imply greater availability of resources not only for governments to electrify rural areas, but also for consumers to purchase electricity. However, this is not necessarily always the case. The Asian region, which has substantially lower per capita incomes than Latin America, nevertheless has a comparable or sometimes superior degree of electrification in rural areas. On the other hand, the relatively low income East and West African countries rate rather poorly, with the great majority of them providing less than 5 percent access to rural electricity. Even within regions, the correlation between economic output and progress in R.E. is not very clearcut. Thus in East Asia, China with about half the per capita income of Indonesia exhibits a much higher rate of rural electrification. Similarly, Costa Rica performs much better than most of its wealthier neighbors in Latin America.

---

[1] Edited version of a paper published in the *Bulletin of Indonesian Economic Studies*, vol.24, August 1988, pp. 87-106.

Table 10.1

Selected Statistics from Some Typical Developing Countries[a]

	Mid-1983 Population (Millions)	1983 GNP Per Capita (US$)	% Pop. Served by Electricity			% Villages Electrified	
			Urban	Rural	Year	Current Year	Projected 2000
East Asia & Pacific							
China	1019.1	300	>80	60	1982	[illegible]5 (1981)	90
Fiji	0.7	2050	50	25	1982	-	-
Indonesia	155.7	560	39	10	1984	25 (1984)	50
Malaysia	14.9	1860	85	55	1983	[illegible]0 (1983)	95
Philippines	52.1	760	40	22	1980	50 (1983)	95
Thailand	49.2	820	78	40	1984	-	-
South Asia							
Bangladesh	99.5	130	20	2	1981	[illegible]2 (1983)	90
India	733.2	260	<25	15	1981	[illegible]5 (1983)	90
Pakistan	89.7	390	<25	15	1981	[illegible]7 (1983)	90
Sri Lanka	15.5	330	35	8	1982	-	-
East & West Africa							
Burkina Faso (Upper Volta)	6.5	180	<15	1	1980	3 (1980)	<10
Ethiopia	40.9	120	10	<1	1982	0 (1978)	<10
Guinea	5.8	300	21	4	1982	-	-
Ivory Coast	9.5	710	93	20	1981	-	-
Liberia	2.1	480	86	4	1982	-	-
Kenya	18.9	340	<10	<2	1982	6	18
Senegal	6.2	440	83	12	1982	-	-

Table 10.1
(continued)

Selected Statistics from Some Typical Developing Countries[a]

	Mid-1983 Population (Millions)	1983 GNP Per Capita (US$)	% Pop. Served by Electricity			% Villages Electrified	
			Urban	Rural	Year	Current Year	Projected 2000
Mediterranean & Middle East							
Algeria	20.6	2320	>95	33	1978	>95 (1980)	>95
Egypt	45.2	700	65	23	1980	90 (1982)	>90
Jordan	3.2	1640	83	60	1982	-	-
Morocco	20.8	760	60	6	1978	-	-
Syria	9.6	1760	80	20	1978	-	-
Tunisia	6.9	1290	90	25	1982	43 (1983)	90
Turkey	47.3	1240	>95	50	1981	-	-
PDR Yemen	2.0	520	45	6	1978	-	-
Yemen AR	7.6	550	74	4	1982	-	-
Latin America							
Argentina	29.6	2770	>95	5	1981	-	-
Bolivia	6.0	510	72	9	1981	-	-
Brazil	129.7	1880	>95	19	1981	3 (1978)	50
Chile	11.7	1870	>95	42	1981	-	-
Colombia	27.5	1430	84	13	1978	-	-
Costa Rica	2.4	1020	>95	>95	1982	-	-
Ecuador	8.2	1420	79	13	1980	-	-
Mexico	75.0	2240	-	<20	1982	25 (1983)	50

[a] The World Bank, Asian Development Bank, USAID, Mason (1985) and author's estimates.

Thus the table shows that wide variations in the degree of rural electrification exist not only between different regions of the world, but also among the individual countries within a given region. Africa provides some of the most striking examples, with the Ivory Coast (and to a lesser extent, Senegal) being well above the regional average. There does appear to be a general link between high levels of access to electricity in rural areas and the period of time for which serious R.E. efforts have been pursued, in a given country. Therefore, nations which have long-standing, well established R.E. programs, with strong government support, tend to have fared better. This point is discussed at greater length below, in the section on institutional frameworks.

Urban access to electricity figures tend to be significantly greater than the corresponding numbers for rural areas, except in a few countries like China. Although the data on villages electrified, shown in the last two columns of Table 10.1, is rather sketchy, it tends to confirm the impression that the percentage of villages electrified is generally significantly greater than the fraction of the rural population served. This most probably implies that bringing power to a rural community does not automatically guarantee that all the residents are both willing and able to connect.

Speaking rather broadly, R.E. schemes would generally cover regions where agricultural activity (including agro-industries) was dominant, the ratio of labor to capital used in production was high, average incomes were small relative to urban areas, and load densities would also be relatively low. Often small or medium sized towns (population up to 50,000) in a rural area may be included, or suburban areas of larger cities. In fact, from the socio-economic and demographic viewpoint, the rural-urban gradation may often be continuous, and the terminology "regional electrification" is sometimes used to indicate programs covering a mix of rural farms, villages and small towns.

Electricity can be introduced into rural areas in several ways, including: isolated generators serving a single consumer (for example a business enterprise); isolated generators serving several consumers who are connected by a local network; and public supplies from a regional or national grid system. Typically, these are not distinct ways to introduce electricity into a rural area, -- rather, they might occur sequentially. This procedure may be thought of as an initial building up of demand for rural electricity, by means of autogeneration, to a point where it becomes economically feasible (in most cases) to connect to a central grid (The World Bank 1975).

Most often, the term "rural electrification" refers to connections to a central grid, since major R.E. programs under consideration are generally those which are sufficiently advanced to make this economically feasible. This is not to say that the most economical method for electrifying every region in every country once the load has built up, is from a central grid. Even in North America and Europe, where the extent of R.E. is almost 100 percent, many areas continue to be served by local autogenerators as this is the least cost way of providing electricity to such areas.

Some specific approaches to R.E. include:

1. Integrated rural development -- electricity is treated as only one component of a wider package of infrastructure.

2. Area coverage -- comprehensive networking seeks to quickly provide electricity supply to as many customers as possible within the designated project area. More modified approaches may first adopt a backbone system, with increasing numbers of consumers being added on over a phased time period.

3. Grid extension -- electrification is based on proximity to existing networks and their relative ease of extension. Each separate branch addition is judged on its own merits.

4. Isolated generation -- electrification in areas remote from the national grid, is based on relative cheapness and abundance of local generating sources.

5. Intensification -- concentrating on increasing the load factor and connection rates in already electrified areas.

In recent years, an increasing number of developing countries have begun to place greater emphasis on the productive uses approach, encouraging electricity consumption for agricultural pumping, agro-industrial and commercial activity, cottage industries, and so on. There are several important reasons underlying this trend, as described later in the Indonesia case study.

Because of the substantial investments required to accelerate the pace of rural electrification, government involvement of some type becomes necessary, and most developing countries are pursuing R.E. programs as a part of their economic development efforts. This in turn implies that there are significant advantages to examining R.E. issues in the context of overall national policy. The broad rationale underlying national level planning and policymaking of all kinds is the need to ensure the best use of scarce resources, in order to further socio-economic development efforts and improve the quality of life of citizens. Planning for energy and electric power, including rural electrification in particular, are all components of national economic planning, and therefore share this same general goal.

Meanwhile, because of the many interactions and non-market forces that shape and affect the energy sectors of every economy, decision makers in an increasing number of countries have realized that energy sector investment planning, pricing and management should be carried out on an integrated basis, e.g., within a integrated national energy planning (INEP) framework which helps analyse the whole range of energy policy options over a long period of time[2]. In summary, R.E. planning must also be closely integrated with overall economic, and energy planning and policy analysis, to meet many inter-related and frequently conflicting national objectives, as effectively as possible.

10.2 Review of Rural Electrification Benefits

In this section, we briefly identify the main potential benefits of R.E., and then review some of impacts based on the actual experience of rural electrification in developing countries.

Some typical benefits (both quantifiable and non-quantifiable) of R.E. schemes have been described already in Chapter 6 and listed in Table 6.1. For convenience, this table is reproduced below as Table 10.2. While the quantifiable benefits of R.E. are self-explanatory, the five broad types of non-quantifiable benefits may be summarized as follows:

---

[2] See the companion volume: *Energy Analysis and Policy*, Chapters 2, 3 and 4.

1. In parallel with the wide-spread modernization and dynamic growth that give rise to quantifiable productivity gains, there are also beneficial catalytic effects due to revitalization of rural communities and attitudinal changes.
2. Social benefits accrue due to overall improvements in the quality of life.
3. Income distribution and social equity might be improved if R.E. benefits are appropriately targeted to low income groups.
4. There could be significant employment impacts following electrification of a rural area that yield not only direct output gains but also intangible improvements in personal satisfaction, family welfare and social stability.
5. General benefits at the national level may also be relaized, including improved political stability and national cohesion, as well as reduced rural to urban migration, interregional tensions and inequalities.

Some results from the economic evaluation of recent R.E. projects were also listed in Table 6.2. The same information is given below in Table 10.3 for convenience and completeness. Incremental revenues provide most of the benefits, except in Bangladesh and India -- which are both low income countries where the ability-to-pay is limited. The internal economic rates of return (IERR) exceed the opportunity costs of capital (8 to 12 percent) in most cases, except in Egypt where rural electricity prices are heavily subsidised.

Table 10.2

Typical Examples of Rural Electrification Benefits

Quantifiable Benefits: Cost Savings and Increased Productivity

Industrial and commercial uses of electricity
i. Motive power - replacing liquid fuel
ii. Lighting - replacing liquid fuel or gas
iii. Space heating, cooling and refrigeration - replacing liquid fuel, coal, gas, biomass or animal waste
iv. Processing food - replacing liquid fuel, coal, gas, biomass or animal waste
v. Transport - replacing liquid fuel

Household Uses of Electricity
i. Lighting - replacing liquid fuel, gas, biomass, or animal waste
ii. Preparing meals - replacing biomass, animal waste, liquid fuel, coal or gas
iii. Space heating, cooling and refrigeration - replacing biomass, animal waste, liquid fuel, coal or gas
iv. Home appliances (fan, iron, radio, television etc) - replacing batteries, biomass or coal
v. Drinking water - replacing liquid fuel (for pumping)

Agricultural Uses of Electricity
i. Water pumping - replacing liquid fuel, coal, gas or muscle power
ii. Parboiling, heating and drying replacing biomass, coal or liquid fuel
iii. Milling, chaff cutting, threshing, etc. - replacing liquid fuel, hydro and muscle power, coal or biomass

Benefits that Are Difficult or Impossible to Quantify

i. Modernization, dynamism and attitude changes - catalytic effects
ii. Qualify of life, community services, and participation
iii. Income redistribution and social equity
iv. Employment creation
v. Other sociopolitical effects

Table 10.3

Benefit Measurement and Internal Economic Rate of Return in Selected Rural Electrification Projects (1975-84)

Country	Incremental Revenues (% of Total Benefits)	Internal Economic Rate of Return (%)
Bangladesh	<5	11
Colombia	55-100	8-13
Egypt	65-85	3-10
India	<10	17-70
Ivory Coast	-	15
Malaysia	100	17
Morocco	100	6
Philippines	50	16-22
Syria	95	8
Thailand	30-60	12-15
Tunisia	60-65	8-12
Yemen (Arab Rep.)	100	16-18
Yemen (People's Dem. Rep.)	40-85	13-15

Sources: World Bank, Asian Development Bank, Inter-American Development Bank, USAID.

The economic basis for claiming additional essentially unrecognised social benefits is based on the argument that the "social" demand curve for electricity lies to the right of the user's private demand curve -- the social surplus benefits are usually not internalised directly within the consumers demand curves, because they are not perceived as benefits by individual users. However, it may be claimed that such benefits ought to be included within the area under the social demand curve which indicates the willingness-to-pay of society as a whole (if they yield identifiable gains to the rural community). Since it is very difficult or almost impossible to quantify these benefits, great care and discipline must be exercised to avoid the temptation of using such claims to justify an R.E. project that otherwise would not have been viable.

While many recent studies of past R.E. efforts in the developing countries are available, the disciplines and viewpoints of the authors, and the rigour with which the data has been collected and analyzed, vary widely. Sometimes, different analysts have provided conflicting interpretations of the same set of facts. Nevertheless, the following broad consensus appears to have emerged concerning the impact of rural electrification in developing countries[3].

[3] Among the many reviews available, the following proved to be the most useful: Barnes (1985), Butler et.al (1980), Cecelski (1982), Fluitman (1983), Kessler et.al (1981), Madigan (1981), Mason (1985), National Electrification Administration (1978), Ross (1972), Samanta and Sundaram (1983), Saunders et al.(1975), Smith (1983), Tendler (1979), USAID (1981), Velez et al. (1983).

1. R.E. promotes agricultural development best when certain complementary inputs such as electric pumps, financial services, and others mentioned earlier, are also available. However, subsidized power prices may conceal the reality that the economic costs of electric pumping often exceeds the costs of the alternative, especially in a period of falling oil prices.

2. R.E. appears to stimulate agro-industrial and commercial activity, although the direction of causality is not completely clear. In other words, rural areas that are ready for sustained growth are likely to be selected for early electrification, and will also exhibit rapid demand growth. There has been a systematic tendency to overestimate productivity gains in the industrial and commercial sectors during the economic appraisal of R.E. schemes.

3. Load growth and beneficial impacts are often hampered by not recognizing the dynamic nature of R.E. Poor planning, inadequate maintenance and follow-up, overloading of systems, and poor supply quality, are some of the undesirable aspects.

4. Domestic consumers perceive a significant improvement in their living conditions, mainly due to the improved quality of lighting, for which they are willing-to-pay far more than their prior expenditures on kerosene might suggest. Access to entertainment via TV and radio is also a contributory factor. However, electricity rarely replaces other fuels for cooking or heating, because of expense or convention.

5. There is a marked tendency for the benefits of R.E. to accrue mainly to the better off and influential persons (wealthy landowners, politicians, and local officials), unless the program is well designed and the poorer beneficiaries carefully targeted. Thus, R.E. is generally not a good method for achieving income re-distributional or social-equity oriented objectives. Furthermore, care must be exercised to ensure that perverse effects do not occur, such as electricity revenues obtained from the urban poor being utilized to subsidize the rural rich.

6. Reading skills, literacy levels, and education are positively correlated with electrified households, but the direction of causality is not clear. Thus, more ambitious and motivated families may be the ones who would acquire an electricity connection first.

7. Farmers and households report greater security and order due to electric lighting, in relation to crime and theft, as well as protection from dangerous animals such as snakes.

8. There is little or no evidence for many other assumed benefits like increased employment opportunities, reduced rural to urban migration, lower birth rates, and so on.

The continuing ex-post analysis of R.E. schemes in the developing countries should continue, in order to resolve some of these ambiguities, and to tailor future programs for greater effectiveness in the widely varying conditions found in different parts of the world.

10.3 Rural Electrification Policy Analysis in Indonesia[4]

The next two sections provide a description of the main rural electrification problems in Indonesia -- the main purpose is to illustrate how some of the principal policy issues discussed earlier, may be identified and analysed. This process would help to provide decisionmakers in Indonesia with a better understanding of the broad R.E. options and strategies available. However, in view of the size and complexity of the country concerned, it is not possible to provide a detailed analysis of all aspects of R.E.

Sector Background

Rural electrification issues have become increasingly important to the Government of Indonesia (GOI), especially since the mid-1970's when the widening gap between the availability of electricity (and other services) in urban and rural areas emerged as a major area of concern. During the past few years, considerable progress has been made in addressing a number of important issues. However, the government still faces significant difficulties in meeting its R.E. goals, given the existing low levels of electrification, low incomes and affordability in rural areas, and the scarcity of resources -- especially during a period when oil export revenues have declined dramatically.

In 1984, Indonesia had a per capita gross national product of about US$550 and a population of 165 million, with nearly 80 percent of this total living in rural areas. Furthermore, almost 100 million people (or 60 percent) were living on Java Island, which comprises only 9 percent of Indonesia's land area. Thus Java supports a high population density of about 700 persons per square kilometer, a value that is much higher than in the other islands. Only about 15 percent of Indonesian households received electricity in 1984, and even this low level of electrification was achieved because the corresponding figure for the capital city of Jakarta was above 40 percent. In the rural areas, barely 10 percent of the homes have electricity connections, and only about 25 percent of the estimated 65,700 villages have been effectively electrified (neglecting those with token electrification, where only a handful of households are connected). Over half these R.E. schemes are supplied by Perusahaan Umum Listrik Negara (PLN), the state owned power utility, while the rest are served by a combination of other government institutions and private (unlicensed) local operators. Given the low rate of electrification, it is not surprising that the national average per capita electricity consumption for 1984 was approximately 100 kWh per annum. Practically all electricity related indicators are low relative to countries at comparable levels of per capita income and development, and especially relative to other ASEAN nations -- a fact that GOI has become increasingly conscious of.

According to the Repelita IV program, or national development plan for the period 1984-89, GOI expects PLN to connect a further 1.6 million rural households, at a cost of almost 1 trillion rupiah (Rp.) in 1985 constant prices (which is 12 percent

---

[4] This case study is based on Munasinghe (1982). The information in the original report has been updated on the basis of more recent data obtained from the Indonesian authorities, US Agency for International Development, Asian Development Bank and World Bank. A useful discussion of policy issues is contained in: McCawley (1978); Sen (1982); and Pabaney (1983).

of PLN's investment program, and 2 percent of total public investment, during the same period). This is expected to be followed by the electrification of another 2.4 million homes during Repelita V (1989-1994), requiring about Rp. 1.4 trillion more. In the present climate of resource scarcity, such a program will strain the government budget significantly, while achieving an overall 1994 target of only 40 percent of all villages (electrified by PLN).

The low income levels and inability to pay for electricity services in rural areas, indicate that most R.E. schemes will not be financially viable for at least a decade. Therefore, the mobilization of both government and non-government resources, as well as the efficiency of rural electrification investments, are critical areas of concern. Rural electrification policy development also needs to be pursued, including further elaboration of the national R.E. master plan, clearer identification and prioritization of specific objectives, and better criteria for selecting target groups and areas. Finally, a significant effort to rationalize the institutional framework for R.E. will yield significant benefits. In particular, the tendency to place the increasing burdens of rural electrification on PLN, without providing the additional financial, technical, and managerial resources, or ensuring adequate leadership, is likely to further erode the capability of PLN in its more critical role as the sole national power utility.

National Objectives of R.E. Policy

Two major areas in which further policy development must be pursued relate to the objectives of rural electrification, and the criteria for selection of areas to be electrified. The broad developmental objectives of GOI which have direct relevance for rural electrification and energy policy are:

1. Achieving a high rate of economic growth; and

2. Realizing a more equitable distribution of income and social welfare.

These two principal goals correspond to the broadly defined objectives: efficiency-growth and equity-income distribution. Many other more specific objectives are also considered important by GOI, including: energy conservation and elimination of wasteful use, reducing dependence on foreign energy sources, balanced regional development (Java versus the other islands, rural versus urban, etc.), ensuring price stability and supply continuity, ensuring financial viability, and preservation of the environment.

As discussed earlier, the economic efficiency objective implies that rural electricity should be produced using the least-cost supply technology, and sold at prices reflecting the long-run marginal costs of supply. Economic growth and financial cost recovery considerations also indicate that consumers who are likely to use electricity most productively should be targeted first. Pursuing R.E. schemes in areas that are relatively close to the existing grid, is also consistent with the cost minimization principle. On the other hand the equity objective suggests that the basic electricity needs of the poorest and most disadvantaged rural groups should receive priority. This in turn implies relatively costly electrification of remote areas and significantly higher subsidies, because consumer affordability would be low. The past approach to R.E. in Indonesia has tended to implicitly favour the distributional viewpoint, although there is increasing acceptance of the growth objective. More importantly, the agencies involved in rural electrification have not been provided a clear enough view of the relative importance of the different goals, and the tradeoffs involved.

There are a number of reasons why greater reliance on the productive use criterion might be beneficial in Indonesia, without necessarily discarding completely, the equity goal. Such an approach is more likely to assure faster load growth and cost recovery, thus facilitating expansion of rural electrification efforts to more remote areas. At the same time, if redistributional considerations are particularly important, other more basic needs such as water supply, health, education and nutrition, may have a greater claim on scarce resources, than electricity. Since potential growth poles where the productive use of electricity is more prevalent, are likely to be better endowed with complementary infrastructure such as transport and other services, the synergistic effects underlying the integrated rural development approach indicate that R.E. schemes will be particularly effective in such areas. Finally, increasing output and employment through electricity induced productivity gains may be a better mechanism for improving welfare in the long-run, rather than subsidized and poorly targeted rural electrification schemes.

Area Selection and Access to Electricity

The relative importance accorded the objectives of R.E. will have a significant impact on the selection of areas and timing of rural electrification projects, as well as cost recovery and pricing policy. Furthermore, from the viewpoint of successful implementation, agreement on, and firm adherence to policy objectives and criteria are important, to avoid ad hoc interpretations and politically motivated interventions in area selection, that can seriously hamper R.E. programs.

The 247 kabupaten (provinces) in Indonesia are convenient administrative units for beginning the process of area selection. They are sufficiently large to be relevant for the laying out of R.E. backbone systems, but also small enough for discriminating on the basis of more localized socioeconomic and demographic factors. These arguments will need to be modified by practical considerations, in certain cases. For example, individual rural electrification networks need not necessarily coincide with administrative boundaries, especially where a kabupaten encompasses several islands outside Java.

Within the kabupaten, GOI has identified about 65,000 desas (villages) falling under village governments and comprising both groups of houses and isolated farms. In 1983, the Ministry of Home Affairs (MHA) classified these villages into three categories: swasembada (modern), swakarya (transitional), and swadaya (traditional), accounting for 16, 41 and 43 percent of the total number of desas, respectively. The classification was carried out according to seven indicators: level of modernization, level of production, sources of income, level of education, availability of infrastructure, existence of village institutions, and community spirit. Within this framework, swasembada villages are entitled to electricity services as early as possible, with the underlying goal that all villages should be eventually brought up to the swasembada level. Java has the highest proportion of swasembada desas (29 percent), while Kalimantan has the lowest (4 percent).

The MHA categorization is a useful starting point for the existing area selection criteria, which give higher priority to swasembada villages nearest the existing grid, and kecamatan (subdistrict) seats. However, this classification is not necessarily an ideal basis for identifying the longer term growth potential of villages, and therefore might be usefully supplemented by other criteria for the selection of areas to be electrified. First and most importantly, a better effort could be made to identify productive uses of electricity, in activities ranging from cottage industries and agricultural pumping, to commercial ventures and agro-industries. Second, household income, especially the ownership of consumer goods (rather than non-cash income) would be a useful indicator of future load growth. In particular, the consumption of energy which may be substituted for by electricity, such as the use of Petromax lamps, ought to be studied more. Finally, the pattern of land ownership needs to be analysed, because this affects the number and ability-to-pay of potential consumers. Much of this data is already available or obtainable without major effort, on the basis of existing surveys or modest extensions thereof. For example, a recent survey of Central Java indicated the following pattern of appliance use in electrified households: lights - 100; cassetes - 35; TV - 34; irons - 31; radios - 27; fans - 4; water pumps - 4; refrigerators - 2; powertools - 2; and sewing machines - 1 percent (US Bureau of Census 1985).

Over 80 percent of productive users, consisting mainly of cottage industries, small shops and commercial enterprises, rice mills, agro-industries, small industries, and restaurants, have obtained connections in electrified areas of Java. The chief uses of electricity were lighting, motive power, and small equipment. Rural businesses tend to use electricity mainly for lighting, and therefore, the average consumption per connection is relatively low (83 kWh per month in Central Java). Small industries (65 percent) and restaurants (14 percent) had unusually low connection rates, probably because they had already purchased significant amounts of equipment using other forms of energy. However, provided that correct energy pricing policies are adopted, over a period of time, connection rates in electrified areas would tend to rise as older equipment was gradually replaced with electric appliances.

Both businesses and industrial establishments have reported increases in output and profits as well as cost savings, following electrification -- especially small industries like garments, textiles, and tool making; certain farm activities like poultry rearing; and agro-processing industries such as soybean factories, and rice mills (Brodman 1982, US Bureau of Census 1985). However, a number of constraints caused by price distortions, poor information, and unavailability of capital, have delayed the switch to electricity. These issues must be analysed and addressed, for example, by establishing specific rural credit programs for electrification loans, publicising other cases of successful productive use more widely, and helping potential industrial and commercial consumers in thinking through the economic and financial implications of the decision to electrify. The information generated through this process would be very helpful in the process of selecting and prioritizing areas for R.E., to promote greater productive use of electricity in the rural regions of Indonesia. While it is clear that dramatic growth and modernization of rural areas will not occur overnight, integrated and coordinated rural development programs (including electrification) could add considerable impetus to ongoing efforts.

Review of the policy on extensions from the grid may be helpful. Present area selection criteria favour villages within 3 km. of the grid. It appears that the rule may appropriately reflect the breakeven distance that equalises costs and benefits for relatively small villages (about 300 homes), of average income. However, too rigid application of this criterion may be unduly conservative. Typical villages with about 700 households, and average or above average incomes, may justify extending the breakeven distance up to 10 km. or more.

Another policy consideration that might be reviewed is the market penetration in villages that are already electrified or earmarked for electrification. Greater emphasis on an intensive approach to rural electrification by targeting specific sites and consumer groups, is likely to yield quicker returns than an extensive R.E. program that would spread increasingly scarce resources ineffectively. Increasing the market penetration is particularly important, because it has a significant impact on both the economic cost-benefit analysis, and the financial cash flow, thus making many otherwise unviable R.E. schemes potentially attractive.

In this context, the currently accepted rate of initial connections (35 to 50 percent of households), may be somewhat conservative. Modest resources devoted to a policy of maximizing connections could yield high marginal returns. Such a program might involve: greater promotional efforts, making rural credit available for meeting initial connection fees as well as the costs of housewiring, and revising the connections policy in the light of the ability-to-pay of poorer households (who may amount to as much as 50 percent of a typical village). Since these aspects are inter-related, any revised policy package must be carefully coordinated. For example, up to three fourths of the homes in an average village could afford the minimum monthly electricity bill for lighting (about Rp 1250 for 10 kWh), but less than a third would be able to meet the connection fees (Rp 30,000) and housewiring costs (Rp 35-70,000). Therefore, the connection rate may be improved by a judicious program of credit facilities and reduced connection charges. Further gains are possible through tariff policy adjustments, but the negative implications (of increasing the subsidy level) on resource mobilization also needs to be considered (see below).

Recent information indicates that many villages are served with electricity, often at a rudimentary level, by unlicensed and unregulated private operators. This phenomenon is also helpful to identify regions in which significant potential demand and willingness-to-pay might exist. For example, a recent survey indicated charges (to rural households consuming under 10 kWh per month) ranging from Rp 200-700 per kWh for private electricity services, compared to the baseline average price of Rp 120-160 for much better quality supply from PLN (Ministry of Mines and Energy 1985). Such information could be usefully integrated into the site selection process.

10.4 R.E. Institutional Framework and Resources in Indonesia

Another major area of concern for the policymaker is the organizational framework that would enhance R.E. implementation. The mobilization of financial resources is a related topic that is also discussed in this section. Both aspects are sensitive, and the policy options are likely to be strongly limited by the existing situation and constraints, most of which have been historically inherited.

Present Organizational and Financing Framework for R.E.

When GOI first considered the promotion of R.E. efforts some years ago, there was no clear consensus as to how the desired results might be best achieved. Consequently, an open-minded attitude prevailed, and several different approaches were encouraged. This process resulted in the existing situation that is characterised by a number of essentially autonomous groups with somewhat different attitudes and mandates to R.E., and having ties to various donor agencies and sources of funds.

There are two principal institutional frameworks for R.E. in Indonesia, that are vertically integrated, with limited cross links:

1. The Ministry of Mines and Energy (MME), its Directorate General for Energy and New Energy (DGENE), and PLN.

2. The Ministry of Cooperatives (MC), its Directorate for the Promotion of Electric Cooperatives (DPEC), and the Project Development Office (PDO).

Several other ministries and provincial governments, as well as private entrepreneurs, are also responsible for a more modest amount of rural electrification.

An important problem to be solved, is how GOI might ensure greater policy coordination and leadership of R.E. work. The diversity of entities involved and fragmentation of overall rural electrification efforts, have made it difficult for a coordinated national R.E. plan to emerge and be successfully implemented. Targets and programs have evolved in an *ad hoc* manner rather than within the framework of a consistent master plan. The involvement of many bilateral and multilateral donors has further complicated the scene, especially with regard to the criteria and standards used in various R.E. projects.

At present, MME through DGENE has the overall responsibility for national rural electrification policy development, coordination, and regulation, including service area allocation, the setting of uniform engineering standards and practices, pricing policy, and performance monitoring. However, its role has not yet been comprehensively defined by GOI, and policies are still in the formative stage. Moreover, the specialized staff and other resources available to formulate, analyze and implement R.E. policy need to be increased.

As indicated earlier, PLN is the chief executing agency for rural electrification schemes, and has a special R.E. subdirectorate (RES) to coordinate this activity. However, while most of the planning and programming is carried out by RES, the actual implementation and operations are done by PLN's decentralised structure: the Wilayah (regional distribution) and special construction project offices. Although PLN has been quite successful, and consistently exceeded its R.E. targets during the last few years, the impact of these efforts on its other operations due to resources diverted, is unclear. During the next decade, over 10 percent of PLN's investment budget will be devoted to R.E., and there is a possibility that the management might increasingly perceive this an unwelcome burden that will drain away manpower and resources from other revenue generating activities -- an attitude that is not uncommon in many developing country utilities. The status accorded to the subdirectorate may also weaken its identity and morale, thereby increasing the risk of urban standards and criteria being applied to rural electrification problems. More importantly, this may make it more difficult for RES, under the present conditions, to deal authoritatively with other national and foreign agencies in key areas like the establishment of engineering standards.

MC functions as the other chief executing agency through DPEC and PDO. Their efforts are aimed at the creation and support of rural electric cooperatives, mainly with USAID assistance. Three major demonstration projects have been implemented in Lampung, Lombok, and Luwu, serving about 12,000 consumers. There is some cooperation between PLN and DPEC in administering a joint MME-MC program by which general village cooperatives (KUD's) are being trained to take over the operation of R.E. schemes constructed by PLN. Meanwhile, the MC-PDO projects establish and support diesel generation based cooperatives, but rely almost exclusively on foreign funding and technical assistance. PLN schemes have their own design standards and criteria, that are not necessarily optimized. All these agencies provide technical assistance and training to their cooperatives, but no support at all to private rural electric operators. Most R.E. schemes are characterised by lengthy waiting periods due to a variety of reasons, including unnecessarily lengthy procedures, lack of materials, and insufficient system capacity.

Several provincial governments have also begun local efforts to supplement the national level programs described earlier. In Central Java, a separate R.E. agency has been established, while the East Java provincial government provides additional funds to enable PLN to increase its existing program. The West Sumatra government is also active in the R.E. area, including the development of mini-hydro sites. DGENE and PLN also have broader, national level new and renewable energy programs. Finally, as noted earlier, there is a significant number of villages served by private entrepreneurs, who have emerged spontaneously.

The sources of funding for the various R.E. activities exhibit an equally complex pattern. PLN has received the bulk of its resources from GOI funds earmarked for R.E. and foreign donors, but has sometimes supplemented these sums with transfers from its non-R.E. budget, and grants from provincial governments. The MC-DPEC-PDO group's activities have been supported through a combination of foreign and GOI financing. The Ministry of Home Affairs administers an independent program of village development grants for R.E., and the Ministry of Finance provides funds to meet the connection costs of PLN consumers through the rural electrification credit program implemented by the Bank Rakyat Indonesia. This complicated web provides little scope for stable and coordinated resource mobilization in the sector, or for the encouragement of rational criteria for monitoring the economic and financial viability of R.E. schemes. At the same time, there is hardly any effort being made to use market forces and tap non-governmental resources that are available at the village level.

Options for Institutional Restructuring

As indicated earlier, one of the principal needs is to strengthen sector leadership. In theory, GOI could restructure its institutional framework in a number of different ways ranging from the creation of a monolithic centralised R.E. entity (e.g., like the Rural Electrification Corporation - REC, of India), to the more or less ad hoc promotion and support of autonomous R.E. cooperatives. Given the existing situation, however, an institutional structure along the lines suggested below would be easiest to achieve, while eliminating most of the present problems.

First, a new body could be set up to facilitate the financing of R.E. schemes. Such an agency for rural electrification (ARE) might report directly to the Minister of Mines and Energy or the Minister of Finance, and would be responsible for finding external and domestic resources for R.E. investments, verifying the economic, financial and technical viability of R.E. schemes submitted to it by the implementing agencies like PLN, monitoring the continued financial health of rural coops, and so on. Thus ARE would enable GOI to oversee the mobilisation and allocation of resources directed towards R.E., while permitting PLN to pursue its goals as a commercially viable entity that is responsible for meeting the overall power needs of the nation.

All major groups involved in R.E. like DGENE, DPEC, PLN, PDO, regional governments, etc., would be represented on ARE's Board of Directors. The staff of this institution would consist mainly of financial analysts and economists, plus some engineers. The success of R.E. efforts in other countries has depended critically on this resource mobilisation issue. For example, REC of India and NEA of the Phillipines have been very active in this regard, although in the proposed Indonesian model, the financing agency may have a more circumscribed role if the establishment of policy and all technical standards and criteria remain the responsibility of MME and PLN.

MME-DGENE would retain its responsibility for establishing national R.E. objectives and broad policy. The views of other interested agencies like AER, MC, DPEE, PDO, the regional governments, MHA and MF, could be incorporated through regular consultations or a committee system, but with the DGENE playing a guiding role. At this level, the early establishment of a definite national R.E. plan (such as in India or the Phillipines) would eliminate many of the existing problems. R.E. policy should clearly define the relative weights to be assigned to different GOI objectives like improving the quality of rural life or providing basic electricity needs, and promoting productive uses of R.E. MME would also continue to licence schemes and allocate to the various agencies involved, the different areas to be electrified, although PLN is likely to maintain its dominant role in implementation, as explained below.

It is logical to retain PLN as the principal agency for implementing R.E. policy. The creation of a separate Directorate for R.E. within PLN may improve the implementation of the R.E. national plan by increasing the importance of rural electrification and giving it a definite identity within PLN, while enabling PLN to take the lead in its dealings with sister agencies. Because of the special characteristics of R.E. work, the financial and technical aspects should be separated as far as possible from PLN's other activities.

For example, at the moment it is not easy to isolate rural electrification investments from the total PLN program. Since PLN has performance criteria to meet based on its overall operations (e.g., GOI requirements, convenants with donors etc.), the R.E. accounting framework should be completely divorced from the rest of PLN's balance sheet. This would virtually eliminate the fears about expanded R.E. activity eroding corporate financial viability, and help to clearly identify rural electrification resource needs as well as any hidden cross-subsidies from urban to rural electricity consumers. On the technical side, R.E. planning should be established as a distinct activity within PLN, while construction, operation and other fieldwork could be more closely integrated with PLN's regular Wilayah-level activities.

MC, DPEC, PDO, the regional governments and private autogenerators would also continue to play a supporting role. This would be helpful to PLN provided there was better policy coordination by MME, and it would also be consistent with GOI's original policy of pursuing a more decentralised program. It was this very flexibility of GOI and the lack of a national R.E. plan that led to different aid donors (US, Canada, Holland, and multilaternal lenders) picking up parts of the R.E. program in a piecemeal manner, as well as the adoption of different criteria and design standards, etc. These difficulties would be resolved under the new arrangement, because PLN would be responsible for establishing all technical standards and interpreting implementation criteria.

As mentioned earlier, there are almost 17,000 rural electric networks that are privately operated, in Indonesia. While hardly any of them are able to match the technical standards, service quality and power prices offered through the official PLN program, the private R.E. schemes nevertheless demonstrate the existence of substantial resources and willingness-to-pay for rural electrification. The mobilization of these resources on a more systematic basis would be extremely helpful in meeting the substantial shortfall between the official Repelita R.E. program targets and actual funds available to the government. In general, village level organizations such as the private rural electric operators and rural cooperatives could play a vital role in this process. The government, through PLN and other relevant agencies could coordinate the activities of such village organizations, and provide the necessary policy framework, and technical and financial support, to enable the latter to more effectively electrify rural areas, and supplement GOI efforts. Since there are a wide variety of private R.E. schemes, and little accurate information on them, it is more helpful to examine the functioning of both rural electric cooperatives and general agricultural cooperatives in the context of their expanded role as suppliers of electricity at the village level.

MC, through PDO, with guidance from USAID and NRECA (National Rural Electrical Cooperative Association of USA) have pursued the approach of setting up autonomous full-fledged R.E. coops in project areas. These cooperatives receive financial and technical support as well as assistance in staff training from GOI. Such a system has been most successful in countries where a strong central agency was first created to provide the coops with initial asistance, as well as continued monitoring of their activities (e.g., REA in the US or the National Electrification Authority or NEA in the Phillipines). Sociological conditions also have to be favourable to ensure the active participation of local consumers in the management and operation of the coop. The inability of PDO to reach its targets may be caused more by the inadequacy of the resources and backing it receives and the limits of it's capability in relation to the original Repelita R.E. program, rather than by any inherent weakness or unsuitability of the cooperative approach to R.E. in Indonesia.

MME and PLN have sought to follow a more gradual method of building up electricity related activity within general agricultural cooperatives in rural areas. For example, in the first stage of the joint DPEC-PLN program, an existing rural coop (KUD) would be requested to undertake simple meter-reading and billing tasks, while all construction and operations activities would be carried out by PLN. Next, the KUD begins to perform basic operation and maintenance (O&M) activities like providing service hookups to new consumers, and line and transformer maintenance. In the final stages, the KUD's R.E. activity would become fully autonomous. The coop would then be responsible for the entire network, either from the distribution substation downwards (if it buys bulk power from the PLN grid), or including the generating plant itself (in the case of isolated systems). PLN would be responsible for training and guiding KUD staff throughout the process, and would continue to provide backup afterwards. The PLN-DPEC program assumes that a KUD would be able to undertake the management and financial responsibilities for R.E. activities rather quickly because of prior familiarity with these aspects through its non-R.E. work, while the more gradual approach described would be more prudent in the context of technical work.

Although it is too early to evaluate the effectiveness of the above approach, it is useful to examine some of the issues. Three criticisms of the KUD approach have been advanced. First, it could become extremely time consuming, especially if the KUD was unwilling to advance towards the final stages. Second, since the KUD and its members are chiefly concerned with agricultural activities, there would be a tendency, at least initially, to downplay or neglect the R.E. effort. Third, the KUD area need not necessarily coincide with the best choice for coverage of a rural electric network, resulting in either arbitrary truncation of the grid, or possible problems of coordination between neighbouring KUD's.

Whatever the outcome of the cooperative approach to R.E., it is clear that they will continue to be a factor in the future.

The final topic to be addressed in this section is the availability of manpower and financial resources, which in turn will affect the extent and timing of any proposed national R.E. plan. It is clear that there is a significant shortage of staff trained in all aspects of R.E. programs, the technical, economic and financial planning and evaluation of projects, their execution and management, continued financial and technical monitoring, and financing of new schemes and upgrading of existing ones.

Rationalisation of the institutional framework will undoubtedly help to better focus existing manpower and minimize duplication of effort. However, this measure alone would be inadequate to meet the needs, and training of new staff should become a high priority. Training targets and timing would require a systematic comparison of the requirements of a national R.E. plan and the available manpower, to identify the resource gap. In the short run, it may be necessary to place greater reliance on consultants and expatriate advice, but with the clear understanding that the training of local counterparts would be an important activity. One major problem to be avoided would be the adoption (often by default) of R.E. standards and practices that were too diverse and/or based on inappropriate foreign or urban criteria.

The inadequacy of financial resources to meet the ambitious rural electrification targets of GOI, are an even more serious issue. While the scope of this case study does not permit a full scale analysis of the sources and uses of funds for R.E., some general conclusions may be drawn. The budgetary consequences of this shortfall for the principal executing agency, PLN, is indicated by the fact that up to 1984, only about two thirds of the total of almost 9,000 villages electrified under the PLN program were financed through R.E. funds earmarked for this purpose -- the remaining one third was covered in a somewhat more *ad hoc* manner, through PLN,s regular distribution extension program. This situation is likely to become even worse during the future, if the R.E. program expands along the lines described at the beginning of this chapter, and GOI's budgetary position deteriorates.

The foregoing discussion indicates that rural electrification policy should seek to bridge the financing gap through a variety of means, including:

1. Mobilizing hitherto untapped resources at the village level (as discussed earlier);
2. Reducing investment costs per consumer, by rationalising the technical standards and parameters;
3. Increasing market penetration by maximizing connections in electrified areas (at relatively modest cost), thereby raising sales revenues;
4. Improving revenues by rationalizing the tariff structure; and
5. Seeking supplementary funding from more conventional sources, including GOI and foreign donors.

10.5 Concluding Observations

The success or failure of rural electrification activities in a developing country invariably depends on the extent to which the relevant issues have been systematically analysed and addressed. The following aspects are some of the essential prerequisite for an effective R.E. effort, based on the experience in Indonesia and other developing countries:

(a) having clear-cut national objectives, on the basis of which the national R.E. Master Plan may be developed -- the Master Plan should take a longer run perspective (of at least 10 years), and would include elements such as the broad R.E. strategy, electrification targets, criteria for selection of areas to be electrified, and so on;

(b) consolidating a viable financing plan for R.E. investments and recurrent costs -- the adequacy of initial financing as well as continued flows of funds, must be verified through government and private equity contributions, local and foreign borrowing, and tariff revenues;

(c) creating an effective institutional framework for rural electrification -- the case for a strong central agency versus a more decentralized approach (for example, cooperatives) must be examined. If a central organization is set up, it should be strong enough to support a long term, nationwide program, or if it is a part of an existing utility or institution, it should have the committment to R.E. and the local organizational infrastructure necessary to implement and follow up the program. The creation of a separate R.E. agency or a R.E. division within an existing utility will have important effects on existing power sector organizations. In particular, competition for scarce financial and skilled manpower resources must be avoided, where this could lead to the further deterioration of already poor urban and industrial electricity services, coupled with unsatisfactory execution of the R.E. program itself;

(d) codifying engineering and technical standards and design criteria, appropriate to local conditions and resource endowment;

(e) devising and establishing procedures for planning, implementation, and follow-up, of R.E. programs and projects at the field level -- some of the more important elements to be explicitly treated include, market surveys and load promotion activities, demand forecasting, least-cost system design and investment planning, project evaluation and cost-benefit analysis, procurement and construction, pricing and connection policy, and supervision and evaluation of schemes; and

(f) maintaining an adequate level of supply quality, and ensuring satisfactory operation and maintenance -- this would also require greater responsiveness to customer complaints and needs.

References

Barnes, D.F., "Electricity's Effect on Rural Life in Developing Nations", Review Paper No. 48, Energy Research Group IDRC, Ottawa, 1985.

Brodman, J., " Rural Electrification and the Commercial Sector in Indonesia", Discussion Paper D73L, USAID, Washington DC, Sept. 1982.

Butler, E., Poe and J. Tendler. "Bolivia: Rural Electrification," Project Impact Evaluation, U.S. Agency for International Development, Washington DC, 1980

Cecelski, E. with S. Glatt. "The Role of Rural Electrification in Development," Discussion Paper D-73E, Resources for the Future, Washington DC, April 1982.

Fluitman, F., "The Socioeconomic Impact of Rural Electrification in Developing Countries, A Review of Evidence," International Labor Organization Working Paper, ILO, Geneva, 1983.

Kessler, J., J. Balantyne, R. Maushammer, and N. R. Simancas, "Ecuador: Rural Electrification," Project Impact Evaluation, U.S.A.I.D., Washington DC, 1981.

Madigan, F.C., "Cooperative Rural Electrification, Income Distribution, Employment, and Fertility: A Case Study from the Southern Philippines," General Conference of the International Union for the Scientific Study of Population, Manila, The Philippines, December 1981.

Mason, M., "Rural Electrification: A Review of Bank-Financed Projects", draft report, Energy Department, The World Bank, Washington DC, Feb 1985.

McCawley, P., "Rural Electrification in Indonesia - Is It Time?", *Bulletin of Indonesian Economic Studies*, vol. 14, July 1978.

Ministry of Mines and Energy, "Survey of R.E. in South Sulawesi", Jakarta, Indonesia, 1985.

Munasinghe, M., "An Integrated Framework for Energy Pricing in Developing Countries", *The Energy Journal*, vol. 1, July 1980, pp. 1-30.

Munasinghe, M., "Integrated National Energy Planning in Developing Countries", *Natural Resources Forum*, vol. 4, 1980, pp. 359-73.

Munasinghe, M., "A Study of the Principal Issues in Rural Electrification in Indonesia", Energy Department, The World Bank, Washington DC, July 1982.

Munasinghe, M., "Practical Application of Integrated National Energy Planning Using Microcomputers", *Natural Resources Forum* Feb. 1986, pp. 17-38.

Munasinghe, M. and C. J. Warren, "Rural Electrification, Energy Economics and National Policy in the Developing Countries", in *Future Energy Concepts*, Institution of Electrical Engineers, London, UK, 1979, pp.44-47.

National Electrification Administration, "Nationwide Survey on Socio-Economic Impact of Rural Electrification," for the U.S. Agency for International Development, Manila, The Philippines, June 1978.

Pabaney, A.D., "Some Thoughts on Developing the Productive Uses of Electricity in the Rural Areas of Indonesia", Jakarta, Indonesia, May 1983.

Ross, J.E., *Cooperative Rural Electrification*, Praeger Press, New York, 1972.

Samanta, B.B. and A. K. Sundaram, "Socioeconomic Impact of Rural Electrification in India," Discussion Paper D-730 by Operations Research Group, U.S. Agency for International Development, Washington DC, January 1983.

Saunders, J., J. M. Davis, G. C. Moses, and J. E. Ross, *Rural Electrification and Development: Social and Economic Impact in Costa Rica and Colombia*, Westview Press, Boulder CO, 1975.

Sen, L.K., "Rural Electrification in Indonesia: Current and Future Perspectives", Discussion Paper No. 131, Harvard Inst. for Int. Dev., Cambridge MA, Feb. 1982.

Smith, D.V., D. B. Mehta, and P. Hayes, "Regional Rural Electrification Survey", draft report, Asian Development Bank, Manila, The Philippines, October 1983.

Tendler, J., "Rural Electrification: Linkages and Justifications," Bureau for Program and Policy Coordination, U.S. Agency for International Development, Washington DC, April 1979.

US Bureau of the Census, "Evaluation of the Indonesian R.E. project: results of 1984 surveys of households, business and public facilities", Washington DC, Dec. 1985.

USAID, "Power to the People: Rural Electrification Sector Summary Report," Draft Summary Report, U.S. AID, Washington DC, 1981.

Velez, E., Becerra and A. Carrasquilla, "Rural Electrification in Colombia," Report by Institution SER de Investigacion, U.S. Agency for International Development, Washington DC, March 1983.

World Bank, "Rural Electrification", Washington DC, Oct. 1975.

Chapter 11

OPTIMIZATION OF ELECTRIC POWER DISTRIBUTION SYSTEMS[1]

11.1 Introduction

The world-wide scarcity of energy resources and high costs of power supply have highlighted the importance of energy conservation and elimination of waste by both producers and users of energy. Power system loss reduction is one of the principal ways of achieving conservation in the electric power sector. Despite the modest (and likely short-term) decline in oil prices in the mid-1980s, power system loss reduction continues to have much scope.

In the process of delivering electricity to consumers, losses are incurred at the generation, transmission, and distribution stages of a power system (see Figure 2.6 in Chapter 2). Generation losses can be decreased by improving plant efficiency and reducing station use, using new technologies like combined cycle thermal plant; by replacing old boilers and upgrading old thermal generators; and by using higher efficiency designs in new hydro installations or replacing older turbines. Leaving generation aside (where acceptable loss norms vary according to plant mix), recent work indicates that average losses in power delivery systems (transmission and distribution) should normally be below 10 percent of gross generation, while economically optimal loss levels might be as low as 5 percent. The corresponding technical losses in many developing country networks approach 20 percent, even after allowing for substantial amounts of theft. Rural electric systems usually tend to have the highest losses.

We recognize that non-technical losses, such as theft or unmetered consumption, might also be significant in some systems and should be reduced. Minimizing such unaccounted-for-energy might involve steps such as replacing broken meters, disconnecting and prosecuting customers who steal electricity, establishing a system of repayment of bills, and having different individuals do meter reading and billing (as an anti-corruption measure).

While recognizing the importance of continued improvements in generation and bulk transmission efficiency, we focus here on reduction of both energy and peak power technical losses in distribution networks, where up to three fourths of total system losses occur.

Two principal reasons account for current high loss levels, especially in rural electric distribution systems in developing countries. First, the selling price of electricity failed to keep pace with rapid increases in supply costs in the decade following the 1973 oil crisis. The consequent decline in many developing country power companies' financial positions has led to reduced investment and system maintenance, while low electricity prices have overstimulated demand. Most utilities have preferred to cut back expenditures on both urban and rural distribution systems

[1] Edited extracts from M. Munasinghe and W.G. Scott, *Energy Efficiency: Optimization of Electric Power Distribution System Losses*, Energy Dept., The World Bank, Washington DC, 1982; and M. Munasinghe, *Rural Electrification for Development*, Westview Press, Boulder CO, 1987.

rather than reduce generation and bulk transmission investments. Supply shortages can lead to major, highly visible power outages, whereas distribution system weaknesses manifest themselves less spectacularly in terms of low voltage and high losses. Investments in new generation projects are also more attractive as a conspicuous sign of "progress," whereas distribution network rehabilitation and improvements are considerably less glamorous.

Second, high distribution system loss levels are related to recent rapid rises in energy costs. Expenditures on system hardware can curtail system losses. Rules of thumb for system design, still used by many developing country utilities, trade off increased system investment costs against reduced losses, on the basis of relative costs more appropriate to the mid-1960s than to the present. As a simple illustration, let generation costs and aluminum (or copper) prices serve as respective proxies for the value of losses and of system hardware (such as conductors and transformers).

Generation costs (in constant terms) have increased several-fold between 1966 and 1986, whereas corresponding aluminum and copper prices have remained constant or even declined. Clearly, an engineer should be willing to use a lot more conductor to reduce system losses now than he or she would have in 1966 (see Figure 11.2). As shown later in this study, this means that acceptable overall target loss levels for the entire distribution network (excluding bulk transmission, but including the distribution substation) should be less than about 5 percent of generation, whereas conventional wisdom has allowed for losses of up to 10 percent or more.

11.2 Economic Loss Optimizing Model

The purpose of this section is to analyze the economic outcome of reducing distribution system technical losses by applying principles of cost-benefit analysis. First, before separating out the distribution network, we consider the net benefit of consumption supplied by the entire electric power system. The electric power system is planned with a time horizon of T periods, each of 1 year in duration.

The net benefits of electricity consumption can be written as:

$$NB(D,R) = TB(D) - SC(D,R) - QC(D,R)$$

where

TB = the total benefits of electricity consumption
SC = costs of supply for the system
QC = costs incurred by consumers due to supply disruptions
R = quality of supply

This formula may be rewritten to indicate the present discounted value of net benefit to society (NB) for the planning period:

$$NB = \sum_{t=0}^{T} [(TB_t(D_t) - SC_t(R_t,D_t) - QC_t(R_t,D_t)] / (1+r)^t \qquad (11.1)$$

We note that the total benefit (TB_t) of electricity consumption in any time period t is a function of the total quantity of electricity demand D_t in the absence of outages. It is assumed that the quality of supply is perfect.

The earlier model sought to maximize net benefit, optimizing reliability through the tradeoff between supply costs (SC_t) and outage costs (QC_t). Here we attempt to maximize net benefits, optimizing supply costs by minimizing technical losses in the distribution system. For this purpose, the SC_t term is broken down into its components.

Total system supply costs consist of generation costs (GSC), transmission costs (TSC), and distribution system cost (DSC):

$$SC = GSC + TSC + DSC \qquad (11.2)$$

Since we are focussing on the distribution network, the generation and transmission systems costs can be represented by the LRMC of bulk supply. LRMC is defined as the ratio of change in system costs associated with a sustained increment of demand in the long run (see Chapter 3 for details).

The expression LRMC = δ(costs)/δ(demand) can be used to calculate the LRMC of bulk supply, generation as well as transmission. This gives the cost per unit of power and energy supplied by the bulk system and entering the distribution network. For example, if a units of energy are entering the distribution network, then the costs of bulk supply are a.MC.

Equation (11.2) can be written as:

$$SC = a.MC + DSC$$

DSC is composed of investment costs and operation and maintenance costs. The technical losses in the distribution network will be reflected in the term a, since more units of electricity will enter the distribution system if losses are higher.

The next step involves giving an economic value to distribution losses. For this purpose, it is necessary to compare net benefits arising from two alternative distribution systems. This approach can be extended to the comparison of many alternative network configurations.

Consider two distribution networks 1 and 2 (Figure 11.1), each supplying different amounts of electricity. Let a_1 be the units of electricity entering distribution system 1 and b_1 be the corresponding units of electricity available to the consumers. Let l_1 be the losses in System 1.

In the distribution network, since a_1 units are entering, with b_1 units leaving and l_1 units being loss, $a_1 = l_1 + b_1$. A similar expression can be derived for System 2.

The net benefit for each power system can be written as:

$$NB_i = \sum_{t=0}^{T} [(TB_{it} - SC_{it} - QC_{it}] / (1+r)^t \qquad ; \text{for } i = 1,2.$$

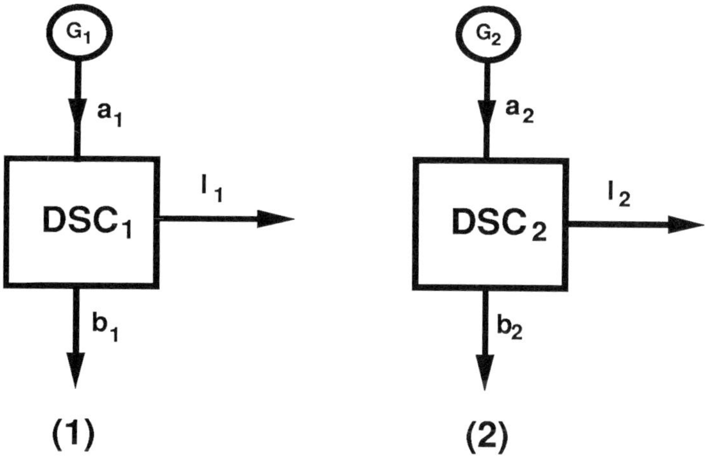

Figure 11.1: Distribution Loss Representation

For each system, the SC term is expanded into its component parts and the net benefit equation (11.1) can be rewritten:

For System 1:

$$NB_1 = \sum_{t=0}^{T} [(TB_{1t} - (a_{1t}.MC_{1t} + DSC_{1t}) - QC_{1t})] / (1+r)^t$$

For System 2:

$$NB_2 = \sum_{t=0}^{T} [(TB_{2t} - (a_{2t}.MC_{2t} + DSC_{2t}) - QC_{2t})] / (1+r)^t$$

Now let us make the simplifying assumption that Systems 1 and 2 are two alternative ways of meeting the same customer load, $b_{1t} = b_{2t}$.

We might imagine that System 1 is an upgraded version of System 2, where distribution costs have been increased to achieve reduced losses. In actual practice, when distribution system rehabilitation is carried out, voltage levels improve together with losses. This is likely to result in increased consumer load. However, this phenomenon will not affect the conclusions of the model described here.

Since TB is a function of b_t, it can be assumed that the total benefits in the two systems are the same, $TB_{1t} = TB_{2t}$. Therefore,

$$NB_1 - NB_2 = \sum_{t=0}^{T} [(TB_{1t}-TB_{2t}) - (a_{1t}.MC_{1t}+DSC_{1t}-a_{2t}.MC_{2t}-DSC_{2t}) - (QC_{1t}-QC_{2t})] / (1+r)^t \quad (11.3)$$

In addition it is assumed that the MC_i are the same for both systems. Since the distribution network is only a small part of a much larger bulk electric supply system, the difference in marginal costs for the two networks at this level will be negligible. Then the above equation (11.3) can be rewritten as:

$$NB_1 - NB_2 = \sum_{t=0}^{T} [(a_{2t}-a_{1t}).MC + (DSC_{2t}-DSC_{1t}) - (QC_{1t}-QC_{2t})] / (1+r)^t \quad (11.4)$$

Since the number of electrical units finally available to the consumers in the two systems is the same, it is easy to show that:

$$a_{1t} - a_{2t} = l_{1t} - l_{2t}$$

Thus the difference in amount of bulk power supplied to the two systems can be replaced by the difference in losses in the two systems. This expression is substituted in equation (11.4):

$$NB_1 - NB_2 = \sum_{t=0}^{T} [(l_{2t}-l_{1t}).MC + (DSC_{2t}-DSC_{1t}) - (QC_{1t}-QC_{2t})] / (1+r)^t \quad (11.5)$$

Let us define the net supply cost (NSC), as follows:

$$NSC_{it} = DSC_{it} + VL_{it}$$

where $VL_{it} = l_{it}.MC$ = Value of Losses.

Next, equation (11.5) may be rewritten in the difference form:

$$\Delta NB = -\Delta NSC - \Delta QC \quad (11.6)$$

where $\Delta NB = NB_1 - NB_2$; $\Delta NSC = NSC_1 - NSC_2$; $\Delta QC = QC_1 - QC_2$;

$NSC_i = \sum_{t=0}^{T} NSC_{it} / (1+r)^t$, and $QC_i = \sum_{t=0}^{T} QC_{it} / (1+r)^t$.

Suppose we assume for the time being that the shortage cost term QC is negligible relative to NSC. We could consider distribution system improvements and loss reduction in systems where the supply quality (in terms of outages) is quite good to begin with. In general, it is assumed that QC would be a small correctional term in the analysis. Neglecting QC, we can rewrite equation (11.6) as $\Delta NB = -\Delta NSC$. In other words $NB_1 > NB_2$ and System 1 provides better net benefits than System 2, if the former has a lower value of net supply costs: $NSC_1 < NSC_2$. Alternatively, we can argue that NB will be maximized when NSC is minimized. It must be stressed here that it is assumed that supply quality and generation and transmission losses are maintained at a constant level. No attempt is made to optimize investment or losses in these higher stages of the electrical supply system, since this is outside the scope of the analysis. This is a partial or localized optimization problem in the electric distribution network.

Writing NSC = VL + DSC and taking derivatives with respect to physical losses L:

$$(\partial NSC/\partial L) = (\partial VL/\partial L) + (\partial DSC/\partial L).$$

The net supply costs in the distribution system are at a minimum with respect to losses when:

$$(\partial NSC/\partial L) = 0 \text{ and therefore } (\partial DSC/\partial L) = -(\partial VL/\partial L)$$

This indicates that to optimize net benefits in the distribution system, marginal distribution supply costs must be increased until they are exactly offset by the marginal decrease in the value of losses.

This is graphically depicted in Figure 11.2 by plotting costs on the vertical axis and losses measured in physical units on the horizontal axis. DSC is a downward sloping curve since, as costs or investments decrease, losses incurred in the distribution system increase. VL (value of losses) is an upward sloping curve since as losses increase, more electricity will have to be supplied by the bulk system to serve the same final demand. The sum of these two components gives net supply costs (NSC). The minimum point of the NSC curve occurs at $L = L^*$, where (the slope of the VL curve) = - (the slope of the DSC curve), assuming negligible differences in the outage costs. The broken curve VL_0 indicates what the value of losses may have been some time ago when generation was cheaper. In this case, the minimum of the corresponding net supply costs (NSC_0) would have occurred at some original optimal loss level L_0, which would be greater than the present optimal level L^* -- as mentioned in Section 11.1.

The Economic Value of kW and kWh Losses

In the engineering studies done until recently, emphasis has often been placed on applying accounting rather than economic principles to loss evaluations. Although concepts such as present worth of annual streams, levelized annual costs, and equivalent investment costs are used, there is inadequate application of rigorous economic theory in these procedures (Nickel 1980).

The principal point is that both kilowatt and kilowatt-hour distribution losses at various time periods should be valued at the long run marginal costs (LRMC) of supply from the bulk supply system. In contrast, some previous authors have significantly underestimated the value of such losses, especially kW losses (Baldwin et.al. 1961, Johnson 1976). The valuation of kWh of energy losses poses no major problems. If distribution losses decrease at any given moment, then the bulk supply LRMC of energy at different times (peak, shoulder, off-peak, or seasonal) provides a measure of the value of kWh lost in the distribution system (see Chapter 3 for details of estimating LRMC).

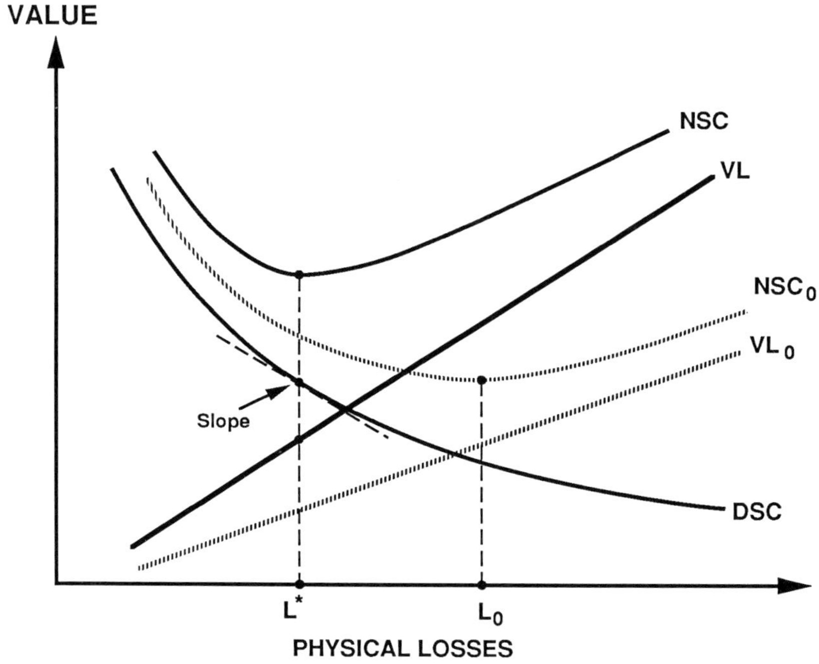

Figure 11.2: Shift in Optimal Economic Loss Levels

However, when distribution system improvements are made, a major change also occurs with respect to kW losses during the peak period. Although the distribution feeder peaks and the bulk system peak might not overlap, any reduction in kW losses during the bulk system peak will lead to a savings in generation and transmission capacity. Even if generation and transmission investments are not actually deferred, the LRMC of bulk kW supply can be used as a proxy for the value of kW distribution system losses at the time of the bulk system peak, as described below.

Thus, losses and customer loads are indistinguishable as far as the bulk supply system is concerned. If losses impose no burdens on bulk capacity, then the incremental costs of serving customers will also be negligible. Furthermore, as described in Chapter 2, in an optimally planned electricity supply system, there are two conditions that must be satisfied: 1) optimal price equals LRMC of supply, and 2) optimal incremental cost of system improvements equals cost of shortages avoided due to improved reliability.

When losses are reduced, it is equivalent to a reduction in demand. Thus bulk system capacity additions can be deferred, yielding cost savings represented by bulk supply LRMC. Alternatively, even if generation and transmission expansion investments continue essentially unchanged (due to lumpiness), then improved bulk supply reliability will provide cost savings (due to averted outage costs at the margin) that are equivalent to marginal savings that could have been realized from deferred generation and transmission investments. We note in the first case that cost savings accrue to the power supplying company whereas, in the second case, the customers gain. Thus from a social viewpoint, both types of cost savings are equivalent, whereas from the power company's viewpoint, the first case might be more desirable.

11.3 Practical Application

Although the loss optimization approach described above requires detailed data and analysis for application, it is possible to reach some practical conclusions of a more general nature, on the basis of some recent studies. Figure 11.3 summarizes some typical power network loss levels. These percentage demand (or kW) losses during the peak load time are usually 1.2 to 1.5 times greater than average energy (or kWh) losses.

Many distribution systems are currently designed and operated mainly on the basis of thermal capabilities of components and quality of supply criteria. The value (usually implicit) assigned to losses is often rather low, when establishing engineering criteria. However, losses should be the most important consideration in establishing design (and operating) criteria. Within practical limitations, in most developing countries it is much less costly to save kilowatts of capacity and kilowatt-hours of energy by reducing distribution losses than to make up these losses by building more generation and transmission facilities.

Loss evaluation and correction are usually far too complex for generalities. However, for the somewhat average conditions of 50 percent load factor, 30 percent loss factor, demand charges in the range of US$130 to 250 per kilowatt per year, energy charges from US$0.011 to 0.034 per kilowatt per year (equivalent to long run marginal supply costs of US$0.04 to .010 per kWh), and a discount rate of 12 percent, recent work indicates that economic loadings at peak should be:

1. Transformers: from 80 to 100 percent of nameplate rating (with appropriate adjustments for specific manufacturing standards used).
2. Primary main conductors: from 15 to 25 percent of rated thermal capacity, and power factor above 95 percent.
3. Secondary conductors: from 10 to 15 percent of capacity (or eliminated entirely).

The above limits are well below criteria used by most developing as well as many developed country utilities. Given these new loading criteria, it is anticipated that voltage drop would be small. Thus neither line voltage, nor even substation regulators would be necessary. The range of US$0.04 to 0.10 per average kWh (in 1984 prices) and the high discount rate of 12 percent (the opportunity cost of capital) are on the conservative side. If energy were valued more highly, or if the discount rate were lower, the conclusions given here regarding loss reduction would be further reinforced.

A rough operational guide for carrying out a quick preliminary assessment of both rural and urban distribution systems is given in Table 11.1. Some specific comments on various items in the table follow.

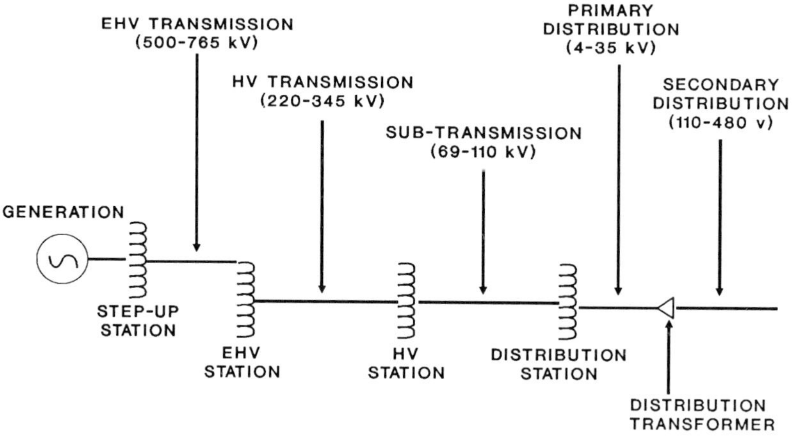

DEMAND LOSSES
(Percent of kW GENERATED)

SYSTEM COMPONENT	TARGET LEVEL		MAXIMUM TO BE TOLERATED	
	Within	Cumulative	Within	Cumulative
Generation	Varies from 0.5-8% depending on type of generation			
Step-up Station	0.25	0.25	0.50	0.50
EHV Trans. & Gen.Station	0.50	0.75	1.00	1.50
HV Trans. & Gen.Station	1.25	2.00	2.50	4.00
Sub Transmission	2.00	4.00	4.00	8.00
Distribution Station	0.25	4.25	0.50	8.50
Distribution Primary St.	2.50	6.75	5.00	13.50
Distribution Transformer & Secondary System	1.50	8.25	3.00	16.50

Source: Munasinghe, M. and W.G. Scott, *Energy Efficiency: Optimization of Electric Power System Losses*, Energy Department, The World Bank, Wash. DC, July 1982 and Ross, W.J., "New Focus on Distribution Levels", *Transmission and Distribution*, Dec. 1981.

Figure 11.3: Normal Power System Demand Loss levels

Table 11.1

Preliminary Checklist for Distribution System Loss Levels

	Item	Good	Fair	Excessive
I.	Demand Loss for the Entire System at Peak	<10%	10-15%	>15%
II.	System Power Factor	95-100%	90-95%	<90%
III.	Impedence of Power Transformers	<5%	5-10%	>10%
IV.	Monitoring the Loading of Distribution Transformers	Annual	Occasional	None
	Maximum Loading on Distribution Transformers	Nameplate	Up to 100% Nameplate	Over 125% Nameplate
V.	Primary Conductor Loading	<20% Thermal Limit	<30% Thermal Limit	>40% Thermal Limit
VI.	Secondary System Maximum Length			
	Urban Areas	1/4 km	1/2 km	> 1/2 km
	Rural Areas	1/2 km	3/4 km	> 3/4 km
VII.	Standards and Specifications	(see text for details)		

I. In most cases, loss reduction should be implemented usually in the following sequence:

1. Power factor correction (to at least 95 percent) by installing capacitors on primary lines
2. Replacing high impedance power transformers
3. Distribution transformer load management
4. Reducing primary conductor loading
5. Reducing secondary conductor loading
6. Reducing subtransmission conductor loading

II. Power factor correction should be accomplished by installing capacitors on the distribution primary system as near to load centers as practical.

1. Install fixed banks to provide 100 percent power factor or slightly leading power factor during off-peak loading periods.
2. Install switched banks to correct power factor during peak loading periods.

III. Power transformer impedance should be checked as follows:

1. The older tap changing under load transformers were often manufactured with impedance values up to 15 percent. These transformers should be removed from service and either scrapped or held for emergency use only.
2. Medium impedance transformers should probably be replaced also, because of their no-load losses.

IV. Monitoring the loading of distribution transformers is essential to reduce losses and burnouts:

1. The lowest cost and most efficient method is to correlate consumers with their transformers and compute loading from energy usage.
2. Install thermal maxi-meters
3. Use clamp-on ammeters at peak time
4. Apply Distribution Transformer Load Management (TLM) techniques, using metered consumer energy requirements to monitor transformer loading. TLM is relatively easy to establish, eliminates most burnouts, reduces losses, and generally produces savings at a high benefit to cost ratio.

V. Primary Conductor loading can be reduced by:

1. Switching loads to other feeders
2. Replacing existing conductors
3. Adding new feeders and sharing the load
4. Raising primary system voltages (for example, 11 kV to 33 kV).

VI. The values for secondary systems in the table are averages (for 240 volt systems) and therefore rather rough. They should be used only as a first check, because specific figures will depend on load densities which are highly variable. The recommended methods for correcting overloaded secondary systems are:

1. Break the secondary system into smaller segments by adding distribution transformers
2. Replace conductors
3. Add more secondary lines.

VII. Standards and specifications should be examined to determine if they are directed towards minimizing losses. The more important areas to be examined are:

1. Power factor correction targets and locations where capacitors are to be installed. The most effective locations are on the primary lines near the load centers, or within customer installations.
2. Specifications for power transformers and distribution transformers, to determine if manufacturers are being informed as to how kW and kWh losses will be evaluated.
3. Initial and normal design electrical loading of transformers and conductors. If thermal capabilities are the basis for electrical loading, then losses are probably excessive.
4. Maximum electrical loading of transformers and conductors before replacement is required.

Summary of Implementation Procedure

Figure 11.4 summarizes the basic procedures for determining economic loss levels for all distribution system components. Specific steps are as follows:

Step 1. Select portion of system to be studied:

Station transformers
Primary
Distribution transformers
Secondary

Step 2. Obtain electrical and physical characteristic of components and form model of system (either manually or on computer).

Step 3. Select a loading cycle (day, week, month, year) and determine the following major parameters:

Peak demand
Load duration
Load factor
Loss factor

Step 4. Compute technical losses for:

kW loss at peak (demand)
Energy losses

Step 5. Select a practical variety of system changes to reduce losses in:

- Transformers
  - Replacement
- Primaries
  - Capacitor Additions
  - Reconductoring
  - New lines
  - Switching
  - Changing voltage levels
- Secondaries
  - Reconductoring (shifting loads)
  - Reduction or elimination

Step 6. Determine costs of the following items, associated with each alternative:

Power (demand and energy)
Investment
Labor
Materials
Other
Operation and maintenance

Step 7. Carry out an economic evaluation of the alternatives, using the methodology described in the model.

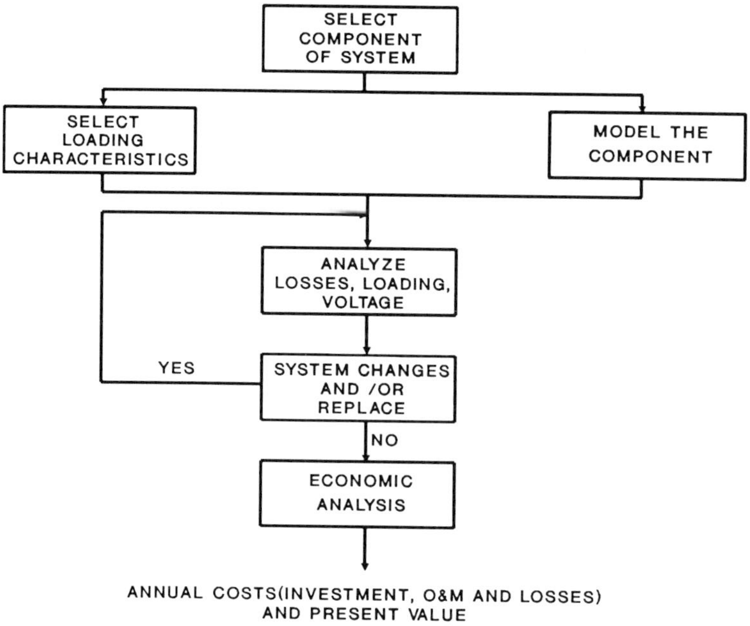

Figure 11.4: Determining Systems Costs and Costs of Losses (transformers, primaries and secondaries)

11.4 Technical Issues in Rural Electric System Optimization

This section provides a general overview and summary for engineering and technical aspects of rural electric systems in developing countries. A detailed checklist that can be used to evaluate the status of a developing country's rural electricity system engineering standards and practices is given in Munasinghe (1987).

Rural electrification projects are generally more dynamic than generation and high voltage transmission projects. For example, a baseload generating plant is designed and constructed to supply a fairly uniform quantity of power over its entire useful life, while a high voltage transmission line is designed and constructed to carry a specific block of power from point A to point B over its useful life. Both are relatively static situations. On the other hand, distribution systems (rural electric systems, in particular) are constantly in flux. With existing customers changing their load patterns and new customers needing service from the existing system, these systems are more dynamic.

The poorest, least developed areas can be the most dynamic because the potential for change is so high. Electrification of a village in such areas might begin with relatively light loads, but the increase in demand is likely to be rapid if the project area has been selected properly on the basis of potential economic growth. Therefore, rural electricity engineering (planning, design, operation, and maintenance) for a developing country must be initiated and organized to meet the dynamic nature of rural requirements. If this is not accomplished, the electrical system will eventually deteriorate to an ineffective level. Under these circumstances, rural electrification funds would probably be better directed to more productive projects.

A good start towards reviewing and obtaining an overview of rural electricity engineering might be made by seeking answers to questions such as:

1. Have material and construction standards been developed to suit rural (rather than urban) conditions, and are they enforced?
2. Do the standards reflect the resource endowment and the economics of the country?
3. Do the standards make use of available resources and provide alternatives and flexibility to meet the forecast needs at least cost?
4. Do design criteria reflect the real economic and electrical requirements of the rural community?
5. Is the quality of service being adequately monitored?
6. Are the resources readily available to correct deficiencies in the quality of service?
7. Is maintenance at a level commensurate with ensuring a full and useful plant life?
8. Do load forecasts realistically reflect the dynamic patterns of demand growth and are they based upon pertinent demographic and socioeconomic data?

Engineering of rural electric systems in developing countries is very difficult due to the number and complexity of the variables. The more important variables are materials, voltages, losses, quality of service criteria, construction standards, planning, design, operation, and maintenance. Each of these must be evaluated in relation to parameters such as costs, availability of resources, skills, training, experience, appropriateness, and originality. Since the scope of this volume does not permit a more comprehensive treatment of rural electricity engineering, attention is focussed upon the following critical areas of concern to the developing countries:

1. Standards - existence, source, and coverage;
2. Materials - poles, conductors, and connectors;
3. Equipment - transformers, regulators, capacitors, and protective devices;
4. Design criteria - phasing, voltage, loading, losses, power factor, and reliability;
5. Operation - control and monitoring of quality; and
6. Maintenance - lines, equipment, and preventive procedures.

Standards

Standardization of materials and construction for rural electrification is important because of the almost infinite combinations of parameters. Conductors provide an excellent example of the need for standardization. There are hundreds of types and sizes of conductors. If all kinds were used, warehoused would soon become overstocked with wires and associated hardware and crews would be bogged with an unnecessarily large variety of items. Even worse, there would rarely be the right connector or splice available when is was needed. On the other hand, standardization must provide a sufficient variety of conductor types and sizes so that engineers have the flexibility to plan and design an effective, economical system.

A developing country without its own rural (and urban) electric distribution standards is likely to become too reliant upon consultants and manufacturers. It might end up with equipment too diverse and inappropriate to its needs. Even in the best of circumstances, the country would find it very difficult to operate and maintain the variety of facilities it would inherit if design and planning were accomplished on a project by project basis.

On the other hand, a country with very rigid, outdated standards would not be able to meet ever changing rural requirements in an economical manner. The checklist given in the Appendix to this chapter indicates ways which a flexible approach to standardization can be developed.

Materials

Poles, conductors and connectors are selected as material items that are representative of the level of technical care that has gone into the creation of engineering practices and standards. A variety of situations exist in developing countries which indicate that economy and usage of natural resources have not been dominant factors in the selection of materials. Some typical cases are summarized below.

Poles

Economic aspects are frequently ignored in the choice of materials for poles. Prestressed concrete poles are often used for rural electrification, although there is good quality wood abundantly available at a fraction of the cost. It is not unusual for a country to use cast concrete poles for rural electrification, while concurrently exporting cheaper treated wooden poles for use in other countries. Steel poles are also used frequently instead of lower cost concrete. In many instances, considerable savings are also possible through the selection of more appropriate pole designs, safety clearances, and span lengths.

Conductors

Many developing countries use Aluminum Clad Steel Reinforced (ACSR) conductors for primary lines in rural areas. They are needed for longer spans, when stresses are high, and their performance can be better under through fault conditions. Some technical disadvantages of ACSR are:

1. Greater care must be exercised during construction because aluminum is soft;
2. Splicing is more difficult because of the steel core; and
3. All connections are critical because aluminum cold-flows under pressure, corrodes when exposed to air, and the corroded coating (aluminum oxide) is an insulator.

ACSR is better restricted to the most important primary feeders, because of high costs.

Few developing countries use all aluminum (AA) wires. Although disadvantages 1 and 3 above also apply to AA, they are lower in cost and easier to handle -- for primaries, in villages where span lengths are short. However, use of AA lines for secondaries is common. Aluminum alloy conductors (AAC) combine the strength of ACSR and the handling qualities of AA; they should be considered where strength requirements are paramount. However, AAC is more expensive and has higher losses than AA. Lighter loading conditions in most developing countries (no ice, moderate wind loads) permit AA to be used for many applications.

Very few developing countries use steel conductors available with zinc, copper, or aluminum coatings to prevent rust. Steel conductors have been used for over 50 years in electrifying the rural areas of North America. The advantages of steel conductors are low cost, simple construction, and high strength. However, their high losses are likely to rule them out in many circumstances where economic optimization is an important determinant of design criteria.

Connectors

Proper splices and connections are essential to the long range successful operation of rural systems using ACSR and AA conductors. Most experienced practitioners agree that only compression splices and connections provide adequate useful life for ACSR and AA conductors. Establishing a good connection and tooling program, and making sure linemen are trained in its use, is one of the most rewarding investments a utility can undertake.

The survey of nine representative developing countries shown in Table 11.2 indicates that, in 1983, only three of them had standardized compression fittings. The other had adopted bolted parallel groove connectors and had left splicing to the ingenuity of the lines crews. Although it is considered standard practice in North America (and practically all well run utilities) to use oxide inhibitors and/or grease to prevent corrosion of aluminum connections, this was not being done in the countries sampled.

In India and Pakistan, parallel groove connectors are included in the standards. However, these connectors are frequently unavailable in these countries, so construction crews make most connections by wrapping the ends of the tap conductors around the line conductor. This practice inevitably leads to an eventual failure of the connection and, thus, a service interruption. It can also cause substantial technical losses, thus increasing the costs of power supply.

In summary, the more common reasons for poor connections are:

1. Wrong size connectors. (If too small, they do not provide sufficient area, or pressure. If too large, they do not grip tightly enough);
2. Loose blades and pressure plates on blade disconnects, gang-operated switches and cutouts;
3. Use of bronze connectors on aluminum conductors, which results in corrosion and eventual failure of the connection;
4. Use of all aluminum connectors on copper conductors, which results in corrosion and eventual failure of the strands of one conductor around the other. This method works for hard drawn copper conductors, but aluminum strands do not have sufficient tensile strength. Invariably, the connection becomes loose, causes losses, begins arcing, and burns down.

Preventing poor connections requires using the right connectors at all times, using compression connectors for aluminum whenever possible, and monitoring of existing connections. The most effective monitoring devices are infrared detectors, which can be used to pinpoint any hot spots on the system.

Table 11.2

Construction Standards & Practices, Materials & Equipment

	Items	Latin America			Africa		Asia			
		Costa Rica	**Brazil**	**Venezuela**	**Ivory Coast**	**Tanzania**	**India**	**Philippines**	**Thailand**	**Pakistan**
1.	Source of Construction Standards									
	a. Developed Solely for Rural Electric									
	b. Modified Urban Standards		x	x	x	x	x		x	x
	c. Adopted from Another Country	x						x		
2.	Primary System									
	a. Three-phase only		x		x	x	x			x
	b. Single & Three-Phase	x		x				x	x	
	c. Single Wire Earth Return (SWER)		p		p					
3.	Voltages									
	a. Primary Distribution Voltages (kV)									
	i) High	24.9	25&35	12	11	11	11	13	11	11
	ii) Low	24.9	15	12	11	11	3&6	13	11	11
	b. Secondary Distribution (Volts)									
	i) Three-Phase	480D	127/220	480	400Y	400Y	415Y	240Y	480	415Y
	ii) Single-Phase	120/240	127	120/240	230	230	240	120/240	240	240
4.	Poles									
	a. Wood	x		x		x		x	x	
	b. Concrete		x		x		x		x	x
5.	Conductors									
	a. ACSR (Alum. Clad Steel Reinforced)	x	x	x	x	x	x	x	x	x
	b. AA All Aluminum			x			x		x	
	c. Copper								x	
	d. Steel (Copper Covered or Galvanized)									
6.	Connectors									
	a. Compression Type	x				x		x	x	
	p. Parallel Groove	x	x	x	x	x	x	x	x	x
	c. Wrapped						x			x

x = usual practice
p = pilot study only
D = delta connected
Y = star connected

Equipment

The availability of a variety of distribution transformers, voltage regulators, capacitors, and protective devices provides flexibility in selecting engineering standards, as indicated below.

Distribution Transformers

Distribution transformers take power from the primary system at the general level between 6 and 35 kilovolts and transform it to a voltage level in the range of 100-400 volts. The standard sizes and costs vary widely from country to country.

The single-phase costs are derived from a blend of the following types of transformers:

1. Primary insulated for line-to-line voltage, two primary bushings;
2. Primary insulated for line-to-neutral voltage, two primary bushings; and
3. Primary insulated for line-to-neutral voltage, one primary bushing.

Distribution transformers can constitute a significant share of rural electrification investment. It is important that a sufficient variety of types and sizes be available to match the needs of consumers. Some poor practices to be avoided are:

1. purchasing only large transformers (e.g., 50 kVA and above);
2. Installation of large transformers for small loads, such as 50 kVA unit for a 5 horsepower motor; and
3. Purchasing only three-phase transformers, especially when loads are light.

Voltage Regulators

These components are installed in substations and out on the primary system to regulate voltage levels. Generally speaking, rural lines are voltage restricted; voltage drop is the major limitation on the length of line and number of consumers that can be served from a specific feeder. For example, without regulation, the typical primary in Figure 11.5 is restricted to about 7 km in length, if a minimum voltage level of 95 percent is maintained. With a station regulator and a line regulator, the coverage of the feeder could be extended an additional 8 km. However, selecting conductor sizes based on the optimal level of losses will often result in circuits having acceptable voltage characteristics without need for regulators under normal load conditions.

The actual application of regulators requires in-depth studies of the economic/engineering balance among losses, service reliability, and power factor control. Station regulators and line regulators are used in many rural electrification systems, but not in systems using primary capacitors for controlling both power factor and voltage.

Capacitors

Capacitors have been used on distribution primary systems to control power factor for over 25 years. The major benefits are:

1. reduced losses
2. reduced demand (which frees capacity), and
3. better voltage levels.

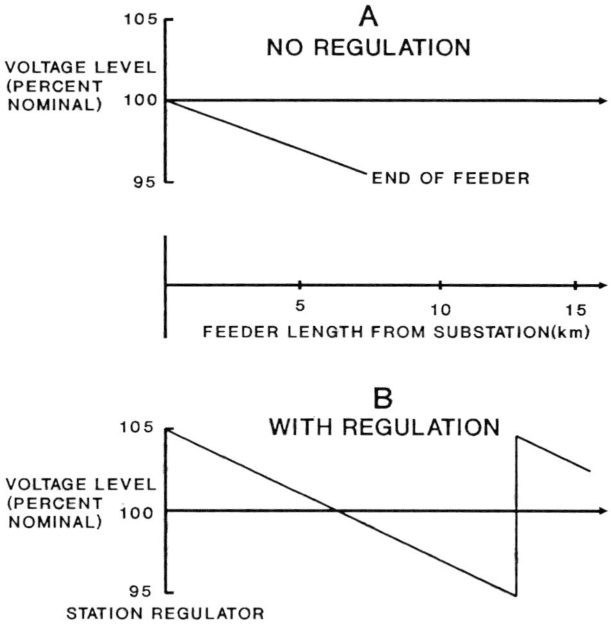

Figure 11.5: Application of Voltage Regulators

Table 11.3 provides an insight into the benefits associated with power factor control through distribution primary capacitors. This simple illustration assumes a concentrated load (3,000 kW, 3,000 kVAR) served through a 12,000 volt single-phase line. The capacitor bank is varied from 0 to 3,000 kVAR and the resulting effects on voltage regulation and losses is shown in the table. The voltage level at the load increases from 82.5 to 87.5 percent, which provides improved but still inadequate service to the consumer. The loss savings will vary from 0 (no capacitors), up to 360 kW and 946,000 kWh at 100 percent power factor. At a very modest assumed value of US$150 per kW per year and US$0.02 per kWh, the capacitors will pay for themselves easily within the first year.

Actual savings on a rural system probably would be less than in the example, but correction of power factors up to 98 percent is economically justified in many cases. Unfortunately, few countries are seriously involved in power factor control. India, for example, has power factors of about 70 percent on many of their rural systems. However, the introduction of primary capacitors is only beginning.

Protective Devices

When properly applied, protective devices assure continuity of service and the isolation of faults from the remainder of the system. It is essential that each developing country's rural electrification program include a major initiative in the protection of lines and equipment. This effort should include periodic fault current studies, coordination studies, and issuance and enforcement of protective guidelines.

Table 11.3

Capacitor Benefits Example

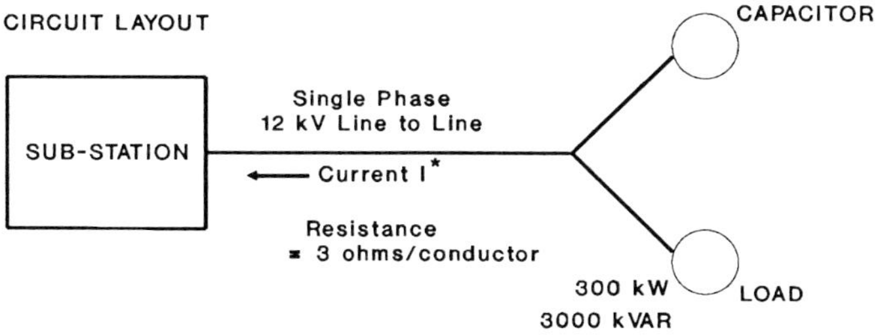

Voltage Drop - Various Capacitor Sizes

Capacitor (kVAR)	Power[b] Factor (%)	Demand (kVA)	Voltage[c] Drop (kV)	Voltage Level: (kV)	(%)
0	70	4.200	2.1	9.9	82.5
1,000	83	3.600	1.8	10.2	85.0
2,000	95	3.222	1.6	10.4	86.7
3,000	100	3.000	1.5	10.5	87.5

Losses - Various Capacitor Sizes[d]

Capacitor (kVAR)	Installed Cost (US$)	Loss Savings Demand (kW)	Loss Savings Energy (MWh)	Loss Savings Value[d] (US$)	Benefit to Cost Ratios (1st Year)
0	0	0	0	0	0
1,000	5,000	195	512	39,490	7.9
2,000	10,000	307	807	62,190	6.2
3,000	15,000	360	946	54,000	4.9

a Current (I) = $\dfrac{[\text{kW} + (\text{kVAR Load} - \text{kVAR Cap})^2]^{1/2}}{12\ \text{kV}}$

b Power Factor = $\dfrac{\text{kW}}{[\text{kW} + (\text{kVAR Load} - \text{kVAR Cap})^2]^{1/2}} \times 100$

c Voltage Drop (kV) = (I x 3 ohms x 2 Conductors)/1000

d Loss Equations:
Energy Losses (kWh) = Demand Loss (kW) x 0.30 (Loss Factor) x 8670 hours
Value ($) = Demand Loss (kW) x $150/kW/year + Energy Loss (MWh) x 1000 x $0.02 /kWh

It should also include the development and enforcement of equipment specifications, and usage of modern equipment -- such as fuses having known characteristics, reclosers, sectionalizers, fault limiting fuses, load break cutouts, air blast breakers, and vacuum switches -- where it is necessary and justified for rural systems. If such an effort is not made, it is likely that service will deteriorate and valuable equipment will be damaged unnecessarily.

Careful selection of lightning arrester specifications is a critical part of developing rural electrification design standards, since lightning arrester characteristics can determine the insulation class to be used for all equipment in the system. Accordingly, price should not be the sole criterion. Judicious use of arrestors, at points such as those near substations and on long secondary circuits, will often prevent expensive damage to other equipment on the system.

11.5 Design Criteria in Power Systems

The more important design criteria for distribution networks are system configuration, voltage levels, loading, losses (demand and energy), power factor, and reliability. These criteria are all interlinked. Determining the proper mix requires intensive and sophisticated economic/engineering studies. However, the reward is a low cost system that meets the needs of the rural consumer. This interaction among criteria will be investigated in the case studies. Some typical results are described below in relation to Table 11.2.

System Configuration

System configuration is the mix of primary voltage levels, phasing, distribution transformers, and secondary systems. There are four basic primary system configurations being applied to the electrification of rural areas in developing countries: Traditional European, Modified European, North American, and Single Wire Earth Return. The major characteristics of the four systems are shown in Figures 11.6 to 11.9 and Table 11.4.

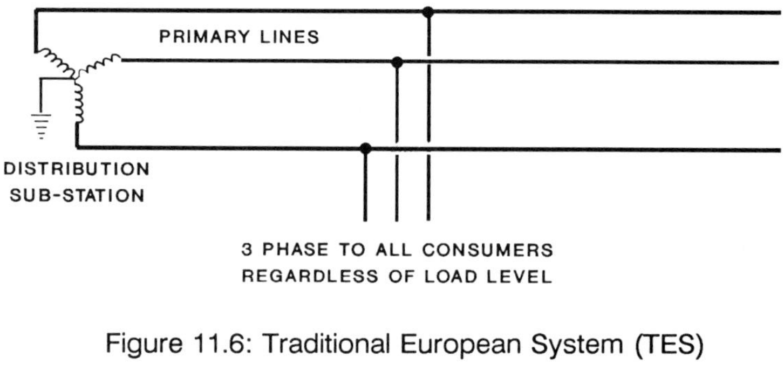

Figure 11.6: Traditional European System (TES)

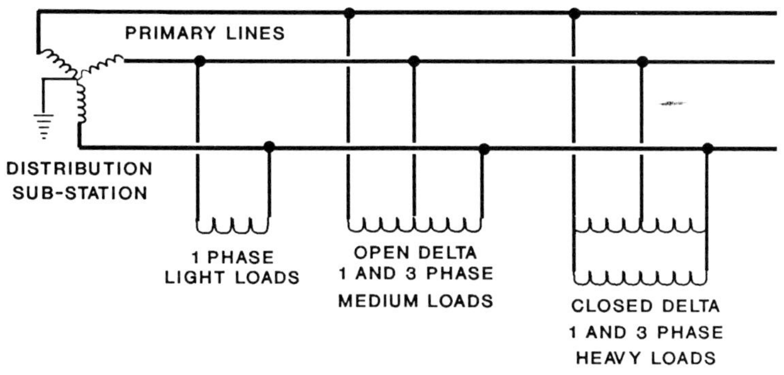

Figure 11.7: Modified European System (MES)

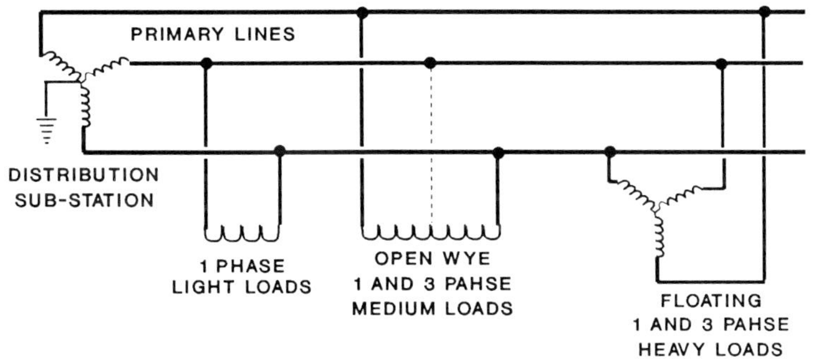

Figure 11.8: North American System (NAS)

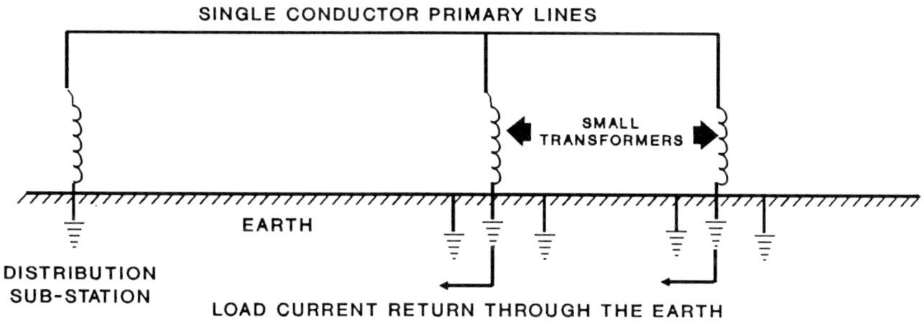

Figure 11.9: Single-wire Earth Return System (SWER)

Table 11.4

Major Characteristics of Rural Electrification Systems

	Primary System			Distribution Transformers				
					Single-Phase			
	Three-Phase	Single-Phase	Typical Voltage	Three-Phase	Line* to Line	Line* to Earth	Coverage of Secondary System	Flexibility
Traditional European	Yes	No	11,000	Yes	No	No	Extensive	Very Little
Modified European	Yes	Yes	11,000	Yes	Yes	No	Medium	Good
North American	Yes	Yes	12,500	Yes	No	Yes	Limited	Very Good
Single Wire Earth Return	No	Yes	10,000	No	No	Yes	Limited	Very Little

* Insulation of primary side of transformers

Traditional European System (TES)

This system was designed to serve the high density towns and cities in Europe. Since the entire system is three-phase, there is little flexibility provided for the engineer to optimize costs. Even the smallest load requires a three-phase primary line, a three-phase transformer, and a three-phase secondary system. This system is not the most appropriate one for electrifying low load density rural areas, because costs are high.

Modified European System (MES)

This modified system includes both three- and single-phase lines (two-phase wires) and transformers (insulated for line-to-line voltage). This provides the engineer with good flexibility in designing a system to meet the early light density loads of rural areas and to expand the system as load density increases. This rural electric system is suitable for countries that use the traditional european system in their urban areas.

North American System (NAS)

The North American System carries the neutral along with the three-phase wires. Single-phase lines consist of a phase wire and an uninsulated neutral conductor, which provides some cost savings cover the two-phase (both insulated) MES, although losses might be higher. The distribution transformer primary is insulated for line-to-earth voltage (versus line-to-line voltage insulation for the European System). Under some circumstances this system is cheaper than the Modified European System, although the difference is generally insufficient to warrant introducing it into a country that has already standardized on the Traditional European System.

Single-Wire Earth Return System (SWER)

This system uses a single conductor with the earth as the return path for the current. It is likely to be economical for very lightly loaded areas with long distances between the station and consumers. It has been used successfully in Australia and Canada, but there are hardly any significant applications in the developing countries. The grounding at each transformer is very important because of potential service problems and danger to the public. Therefore, detailed and in-depth engineering/economic studies are essential in determining feasibility.

Secondary Systems

There are two basic types of secondary distribution systems:

1. Centralized (originating mainly in Europe) - based on larger, centrally located distribution transformers and lengthy secondary lines, serving from 10 to 200 consumers (per transformer); and

2. Decentralized (originating mainly in North America) - based on smaller transformers installed near the load centers, with short secondary systems. Each transformer serves 1 to 15 consumers, depending on load density.

Voltage Levels

Primary system levels have been steadily increased in the electrification of rural areas in North America and Latin America to obtain the benefits of wider coverage and lower losses per dollar of investment. One of the most significant developments in the past twenty years has been the availability of low cost equipment, permitting utilization of relatively high voltages (25-35kV) for primary distribution circuits.

Loading

The loading of facilities affects both useful life and losses. However, few utilities are making the studies required to determine the economic loading levels of lines and equipment for their system.

Power Factor

This parameter is being largely ignored in most developing countries, although as discussed previously, a poor power factor is a major source of both losses and voltage drop. Power factor at low levels can be corrected with a very high benefit to cost ratio.

Reliability

This aspect should be studied more seriously by many developing country utilities, although there are savings to be made through optimizing design according to consumer needs and their outage costs.

11.6 Operation and Maintenance of Equipment

Operations

Specific comments with regard to operation include:

1. Operations staff may inherit a poorly constructed system to begin with;
2. New consumers are often added ad hoc to an already overloaded system;
3. In general, operations suffer due to lack of trained personnel, clear cut procedures, communications, vehicles, tools, and equipment; and
4. Where records are kept of outages, voltages and loading, there usually are insufficient staff or incentives for processing and analyzing the data to improve operations.

Maintenance

Maintenance is rarely adequate in the power systems of developing countries. There is a poor response to actual failures, and more importantly, very little in the all important area of preventive maintenance. As with operations, maintenance is also affected by a shortage of manpower, procedures, workshop facilities, vehicles, tools, and equipment.

Quality of Supply (Reliability)

The quality of service of rural electric systems is rarely monitored or maintained. Most of these systems in the countries surveyed in Table 11.2 were planned, designed, and constructed to provide service to a large area (500 to 1,000 square miles, 5 to 100 villages), with sufficient capacity to meet five to eight years of load growth.

Future loads are often connected with little or no regard to their effect on the supply quality to existing customers, and little attention to system upgrading. Consequently, the following problems have arisen in many rural areas:

1. Voltage levels of 70 to 80 percent of the nominal service voltage at peak. One typical area in India reported that 80 to 90 percent of all irrigation pump motors had suffered some damage each year due to poor quality of service;
2. Losses in the general range of 20 to 30 percent (10 percent is considered high for a well-managed system);
3. High failure rates for expensive (often imported) equipment, such as transformers, reclosers, breakers, switches, and fuse holders; and
4. Low reliability levels because of frequent failures.

11.7 Technoeconomic Parameters and Scope for Cost Reduction

The actual technical and economic parameters adopted in different developing countries, and the scope for reduction of unit costs, are described in this section. The data summarized in Table 11.5 are derived from a number of rural electrification projects implemented over the period 1975-84. It shows that distribution system parameters vary widely. In particular, the reliance on voltages of 11kV and below for primary feeders could indicate insufficient attention to reducing costs and losses by moving to higher voltages, as explained earlier in this chapter. Similarly, the use of three-phase, four-wire secondary systems is also likely to raise cost per connection. Most of the loss figures quoted are unacceptably high, except for Thailand.

The number of connections per kilometer of primary feeder varies widely, and the evolution of connections over time also exhibits interesting contrasts. Thus, the Bangladesh rural electrification project starts at a very low level of five consumers per km and increases in value nineteen-fold (to saturation), whereas the Yemen Arab Republic has a much higher initial year connection rate that rises very little over time. Average distribution costs (in 1983 equivalent prices) range from US$200 to 3,680 per connection, but the costs per unit of electricity consumed vary somewhat less (US$0.03 to 0.29 per kWh). At the same time, there is no obvious direct correlation between costs per connection and costs per kWh, although higher connection costs tend to increase unit costs. Both the above points might be explained by the fact that the more expensive systems tend to be associated with rural electrification projects where electricity consumption per connection is higher.

Nevertheless, these cost figures suggest that countries with relatively high investment costs per customer should make greater efforts to actively seek methods of reducing such costs to approach levels in some of the low- and middle-income Asian countries like Bangladesh, India, Philippines, and Thailand. A recent study (Smith et al. 1983) indicates that, even in Asia, overall rural electrification costs (especially initial investments) might be reduced by up to 30 percent, through measures such as:

1. improved design of distribution substations;
2. selection of higher and more appropriate primary system voltages (in relation to subtransmission and bulk supply voltages);
3. improved design of line supports;
4. revised specification of distribution transformers, switching, and protection;
5. improved metering and billing systems;
6. improved operation and maintenance;
7. revised service quality criteria (such as voltage variation standards, and elimination of sectionalizers).

Table 11.5

Technical and Economic Data from Selected Rural Electrification Projects (1975-84)

Country	Project Year	Distribution System: Primary Voltage (kV)	Distribution System: Second.[1] Config.	Distribution System: Losses[2] (% of Generation)	Conn. per km of Primary Feeder: Year 1	Conn. per km of Primary Feeder: Saturation	Distribution Costs (1983): Per Av.[3] Connection (US$)	Distribution Costs (1983): Per Av.[4] kWh (USc)
Bangladesh	1981	11	1p/2w 3p/4w	22	5	95	295	4
Brazil	1982	13.8	1p,2p and 3p	-	-	-	3680	7.5
Colombia	1979-80	13.8	1p,3p	>30	-	-	1120-1220	5-10
Egypt	1976	11	3p/4w	-	-	-	-	3.5
India	1975-81	5-7	1p/2w 3p/4w	21	10-15	-	200-340	3-4
Malaysia	1981-83	11,22 & 33	1p,3p	-	90-110	105-125	960-1050	3-5
Morocco	1978	22	3p/4w	18	30	50	1220	29
Philippines	1979	13.2	1p/2w	10-17	30-50	-	180-430	3

Table 11.5 (contd.)

Technical and Economic Data from Selected Rural Electrification Projects (1975-84)

Country	Project Year	Distribution System			Conn.per km of Primary Feeder		Distribution Costs (1983)	
		Primary Voltage (kV)	Second.[1] Config.	Losses[2] (% of Generation)	Year 1	Saturation	Per Av.[3] Connection (US$)	Per Av.[4] kWh (USc)
Syria	1977	20	3p/4w	20	754	100	640	6
Thailand	1976-83	22,23	1p/2w 1p/3w 3p/4w	7-12	35-120	>250	200	5-8
Tunisia	1981-83	10,15&30	1p,3p	14	30-35	-	-	8
Yemen (Arab Republic)	1981-84	11	1p,3p	-	35-100	40-140	-	20
Yemen People's Dem. Rep.	1978-81	11	3p	23	105	160	1870-2930	-

Sources: World Bank, Asian Development Bank, Inter American Bank, Mason (1985), and author's estimates.

1 p = number of phases
c = number of conductors
2 Existing rural electrification networks.
3 Present value of investments, adjusted to 1983 prices.
4 Average incremental cost, including O&M costs, but excluding generation or bulk supply costs.

References

Baldwin, C. J., et al., "A Further Look at Losses," AIEE Fall General Meeting, Oct. 1961.

Johnson, B. F., "Evaluation of the Costs of Losses in Power Systems," Consumer's Power Company, April 1976.

Munasinghe, M., *Rural Electricity for Development*, Westview Press, Boulder CO, 1987.

Nickel, D. L., "Distribution Transformer Loss Evaluation: Proposed Techniques," Proceedings of the IEEE, PES Winter Meeting, New York, Feb. 1980.

Smith, D. V., D. B. Mehta, and P. J. Hayes, "Regional Rural Electrification Survey", draft report, Asian Development Bank, Manila, Philippines, Oct.1983.

BIBLIOGRAPHY

Allan, R.N. and F.N. Takieddin, "Network Limitations on Generating Systems Reliability Evaluation Techniques", Paper A77-8070-5, in Proc. IEEE Power Eng. Soc. Winter Meet., New York, Jan.-Feb. 1978.

Allan, R.N. and F.N. Takieddin, "Generation Modelling in Power System Reliability Evaluation," in *Reliability of Power Supply Systems* (IEEE Conf. Publ. 148) London, England, Feb. 1977, pp. 47-50.

Anderson, D., "Models for Determining Least-Cost Investments in Electricity Supply," *Bell Journal of Economics*, vol. 3, Spring 1972, pp. 267-301.

Andersson, R. and L.Taylor,"The Social Cost of Unsupplied Electricity: A Critical Review", *Energy Economics*, July 1986, pp.139-46.

Arvanitidis, N.V. and J. Rosing, "Composite Representation of Multi-Reservoir Hydroelectric Power System", *IEEE Transactions on Power Apparatus and Systems*, vol. PAS-90, February 1970a.

Arvanitidis, N.V. and J. Rosing, "Optimal Operation of Multi-Reservoir System Using Composite Representation," *IEEE Transactions on Power Apparatus and Systems*, vol. PAS-90, February 1970b.

Baldwin, C.J., C.H. Hoffman, and P.H. Jeynes, "A Further Look at Cost of Losses," *AIEE Trans. (Power App. Syst.)*, vol. 80, 1961.

Banco Central de Costa Rica, *Cifras Sobre Produccion Agropecuaria*, San Jose, Costa Rica, July 1978.

Banco Central de Costa Rica, *Cifras de Cuentas Nacionales de Costa Rica*, Serie 1957-77, San Jose, Costa Rica, July 1978.

Barnes, D.F., "Electricity's Effect on Rural Life in Developing Nations", Review Paper No. 48, Energy Research Group IDRC, Ottawa, 1985.

Baum, W., "The Project Cycle", *Finance and Development*, vol.15, Dec. 1978.

Baumol, W.J. and D.F. Bradford, "Optimal Departures From Marginal Cost Pricing," *Amer. Econ. Rev.*, pp. 265-283, June 1970.

Becker, G., "A Theory of the Allocation of Time", *Economic Journal*, vol.75, Sept.1965, pp.493-517.

Bental, B. and S.Ravid,"A Simple Method for Evaluating the Marginal Cost of Unsupplied Electricity", *The Bell Journal of Economics*, vol.13, 1982, pp.249-53.

Bernstein, M. and Y. Hegazy, "Economic Costs of Electricity Shortages: Case Study of Egypt",*The Energy Journal*, Special Electricity Reliability Issue, vol. 9, 1988, pp.173-188.

Berrie, T.W., "Power System Planning under Electricity Spot Pricing", *IEEE Transactions on Power Systems*, Vol.PWRS-2, No.3, August 1987.

Betancourt, T., "An Econometric Analysis of Peak Electricity Demand in the Short Run", *Energy Economics*, Jan. 1981, pp.14-29.

Bodlund, B., et al., "Technology Options for the Swedish Power Sector", in Johansson, T.B., et al.(eds), *Electricity*, Lund University Press, Stockholm, Sweden, 1989.

Boiteux, M., "La Tarification des Demandes en Pointe", *Revue Generale de l'Electricite*, vol.58, 1949.

Boiteux, M. and P. Stasi, "The Determination of Costs of Expansion of an Interconnected System of Production and Distribution of Electricity," in J. Nelson (Ed.), *Marginal Cost Pricing in Practice*, Prentice-Hall, Englewood Cliffs NJ, 1964.

Brodman, J., " Rural Electrification and the Commercial Sector in Indonesia", Discussion Paper D73L, USAID, Washington DC, Sept. 1982.

Brown and M.Johnson,"Public Utility Pricing and Output Under Risk", *American Economic Review*, vol. 59, March 1969, pp.119-28.

Burchnall, J.A. "A Review of Off-Peak Tariff Incentives and Their Effect in England and Wales," Economic Commission for Europe- Committee on Electric Power, Brussels, Belgium, Report No. EP/SEM. 5/R17, Mar. 1977.

Butler, E., Poe and J. Tendler. "Bolivia: Rural Electrification," Project Impact Evaluation, U.S. Agency for International Development, Washington DC, 1980

Canadian International Development Agency, Proc. Regional Power systems Engineering Seminar, Kathmandu, Nepal, Feb.-Mar. 1977.

Caves, D.W. and L.R. Christensen, "Residential Substitution of Off-Peak for Peak Electricity Usage Under Time of Use Pricing," *The Energy Journal*, vol. 1, Apr. 1980, pp. 85-142.

Cecelski, E. with S. Glatt. "The Role of Rural Electrification in Development," Discussion Paper D-73E, Resources for the Future, Washington DC, April 1982.

Chao, H.P., S. Oren, S. Smith and R. Wilson, "Priority Service: Market Structure and Competition", *The Energy Journal*, Special Electricity Reliability Issue, vol. 9, 1988, pp.77-104.

Chao, H.P. and R.Wilson, "Priority Service: Pricing, Investment and Market Organization", *American Economic Review*, vol. 77, December 1987, pp.899-916.

Chow, G.C., "Tests of Equality Between Subsets of Coefficients in Two Linear Regressions", *Econometrica* vol.28, July 1960, pp.591-605.

Cicchetti, C.J., W.J. Gillen and P. Smolensky, *The Marginal Cost and Pricing of Electricity*, Ballinger Publ., Cambridge MA, 1977.

Crew, M.A. and Kleindorfer, P.R., "Peak Load Pricing with a Diverse Technology", *The Bell Journal of Economics*, vol. 7, No. 1 ,Spring 1976, pp. 207-231.

Crew, M.A. and P.R. Kleindorfer, "Reliability and Public Utility Pricing," *Amer. Economic Rev.*, vol. 68, Mar. 1978.

Crew, M.A. and P.R. Kleindorfer, *Public Utility Economics*, St. Martins Press, New York, 1979.

Cummings, R.G., D.S.Brookshire and W.D.Schulze, *Valuing Environmental Goods: An Assessment of the Contingent Valuation Method*, Rowman and Allanheld, Totowa NJ, 1986.

de Grasse, R.V., "Electric Storage Heating After Two Years," *Public Utilities Fortnightly*, vol. 23, Jan. 6, 1977.

Dias-Bandaranaike, R. and M. Munasinghe, "The Demand for Electricity Services and the Quality of Supply," *The Energy Journal*, vol. 4, Apr. 1983.

Doane, M., R. Hartman and C-K. Woo, "Household's Perceived Value of Service Reliability: An Analysis of Contingent Valuation Data", *The Energy Journal*, Special Electricity Reliability Issue, vol. 9, 1988, pp.135-150.

Doane, M., R. Hartman and C-K. Woo, "Household Preference for Interruptible Rate Options and the Revealed Value of Service Reliability", *The Energy Journal*, Special Electricity Reliability Issue, vol. 9, 1988, pp.121-134.

Donziger, A., "The Underground Economy and the Theft of Utility Services," *Public Utilities Fortnightly*, Nov. 22, 1979, pp. 23-27.

Dupuit, P., "De l'Utilite et de sa Mesuer," *La Reforma Soziale*, Turin, Italy, 1932.

Electric Power Research Institute (EPRI), "Power Shortage Costs: Estimates and Applications," report prepared for EPRI by Faucett Associates, EA-1215, California, December 1981.

Electric Power Research Institute, "Utility Fuel Inventory Model: Version 1-0," ER-4079-CCM, California, August 1985.

Electric Power Research Institute, "Value of Service: Scoping Study", EM-4389, California, January 1986.

Electricity Council, "Domestic Tariffs Experiment", Load and Market Res. Rep. 121, London, U.K., 1978.

Endrenyi, J., *Reliability Modeling in Electric Power Systems*, Wiley, New York, 1978.

Energy Journal, Special Issue on Electricity Reliability, 1988.

Federal Energy Regulatory Commission, "Procedures Governing the Collection and Reporting of Information Associated with the Cost of Providing Electric Service," Final Regulation, Docket No. RM79-6d, Washington, DC, June 5, 1979.

Finnish Power Producers Coordinating Council (STYV), "Report on the Value of Non-distributed Energy", August 1979.

Fisher, F.M., "Tests of Equality Between Sets of Coefficients in Two Linear Regressions", *Econometrica*, vol.38, March 1970, pp.361-6.

Florida Public Service Commission, "Need Determination for Proposed Lake Tarpon-Kathleen 500 kV Line", Florida Power Corporation, Docket No.840044-EI, Order No.13676, Issued September 13, 1984.

Fluitman, F., "The Socioeconomic Impact of Rural Electrification in Developing Countries, A Review of Evidence," International Labor Organization Working Paper, ILO, Geneva, 1983.

Ford, L.R. and D.R. Fulkerson, *Flows in Networks*, Princeton Univ. Press, Princeton NJ, 1962.

Francfort, A. and P. Woo, "Lifeline and Incremental Cost Residential Electric Rates," *Public Utilities Fortnightly*, Feb. 17, 1977, pp. 15-20.

Fulkerson, W., et al., *Energy Technology R&D*, Oak Ridge National Lab., Tenn., USA, 1988.

Garfield, P.J. and W.F. Lovejoy, *Public Utility Economics*, Prentice-Hall, Englewood Cliffs NJ, 1964.

Goett, A., D. McFadden and C-K. Woo, "Estimating Household Value of Electrical Service Reliability with Market Research Data", *The Energy Journal*, Special Electricity Reliability Issue, vol. 9, 1988, pp.105-120.

Green, H.A.J., *Aggregation in Economic Analysis: An Introductory Survey*, Princeton Univ. Press, Princeton, NJ, 1964.

Gronau, R., "The Intrafamily Allocation of Time: The Value of Housewives Time", *American Economic Review*, vol.63, Sept.1973, pp.634-50.

Hall, G., M.Healy and W.Poland, "PG&E's New Methodology for Value Based Generation Reliability Planning", paper presented to the System Planning Committee, Edison Electric Institute, Minneapolis, June 1986.

Hastings, B.F., "Ten Years of Operating Experience with a Remote Controlled Water Heater Load Management System at Detroit Edison," *IEEE Trans. Power App. Syst.*, vol. PAS-99, July/Aug. 1980, pp. 1437-1441.

Herman, G.F.K., "Electricity Tariffs," Economics Dept., Univ. Stellenbosch, South Africa, 1974.

Holt, S.B., Jr. and D.M. Crawford, "Distribution Substation Planning Using Optimization Methods," in *Application of Optimization Methods in Power System Engineering* (IEEE Tutorial Course), pp. 69-76.

Hotelling, H., "The General Welfare in Relation to Problems of Taxation and of Railway and Utility Rates," *Econometrica*, vol. 6, July 1938, pp. 242-269.

Houthakker, H.S., "Some Calculations of Electricity Consumption in Great Britain", *J. Royal Statistical Society*, vol.114, 1951, pp. 351-71.

IAEA, *Expansion Planning for Electrical Generating Systems: A Guidebook*, International Atomic Energy Agency, Vienna, 1984.

IEEE Standard 346-1973, "IEEE Standard Definitions in Power Operations Terminology Including Terms for Reporting and Analyzing Outages of Electrical Transmission and Distribution Facilities and Interruptions to Customer Service," New York, IEEE, 1973.

IEEE Tutorial Course, no. 76CH1107-2-PWR, Application of Optimization Methods in Power System Engineering, IEEE, New York, 1976.

Interconexion Electrica S.A., "Costos De Racionamiento De Energia Electrica Sectores Industrial, Residencial, Comercio Y Servicios Sistema Interconectado Colombiano", Medellin, Colombia, Oct.1986.

Jamaica Public Service Company, "Reliability Dollar Value Analysis" and "The Selection of a Generation Reliability Criterion for JSPC", November 1975.

Jaramillo,P. and E. Skoknic, "Costo Social de las Restricciones Energia Electrica," Oficina de Planificacion, Santiago, Chile, Aug. 1973.

Johnson, C.R. (FEA), prepared testimony before the Connecticut Public Utilities Control Authority, Docket No. 760204, 1976.

Johnson, B. F., "Evaluation of the Costs of Losses in Power Systems," Consumer's Power Company, April 1976.

Johnston, J., *Econometric Methods*, McGraw-Hill, New York, 1972.

Julius, D., "The Economic Cost of Power Outages: Methodology and Application to Jamaica", EW&T Department, World Bank, August 1976.

Kaufman, A., "Reliability Criterion - A Cost-Benefit Analysis", Office of Research, N.Y. State Public Service Commission, August 1975.

Keane, D., L. McDonald and C-K. Woo, "Estimating Residential Partial Outage Cost with Market Research Data", *The Energy Journal*, Special Electricity Reliability Issue, vol. 9, 1988, pp.151-160.

Kessler, J., J. Balantyne, R. Maushammer, and N. R. Simancas, "Ecuador: Rural Electrification," Project Impact Evaluation, U.S.A.I.D., Washington DC, l981.

Knight, U.G., *Power System Engineering and Mathematics*, Pergamon, New York, 1972.

Lancaster, K., "A New Theory of Consumer Behaviour", *Journal of Political Economy*, vol.74, April 1966, pp.132-57.

Little I.M.D., and J.A. Mirrlees, *Project Appraisal and Planning for Developing Countries*, Basic Books, New York, 1974.

Lolander, D., "Memorandum Concerning Estimation of the Costs of the National Economy in the Event of a Loss of Supply", Vattenfall, Sweden, 1948.

Madigan, F.C., "Cooperative Rural Electrification, Income Distribution, Employment, and Fertility: A Case Study from the Southern Philippines," General Conference of the International Union for the Scientific Study of Population, Manila, The Philippines, December 1981.

Malko, J.R. and A. Faruqui, "Implementing Time-of-Day Pricing of Electricity: Some Current Challenges and Activities," in M.A. Crew (Ed.), *Issues in Public Utility Pricing and Regulation*, Lexington Books, New York, 1980.

Marginal Costing and Pricing of Electrical Energy (Proc. of the State of the Art Conf.), Canadian Electrical Association, Montreal, Que., May 1978.

Marko, G.E., "A Method of Combining High Speed Contingency Load Flow Analysis with Stochastic Probability Methods to Calculate a Quantitative Measure of Overall Power System Reliability," paper No. A780-53-1 in Proc. IEEE Power Eng. Soc. Summer Meet, Los Angeles, July 1978.

Mason, M., "Rural Electrification: A Review of Bank-Financed Projects", draft report, Energy Department, The World Bank, Washington DC, Feb 1985.

McCawley, P., "Rural Electrification in Indonesia - Is It Time?", *Bulletin of Indonesian Economic Studies*, vol. 14, July 1978.

Miedema, A.K., S.B. White, C.A. Clayton, and D.P. Lifson, "Analysis of Experimental Time-of-Use Electricity Prices," Research Triangle Institute, NC 27709, Dec. 1978.

Miedema, A.K. "Overview of DOE Analyses of Residential TOU Data," Research Triangle Institute NC, Apr. 1980.

Ministerio de Economia, Industria y Comercio, Censos Nacionales de 1973 Agropecuaria, San Jose, Costa Rica, July 1974.

Ministry of Mines and Energy, "Survey of R.E. in South Sulawesi", Jakarta, Indonesia, 1985.

MIT Energy Lab., "New Utility Managment and Control Systems", Cambridge, MA, 1979.

Mitchell, B.M., W.G. Manning, Jr. and J.P. Acton, *Peak Load Pricing*, Ballinger Publ., Cambridge MA, 1978.

Munasinghe, M., *The Leisure Cost of Electric Power Failures*, World Bank Staff Working Paper No. 285, World Bank, Washington DC, June 1978.

Munasinghe, M. *The Economics of Power System Reliability and Planning*, Johns Hopkins Univ. Press, Baltimore, MD, 1979.

Munasinghe, M., "Costs Incurred by Residential Electricity Consumers Due to Power Failures", *Journal of Consumer Research*, vol.6, March 1980, pp.361-69.

Munasinghe, M., "System Planning to Achieve Optimal Reliability Levels", Proceedings of the Third International Conference on Analysis, Forecasting and Planning of Public Utilities, INSEAD, Paris, June 1980, pp. 303-24.

Munasinghe, M., "A New Approach to System Planning," *IEEE Trans. Power App. Syst.*, vol. PAS-99, May-June 1980.

Munasinghe, M., "Integrated National Energy Planning (INEP) in Developing Countries", *Natural Resources Forum*, vol. 4, 1980, pp. 359-73; also available as reprint no. 165, The World Bank, Washington DC.

Munasinghe, M., "An Integrated Framework for Energy Pricing in Developing Countries", *The Energy Journal*, vol.1, July 1980, pp.1-30.

Munasinghe, M., "World Bank Research in Electric Power", *Research News*, The World Bank, Washington DC, Spring 1981, pp. 3-10.

Munasinghe, M., "Principles of Modern Electricity Pricing," *Proc. IEEE*, vol. 69, Mar. 1981, pp. 332-348.

Munasinghe, M. "Optimal Electricity Supply: Reliability, Pricing and System Planning," *Energy Economics*, vol. 3, July 1981, pp. 140-152.

Munasinghe, M., "A Study of the Principal Issues in Rural Electrification in Indonesia", Energy Department, The World Bank, Washington DC, July 1982.

Munasinghe, M., "Engineering-Economic Analysis of Electric Power Systems", *Proc. of the IEEE*, vol. 72, April 1984, pp.424-61.

Munasinghe, M., "Practical Application of Integrated National Energy Planning Using Microcomputers", *Natural Resources Forum*, Feb. 1986, pp. 17-38.

Munasinghe, M., *Rural Electrification for Development: Policy Analysis and Applications*, Westview Press, Boulder CO, 1987.

Munasinghe, M. and V. Corbo, "The Demand for CATV Services in Canada," *Canadian J. of Economics*, Vol. 11, August 1978, pp. 506-520.

Munasinghe, M. and M. Gellerson, "Economic Criteria for Optimizing Power System Reliability Levels", *The Bell Journal of Economics*, vol.10, no.1, Spring 1979, pp.353-365.

Munasinghe, M., J. Gilling and M. Mason, *A Review of World Bank Lending For Electric Power*, Energy Series Paper No. 2, The World Bank, Wash. DC, March 1988.

Munasinghe, M. and S. Rungta, Eds., *Costing and Pricing Electricity in Developing Countries*, Asian Development Bank, Manila, Philippines, 1984.

Munasinghe, M. and G. Schramm, *Energy Economics, Demand Management and Conservation Policy*, Van Nostrand Rheinhold, New York, 1983.

Munasinghe, M. and W. Scott, "Long-Range Distribution System Planning Based on Optimum Economic Reliability Levels", Paper No. A78576-1, *Proceedings of the IEEE Power Engineering Society*, Summer Meeting, Los Angeles, August 1978.

Munasinghe, M. and W. Scott, *Energy Efficiency: Optimization of Electric Power Distribution System Losses*, Energy Department Paper No. 6, The World Bank, Washington, DC, July 1982.

Munasinghe, M., W.G. Scott, and R. Dias-Bandaranaike, *Optimizing Rural Electricity Supply*, Energy Dept. Report, The World Bank, Washington DC, 1982.

Munasinghe, M., and J. Warford, "Shadow Pricing and Evaluation of Public Utility Projects", Report No. GAS 14, Energy Department, World Bank, Washington, D.C., 1977.

Munasinghe, M. and J. Warford, "Shadow Pricing and Power Tariff Policy", in *Marginal Costing and Pricing of Electricity*, State of the Art Conference on Marginal Cost Pricing, Montreal, April-May 1978.

Munasinghe, M. and J. Warford, *Electricity Pricing*, Johns Hopkins University Press, Baltimore MD, 1982.

Munasinghe, M., and C. J. Warren, "Rural Electrification, Energy Economics and National Policy in Developing Countries", in *Future Energy Concepts*, Institution of Electrical Engineers, London, UK, 1979.

NARUC Electric Utility Rate Design Study, "Rate Design and Load Control: Issues and Directions," EPRI, Palo Alto, CA, Nov. 1977.

NARUC Electric Utility Rate Design Study, "1978 Plan of Study", EPRI, Palo Alto CA, Dec. 1977.

NARUC Electric Utility Rate Design Study, "Guide to Study Reports," EPRI, Palo Alto CA, Aug. 1979.

NARUC Electric Utility Rate Design Study, Rep. 69, EPRI, Palo Alto CA, Nov. 1979.

NARUC Electric Utility Rate Design Study, Rep. 72, EPRI, Palo Alto CA, Nov. 1979.

NARUC Electric Utility Rate Design Study, Rep. 79, EPRI, Palo Alto CA, Apr. 1980.

NARUC Electric Utility Rate Design Study, Rep. 66, EPRI, Palo Alto CA, June 1980.

NARUC Electric Utility Rate Design Study, Rep. 91, EPRI, Palo Alto CA, June 1980.

NARUC Electric Utility Rate Design Study, Rep. 67, EPRI, Palo Alto CA, July 1980.

National Electrification Administration, "Nationwide Survey on Socio-Economic Impact of Rural Electrification," for the U.S. Agency for International Development, Manila, The Philippines, June 1978.

Nicholson, W., *Microeconomic Theory*, Second Edition, Dryden Press, Hinsdale IL, 1978, Chap.4.

Nickel, D. L., "Distribution Transformer Loss Evaluation: Proposed Techniques," Proceedings of the IEEE, PES Winter Meeting, New York, Feb. 1980.

Noferi, P.L., L. Paris, and L. Salvaderi, "Monte Carlo Methods for Power System Reliability Evaluation in Transmission and Generation Planning," in Proc. Annual Reliability and Maintainability Symp., Washington, DC, Jan. 1975, pp. 460-469.

Office of Technology Assessment (OTA), *New Electric Power Technologies*, US Congress, Washington DC, July 1985.

Ontario-Hydro, "Survey on Power System Reliability: Viewpoint of Large Users," Rep. PMA 76--5, Toronto, Ont., Canada, 1976.

Ontario Hydro, "Survey on Power System Reliability - Summary of Customer Viewpoints", Report RU80-12, 1980.

Oren, S., S.Smith, R.Wilson and H.Chao, "Priority Service: Unbundling the Quality Attributes of Electric Power", Electric Power Research Institute, Report No. EPRI EA-4851, Palo Alto, CA, 1986.

Organization for Economic Cooperation and Development (OECD), *Environmental Effects of Electricity Generation*, Paris, 1985.

Pabaney, A.D., "Some Thoughts on Developing the Productive Uses of Electricity in the Rural Areas of Indonesia", Jakarta, Indonesia, May 1983.

Panzar, J.C., "A Neo-Classical Approach to Peak Load Pricing," *Bell J. Economics*, vol. 7, Autumn 1976, pp. 521-30.

Pelletier, Y., "Application des Methodes d'Analyse de Donnees a l'Etude d'Un Echantillon de Courbes de Chargemoyenne Tension," Electricite de France, Paris, Apr. 1977.

Pindyck, Robert S., *The Structure of World Energy Demand*, MIT Press, Cambridge MA, 1979.

Pioger, Y. "Elements de Reflexion Sur la Deformation des Courbes de Chage Clobales et Sectorielles: Evaluation de l'effet induit par la tarification," Electricite de France, Paris, July 1977.

Plano Diretor Do Desenvolvimento, Prefeitura Municipal de Cascavel, Cascavel, Parana, Brazil 1975.

Poland, W., "The Importance of Including Uncertainties in Economic Generation Reliability Planning", *The Energy Journal*, Special Electricity Reliability Issue, vol. 9, 1988, pp.19-32.

Priestman, D., "The Cost of a Long Term Electrical Energy Shortage Due to Underplanning", paper presented at the Eleventh Annual Pacific Northwest Regional Economic Conference, Eugene, Oregon, May 1977.

Public Utilities Regulatory Policies Act (PURPA), Rep. 95.1750, House of Representatives, 95th Congress-2nd Session, Oct. 10, 1978.

Ray, A., *Cost-Benefit Analysis*, Johns Hopkins Univ. Press, Baltimore, MD, 1984.

Reynolds, S.P. and T.E. Creighton Jr., "Time of Use Rates for Very Large Customers on the Pacific Gas and Electric Company System," *IEEE Trans. Power App. Syst.*, vol. PAS-99, Jan./Feb. 1980, pp. 147-151.

Rischl, R., "Optimal System Expansion Planning. A Critical Review", *Systems Engineering for Power: Status and Prospects*, Proc. USERDA Conf. (Henniker, NH, Aug. 1975) NTIS CONF-750867, pp. 233-260.

Ross, J.E., *Cooperative Rural Electrification*, Praeger Press, New York, 1972.

Ruggles, N.,"The Welfare Basis of the Marginal Cost Pricing Principle," *Rev. Economic Studies*, vol. 17, 1949-1950, pp. 29-46.

Ruggles, N.,"Recent Developments in the Theory of Marginal Cost Pricing," *Rev. Economic Studies*, vol. 17, 1949-1950, pp. 107-126.

Samanta, B.B. and A. K. Sundaram, "Socioeconomic Impact of Rural Electrification in India," Discussion Paper D-730 by Operations Research Group, U.S. Agency for International Development, Washington DC, January 1983.

Sanghvi, S., "Economic Costs of Electricity Supply Interruptions: US and Foreign Experience", *Energy Economics*, vol 4, no.3, July 1982, pp.180-198.

Sanghvi, S., "Household Welfare Loss Due to Electricity Supply Disruptions", *The Energy Journal*, vol.4, 1983, pp.33-54.

Sanghvi, S., "Optimal Electric Supply Reliability Using Customer Shortage Costs", *Energy Economics*, vol.5, April 1983.

Sanghvi, S., "Optimum Generation Reserve Study", Final report to Public Service Commission and Office of the Attorney General of New Mexico, 1985a.

Sanghvi, S., "Survey of Reliability Planning Practices at Major U.S.Utilities", Report prepared for Pacific Gas and Electric Co. of California, October 1985b.

Sanghvi, S., "Incorporating Risk Preference in Generation Reliability Planning Using the EUE Criterion", report prepared for Pacific Gas and Electric Co. of California, July 1985c.

Sanghvi, S., "Residential Outage Cost Estimation", final report to Rate Department, Pacific Gas and Electric Company (PG&E) of California, December 1986.

Sato, K., "Additive Utility Function with Double-Log Consumer Demand Functions", *Journal of Political Economy*, 1982, pp. 102-24.

Saunders, J., J. M. Davis, G. C. Moses, and J. E. Ross, *Rural Electrification and Development: Social and Economic Impact in Costa Rica and Colombia*, Westview Press, Boulder CO, 1975.

Schweppe, Fred C., M. Caramanis, R. Tabors and R. Bohn, *Spot Pricing of Electricity*, Kluwer Publishing, New York, 1988.

Sen, L.K., "Rural Electrification in Indonesia: Current and Future Perspectives", Discussion Paper No. 131, Harvard Inst. for Int. Dev., Cambridge MA, Feb. 1982.

Shepard, R.W., *Cost and Production Functions*, Princeton Univ. Press, Princeton NJ, 1953.

Sherman, R. and Visscher, M., "Second-Best Pricing with Stochastic Demand", *The American Economic Review*, vol. 63, No. 1, March 1973, pp. 41-53.

Shipley, B., A.Patton and J.Denison, "Power Reliability Cost vs. Worth", *IEEE Transactions on Power Apparatus and Systems*, vol.PAS-91, Sept/Oct. 1972, pp.2204-12.

Smith, D.V., D.B. Mehta, and P. Hayes, "Regional Rural Electrification Survey", draft report, Asian Development Bank, Manila, Philippines, Oct. 1983.

Soares, S., C. Lyra and H. Tavares, "Optimal Generation Scheduling of Hydro-thermal Power Systems", *IEEE Transactions on Power Apparatus and Systems*, vol. PAS-99, May-June 1980.

Squire, L. and H. van der Tak, *Economic Analysis of Projects*, Johns Hopkins Univ. Press, Baltimore MD, 1975.

Steiner, P., "Peak Loads and Efficient Pricing," *Quart. J. Economics*, Nov. 1957, pp. 585-610.

Stott, B., "Review of Load-Flow Calculation Methods," *Proc IEEE*, vol. 62, July 1974, pp. 916-929.

Suelflow, J., *Public Utility Accounting: Theory and Applications*, Michigan State Univ. Press, East Lansing MI, 1973.

Sullivan, J.B., "The A.I.D. Experience with Independent Power Generation", paper presented at the Central American and Caribbean Workshop on Electric Power, San Jose, Costa Rica, August 1988.

Sullivan, R.L., *Power System Planning*, McGraw-Hill, New York, 1977.

Swedish Association of Electricity Supply Undertakings, "Evaluating Electricity Interruptions", Stockholm, Sweden, 1981.

Swedish Joint Committee for the Electricity Council, "Cost of Interruptions in Electricity Supply", Stockholm, Sweden, Sept. 1969.

Swedish Committee on Supply Interruption Costs, "Costs of Interruptions in Electricity Supply", A.O. Trans. 450, The Electricity Council, UK, September 1968.

Symposium on Peak Load Pricing, *Bell J. Economics*, vol. 7, Spring 1976, pp. 197-250.

System Control, Inc., "Analysis of Electric Power System Reliability", prepared for the California Energy Resources Conservation and Development Commission, Palo Alto, CA, 1978.

Taiwan Power Company, "Evaluation of Generating Capacity Reserve", Taiwan, 1975.

Takayama, A., *Mathematical Economics*, Cambridge University Press, Cambridge, UK, 1985.

Taylor, Lester D., "The Demand for Energy: A Survey of Price and Income Elasticities," in: W.D. Nordhaus (ed), *International Studies of the Demand for Energy*, North Holland, Amsterdam, 1977.

Telson, M., "The Economics of Alternative Levels of Reliability for Electric Generation Systems", *Bell Journal of Economics*, vol. 6, no.2, 1975, pp.679-94.

Tendler, J., "Rural Electrification: Linkages and Justifications," Bureau for Program and Policy Coordination, U.S. Agency for International Development, Washington DC, April 1979.

The Energy Journal, Special Electricity Issue, 1983.

Tietenberg, T., *Environmental and Natural Resource Economics*, Scott Foresman Co., Glenville, IL, 1988.

Turvey, R. and D. Anderson, *Electricity Economics*, Johns Hopkins Univ. Press, Baltimore MD, 1977.

Turvey, R., *Optimal Pricing and Investment in Electricity Supply*, MIT Press, Cambridge, MA, 1968.

U.S.Agency for International Development (USAID), *Power Shortages in Developing Countries*, Washington, DC, March 1988.

US Bureau of the Census, "Evaluation of the Indonesian R.E. project: results of 1984 surveys of households, business and public facilities", Washington DC, Dec. 1985.

USAID, "Power to the People: Rural Electrification Sector Summary Report," Draft Summary Report, U.S. AID, Washington DC, 1981.

Vardi, J. and B. Avi-Itzhak, *Electric Energy Generation: Economics, Reliability and Rates*, MIT Press, Cambridge, MA, 1981.

Velez, E., Becerra and A. Carrasquilla, "Rural Electrification in Colombia," Report by Institution SER de Investigacion, U.S. Agency for International Development, Washington DC, March 1983.

Viswanathan, "Service-Reliability Differentiated Pricing with Special Reference to Electric Utilities", Unpublished Doctoral Dissertation, Stanford University, CA, 1985.

Webb, M., "The Determination of Reserve Generating Capacity Criteria in Electricity Supply", *Applied Economics*, vol.22, March 1977, pp.19-31.

Wenders, J.T., "Peak Load Pricing in the Electric Utility Industry," *Bell J. Economics*, vol. 7, Spring 1976, pp. 232-241.

Williamson, O.E., "Peak Load Pricing and Optimal Capacity under Indivisibility Constraints," *Amer. Economic Rev.*, vol. 56, no. 4, Sept. 1966, pp. 810-827.

Willig, R.D., "Consumers Surplus Without Apology", *American Economic Review*, vol.66, Sept.1976, pp.589-97.

Wilson, J.W., "Residential Demand for Electricity", *Quarterly Review of Economics and Business*, vol.11, Spring 1971, pp. 7-22.

Wismer, D.A., *Optimization Methods for Large Scale Systems*, McGraw-Hill, New York, 1971.

Woo, C.K. and K. Train, "Cost of Electric Power Interruptions to Commercial Firms", *The Energy Journal*, Special Electricity Reliability Issue, vol. 9, 1988, pp.161-172.

Woo, C.K. and N.Toyama, "Service Reliability and the Optimal Interruptible Rate Option in Residential Electricity Pricing", *The Energy Journal*, vol. 7, July 1986, pp.123-36.

World Bank, *Rural Electrification*, Washington DC, 1975.

World Bank, *World Development Report*, Washington DC, August 1989.